Remote Sensing and Hydrology

Remote Sensing and Hydrology

Edited by Sarah Jackson

SYRAWOOD
PUBLISHING HOUSE

New York

Published by Syrawood Publishing House,
750 Third Avenue, 9th Floor,
New York, NY 10017, USA
www.syrawoodpublishinghouse.com

Remote Sensing and Hydrology
Edited by Sarah Jackson

International Standard Book Number: 978-1-64740-436-9 (Hardback)

Cataloging-in-publication Data

Remote sensing and hydrology / edited by Sarah Jackson.
    p. cm.
Includes bibliographical references and index.
ISBN 978-1-64740-436-9
1. Hydrology--Remote sensing. 2. Remote sensing.
3. Water resources development--Remote sensing.
I. Jackson, Sarah.
GB656.2.R44 R46 2023
551.48--dc23

TABLE OF CONTENTS

PREFACE

Remote sensing refers to a process involving the identification and observation of the topographies of a region by measuring its reflected and emitted radiation from a distance. It involves gathering information about objects or regions from distant objects such as aircrafts, satellites and drones. It is commonly used for gathering information about the Earth's surface from space. Weather forecasting, natural disaster research, land-use mapping, resource exploration, and environmental research are some of the prominent applications of remote sensing. One of the significant applications of remote sensing is in hydrology, where it can be used for measuring several hydrological variables over a region repeatedly. The hydrological variables measured by remote sensing include surface temperature, snow water equivalent, surface soil moisture, snow covered area and land cover. This book is a valuable compilation of topics, ranging from the basic to the most complex studies on the role of remote sensing in hydrology. Scientists and students actively engaged in this field will find it full of crucial and unexplored concepts.

Various studies have approached the subject by analyzing it with a single perspective, but the present book provides diverse methodologies and techniques to address this field. This book contains theories and applications needed for understanding the subject from different perspectives. The aim is to keep the readers informed about the progresses in the field; therefore, the contributions were carefully examined to compile novel researches by specialists from across the globe.

Indeed, the job of the editor is the most crucial and challenging in compiling all chapters into a single book. In the end, I would extend my sincere thanks to the chapter authors for their profound work. I am also thankful for the support provided by my family and colleagues during the compilation of this book.

Editor

An Assessment of the Hydrological Trends using Synergistic Approaches of Remote Sensing and Model Evaluations over Global Arid and Semi-Arid Regions

Wenzhao Li [1], Hesham El-Askary [1,2,3,*], Rejoice Thomas [4], Surya Prakash Tiwari [5], Karuppasamy P. Manikandan [5], Thomas Piechota [1] and Daniele Struppa [1]

[1] Schmid College of Science and Technology, Chapman University, Orange, CA 92866, USA; li276@mail.chapman.edu (W.L.); piechota@chapman.edu (T.P.); struppa@chapman.edu (D.S.)

[2] Center of Excellence in Earth Systems Modeling and Observations, Chapman University, CA 92866, USA

[3] Department of Environmental Sciences, Faculty of Science, Alexandria University, Moharem Bek, Alexandria 21522, Egypt

[4] Computational and Data Sciences Graduate Program, Schmid College of Science and Technology, Chapman University, Orange, CA 92866, USA; rejthomas@chapman.edu

[5] Center for Environment and Water, The Research Institute, King Fahd University of Petroleum & Minerals (KFUPM), Dhahran 31261, Saudi Arabia; surya.tiwari@kfupm.edu.sa (S.P.T.); manikand@kfupm.edu.sa (K.P.M.)

*   Correspondence: elaskary@chapman.edu

Abstract: Drylands cover about 40% of the world's land area and support two billion people, most of them living in developing countries that are at risk due to land degradation. Over the last few decades, there has been warming, with an escalation of drought and rapid population growth. This will further intensify the risk of desertification, which will seriously affect the local ecological environment, food security and people's lives. The goal of this research is to analyze the hydrological and land cover characteristics and variability over global arid and semi-arid regions over the last decade (2010–2019) using an integrative approach of remotely sensed and physical process-based numerical modeling (e.g., Global Land Data Assimilation System (GLDAS) and Famine Early Warning Systems Network (FEWS NET) Land Data Assimilation System (FLDAS) models) data. Interaction between hydrological and ecological indicators including precipitation, evapotranspiration, surface soil moisture and vegetation indices are presented in the global four types of arid and semi-arid areas. The trends followed by precipitation, evapotranspiration and surface soil moisture over the decade are also mapped using harmonic analysis. This study also shows that some hotspots in these global drylands, which exhibit different processes of land cover change, demonstrate strong coherency with noted groundwater variations. Various types of statistical measures are computed using the satellite and model derived values over global arid and semi-arid regions. Comparisons between satellite- (NASA-USDA Surface Soil Moisture and MODIS Evapotranspiration data) and model (FLDAS and GLDAS)-derived values over arid regions (BSh, BSk, BWh and BWk) have shown the over and underestimation with low accuracy. Moreover, general consistency is apparent in most of the regions between GLDAS and FLDAS model, while a strong discrepancy is also observed in some regions, especially appearing in the Nile Basin downstream hyper-arid region. Data-driven modelling approaches are thus used to enhance the models' performance in this region, which shows improved results in multiple statistical measures ((RMSE), bias (ψ), the mean absolute percentage difference ($|\psi|$)) and the linear regression coefficients (i.e., slope, intercept, and coefficient of determination (R^2)).

Keywords: drylands; climate classification; GLDAS; FLDAS; machine learning; Google Earth Engine

1. Introduction

Arid and semi-arid regions, also known as drylands, are areas where the annual total surface evaporation and vegetation transpiration substantially exceeds precipitation. Recent studies show that they account for about 40% of the Earth's land surface [1] and play an important role in the process of global climate change as a regulator of trends and variabilities of atmospheric carbon dioxide (CO_2) concentrations [2–5]. Anders et al. [2] showed that the mean sink, trend, and interannual variability in CO_2 uptake by terrestrial ecosystems are influenced by diverse biogeographic regions where the trend and interannual variability of the sink are largely influenced by semi-arid ecosystems. In fact, Joel et al. [3] suggested that the contribution of the dryland regions towards the changing regional and global CO_2 exchange may be three to five times larger than current estimates. Benjamin et al. [5] found that the global carbon cycle inter-annual variability is driven by the carbon pools in semi-arid biomes. Due to the lack of water supplies, low vegetation coverage, and fragile ecological environment in these areas, these regions are suitable for activities with lower annual productivity and more sensitive and rapid in response to external forces. In the past 100 years, the arid and semi-arid regions have been the areas with the most significant land temperature increase and area expansion [6,7]. The situation is projected to intensify, as indicated by climate models [8,9], likely resulting in an increase in aridity, warming, along with land degradation and desertification in the drylands of developing countries [7]. The drylands would also undergo the consequences of climate change through emissions from humid lands [7,10].

Nowadays, more than two billion people live in these regions, with about 90% of them from developing countries where people's living standards and technological capabilities lag far behind developed countries [1]. The interaction of anthropogenic activities (e.g., farming, grazing and irrigation) and natural processes is suggested to have a strong influence on the water balance. The increasing demand for water resources in support of these activities can exacerbate the severity of existing drought conditions. Mitigation actions such as groundwater exploitation for irrigation and industry would alleviate the ongoing situation yet increase the vulnerability for upcoming droughts. Therefore, moderate to severe droughts, combined with the depletion of surface and groundwater resources, pose significant risks to water and food security, leading to wide-ranging and long-term impacts, including conflicts among dryland countries over shared transboundary river basins. As a result, there is a pressing need to improve our understanding of the ongoing situations and the future trends of hydrological parameters in these global arid and semi-arid regions to perform proper mitigation assessments for these important and vulnerable ecosystems.

Dryland vegetation is strongly governed by the spatial and temporal variation of water availability. These variables show apparent links between hydrology and ecology in these regions, where the temporal variation of precipitation is more substantial than other areas at different scales (diurnal, seasonal and interannual). Generally, precipitation pulses trigger a cascade of pulsed ecosystem responses to end dry periods. Meanwhile, they can also lead to a rapid increase in evapotranspiration (ET), which consists mainly of evaporation (E) from plant and soil surfaces and plant transpiration (T) through leaf stomata. Consequently, transpiration can contribute significantly to total ET if the water added is adequate to wet the root zone, indicated by the surface/sub-surface soil moistures [3,11–14]. In drylands, ET is an effective indicator of vegetation resilience and resistance to drought conditions, since water is the major constraint on plant productivity [15,16]. In the hydrological cycle, ET also serves as a key process that links energy, carbon and water cycles [17].

Ground-based observation networks, mainly from eddy covariance flux towers, have been established to estimate various variables such as ET and gross primary productivity (GPP). Some typical networks include FLUXNET [18], AmeriFlux [19], ILTER [20] and NEON [21]. These sites are mostly situated in more humid and developed areas, while large drylands in Africa, South America, Middle East, and central Asia have limited representation. Remote sensing has been functional in demonstrating the role of GPP and ET, as well as ecosystem structure within the context of the broader Earth system [4,5]. Favorable atmospheric conditions (lower cloud coverage) in drylands offer advantages for optical remote sensing and increase the chances of high-quality observations. In addition, pioneering technologies were developed in drylands, including the retrieval of surface reflectance [22,23], as well as quantitative estimates of vegetation conditions and photosynthetic capacity in the rangelands of the Great Plains [24]. The later research afterward became the Normalized Difference Vegetation Index (NDVI) [25], calculated as NDVI = (NIR − R)/(NIR + R), which is generally based on the high reflectivity of plants in the near-infrared (NIR) wavelength range and the high absorption by plants in the red (R) wavelength range for photosynthesis. With spectral normalization, NDVI minimizes effects from topographic variation, sun-sensor geometry and shadows [26], thus becoming one of the most widely used vegetation indices [26–28]. The long-term and high-frequency NDVI time series retrievals, like the 23 year time series generated for long-term studies of the Sahel region [26], enhanced our understanding of the long-term trends and driving forces behind regional dryland dynamics. In Australia, an analysis to check whether vegetation cover increased was conducted using calibrated advanced very high-resolution radiometer data spanning 1981–2006 [27]. In the same region, an analysis of the decadal time-scale changes was conducted using AVHRR GIMMS NDVI dataset to check the relationship between a proxy for vegetation productivity and annual rainfall [28]. Subsequently, Randall et al. [29] observed a greening of the globe over recent decades which they attributed to CO2 fertilization effect. A study by Ramus et al. [30] to analyze trends in vegetation greenness of semi-arid areas showed an increase in greenness found both in semi-arid areas where precipitation is the dominating limiting factor for plant production and where air temperature is the primary growth constraint. Similarly, Ulf et al. [31] in their research using NOAA AVHRR data in Mediterranean basin, the Sahel from the Atlantic to the Red Sea, major parts of the drylands of Southern Africa, China–Mongolia and the drylands of South America showed a greening-up. Other studies conducted by Kolby et al. [32], which showed an increase in net primary productivity (NPP), and by Zaichun et al. [33], which discussed the greening of earth and its drivers, attributed the major cause to be CO_2 fertilization. These investigations recently elucidated that global drylands greening is largely due to CO_2 fertilization's effects in the context of global climate change [29–33]. Since the 2000s, there has been an exponential increase in the volume of remote sensing data, resulting from newly launched sensors and products, as well as higher resolution domains (spatial, temporal, and spectral). Such expansion of big data revolutionized our understanding of the Earth system, while posing a higher request for advances in data management and processing capability. Accordingly, increasing the availability of cloud-based computing and analytical platforms, such as Amazon Web Service (AWS) and Google Earth Engine [34], is being widely used for dryland research [35–40]. Using the tools within Google Earth Engine, it was found that global dryland tree cover was significantly underrated and recent estimates exceeded previous ones by over 40% [41].

The goal of this paper is to evaluate hydrologic and land cover changes in global arid/semi-arid regions over the last decade (2010–2019), through time series and trends analysis, correlation and interactions of multiple hydrological indicators and vegetation indices. Another key point of this research is to evaluate two major process-based land surface modeling and data assimilation models (GLDAS and FLDAS) by satellite-based hydrological products. Additional analysis displayed an overestimation of surface soil moisture (SSM) and an underestimation of evapotranspiration (ET), as well as a discovery of a strong discrepancy observed in Egypt by the aforementioned models. To mitigate this discrepancy, machine learning models using extra data from satellites (Moderate Resolution Imaging Spectroradiometer (MODIS)) were developed which increased the process-based model performance significantly and provided results with lower RMSE and bias. This optimization effort highlights the promising applications of the synergistic approach of data-driven and process-based modeling for hydrological studies.

This paper is organized into four sections. The information regarding the study area, data sources and methods are described in Section 2. Section 3.1 illustrates the time series boxplots of multiple hydrological parameters used over arid regions. In Section 3.2, the trends of ET, soil moisture and precipitation during the period of 2010–2019 over global drylands are presented. Section 3.3 shows the landcover and water storage changes of selected areas. In Section 3.4, the comparison between FLDAS and GLDAS surface soil moisture and ET products is shown by correlation maps, followed by the approach of data-driven modelling of the aforementioned products. Finally, the results are discussed in detail in Section 4 and a conclusion in Section 5 is provided for the paper.

2. Materials and Methods

2.1. Study Area

The global arid and semi-arid regions were selected from the updated version of the Köppen–Geiger climate map [42], which shows five main climate groups, with sub-division of each group based on seasonal precipitation and temperature patterns. The five main groups are A (tropical), B (dry), C (temperate), D (continental), and E (polar). In this research, the subgroups starting with the letter "B" (dry) were identified as arid and semi-arid classes, namely, BWh (hot desert climates), BWk (cold desert climates), BSh (hot steppe climates) and BSk (cold steppe climates) (Figure 1a). The corresponding biome type distribution over these regions is also presented in Figure 1b using OpenLandMap Potential distribution of biomes dataset [43].

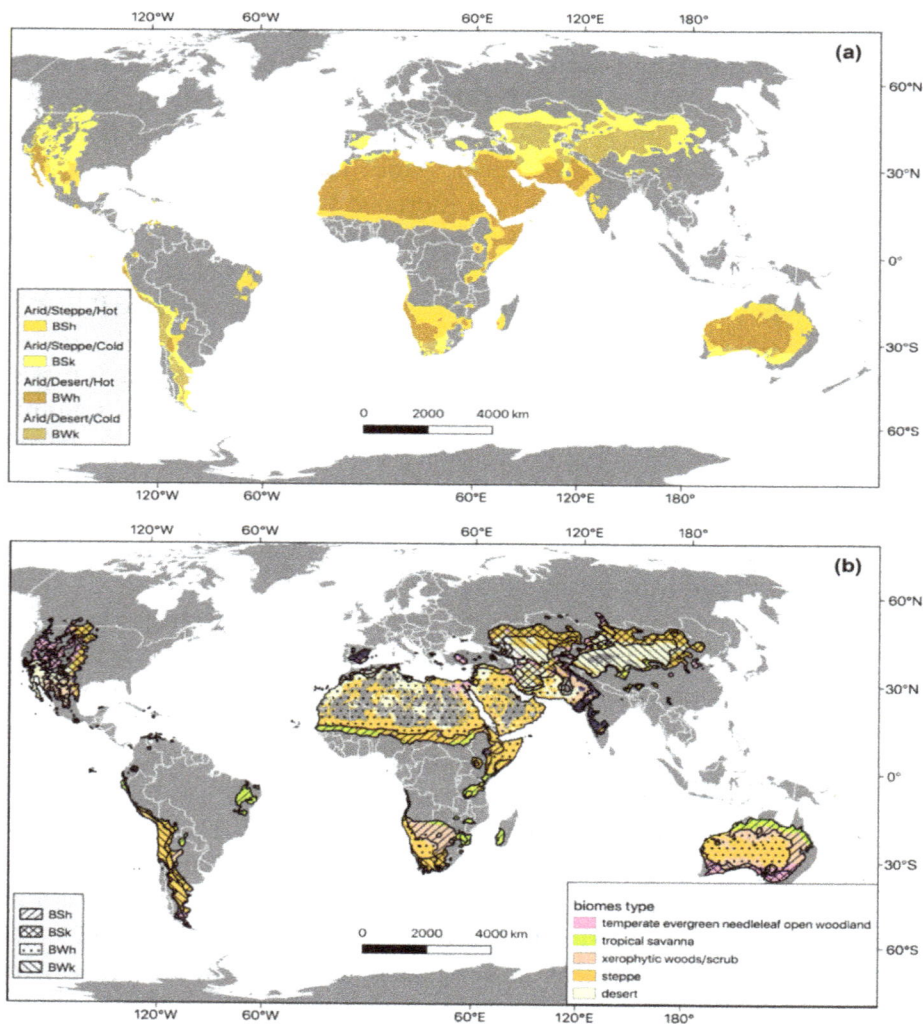

Figure 1. (**a**) Global arid regions in Köppen–Geiger climate classification, including four arid types: BSh (hot semi-arid climate), BSk (cold semi-arid climate), BWh (hot desert climate) and BWk (cold desert climate); (**b**) biomes types distribution map over the global arid region.

2.2. Data

2.2.1. Data from Remote Sensing Observations

- Precipitation Data

The precipitation data used in this study were obtained from the Climate Hazards Group Infrared Precipitation with Station (CHIRPS) dataset (version 2.0) [44], with 0.05° spatial resolution satellite imagery, and were converted into monthly temporal resolution. Developed by Earth Resource Observation and Science Center and the United States Geological Survey (USGS), CHIRPS is a 30+ year quasi-global rainfall dataset incorporating satellite and station data and to create gridded precipitation time series for hydrology monitoring and correlational and trend studies.

- Surface Soil Moisture Data

The surface soil moisture data were obtained from the National Aeronautics and Space Administration (NASA)–United States Department of Agriculture (USDA) Global Soil Moisture Dataset [45–48], with 0.25° spatial resolution and monthly temporal resolution. This dataset is generated through the data assimilation approach using a modified two-layer Palmer model using a 1-D ensemble Kalman filter (EnKF) to integrate the satellite-derived soil moisture observations. This dataset includes products of soil moisture profile (%), surface and subsurface soil moisture, as well as their standardized anomalies values computed using a 31 day moving window.

- Water Storage Data

The changes in groundwater storage are represented by the variations in the water equivalent height (in cm) observed from the Gravity Recovery and Climate Experiment (GRACE) and GRACE Follow-on mission monthly gravitational anomalies datasets [49–51] in 1° spatial resolution. GRACE is a twin satellite setup that measures global changes in water anomalies. These anomalies depict the changes in the total water column represented by water equivalent height, which includes surface, sub-surface, and groundwater components.

- Vegetation Indices Data

The Moderate Resolution Imaging Spectroradiometer (MODIS) sensor onboard of Terra satellite provides three vegetation index products: evapotranspiration (ET) [52], NDVI [53], Leaf Area Index (LAI) and fraction of absorbed photosynthetically active radiation (FPAR) [54]. All the products are regridded and converted into monthly data over 1 km spatial resolution. The calculation of MODIS ET is affected by other variables, including albedo, biomes type and LAI [52]. LAI is defined as half the total leaf area per unit ground area [55]. FPAR is defined as the fraction of the incoming solar radiation that is absorbed by a photosynthetic organism in the photosynthetically active radiation spectral region. The MODIS products are proved to effectively capture the dominant seasonal LAI and FPAR patterns in many arid and semi-arid ecosystems [56].

2.2.2. Data from Process-Based Models

The Land Data Assimilation Systems (LDAS) use data from multiple ground and space observations that are integrated with advanced numerical models of physical processes to simulate fields of water and energy states and fluxes, which helps in filling gaps and minimizing errors in the observations. Different projects (e.g., NLDAS, GLDAS, FLDAS) of LDAS have been configured for specific domains and purposes [57]. There are some known issues with GLDAS version 1 data, which might present some challenges in the runoff, soil moisture and ET analysis performed with the GLDAS model. These issues are suggested to be solved in the newer version (GLDAS version 2.1) used in this study. The GLDAS 2.1 simulation started on 1 January 2000 using the conditions from the GLDAS version 2.0

simulation. The GLDAS-2.1 is revised with upgraded models using National Oceanic and Atmospheric Administration (NOAA)/Global Data Assimilation System (GDAS) atmospheric analysis fields [58], the disaggregated Global Precipitation Climatology Project (GPCP) precipitation fields [59] and the Air Force Weather Agency's AGRiculturalMETeorological modeling system (AGRMET) radiation fields. In this study, the surface soil moisture and evapotranspiration were selected from the GLDAS-2.1 Noah model from 2010 to 2019, with its 0.25 degree spatial resolution and monthly temporal resolution.

FLDAS [60] is designed for semi-arid, food-insecure regions of Africa. Unlike GLDAS, which relies on global rainfall products (e.g., CMAP [61] and Princeton [62]), FLDAS uses a combination of the forcing data from Modern-Era Retrospective analysis for Research and Applications version 2 (MERRA-2) data and CHIRPS [44], which is designed for seasonal drought monitoring and trend analysis. In this study, the surface soil moisture and evapotranspiration were selected from the FLDAS dataset in Noah version 3.6.1 from 2010 to 2019, with its 0.1 degree spatial resolution and monthly temporal resolution.

2.3. Methods

2.3.1. Harmonic Analysis

Harmonic analysis is a method of expressing a function or signal as a superposition of fundamental waves as Equation (1). In this study, to estimate the variation of multiple variables (e.g., SSM, precipitation), a harmonic model $H(t)$ was built with elements of a constant band (β_0), a linear term of slope (β_1) and harmonic terms of amplitudes (β_2, β_3, β_4 and β_5). The constant band β_0 represents the extent of consistency of the time series. β_1 shows linear trend of a time series in spite of seasonal variations. Therefore, the term β_1 represents the annually increasing/decreasing trend. Additionally, f represents the fundamental frequency and harmonic terms of amplitudes (β_2, β_3, β_4 and β_5) are used to simulate seasonal variations.

$$H(t) = \beta_0 + \beta_1 t + \beta_2 \cos(2\pi f t) + \beta_3 \sin(2\pi f t) + \beta_4 \cos(4\pi f t) + \beta_5 \sin(4\pi f t) \tag{1}$$

Following the methods in [38,39], harmonic analysis was applied to show a yearly trend followed by precipitation, ET and surface soil moisture in global drylands from 1 January 2000 to 31 December 2019 using monthly values (units in mm). The units of SSM/ET outputs from FLDAS and GLDAS products were also transformed into mm to compare with MODIS products.

2.3.2. Correlational Analysis

For the purpose of finding the correlational relationship between the multiple variables, correlation maps were developed to show the standard correlation using the Pearson correlation coefficient (r) in the range from -1 (anti-correlation) to $+1$ (perfect correlation), between these two monthly time series x and y, with N elements as Equation (2):

$$r = \frac{\sum_{i=1}^{N}(x_i - \overline{x})(y_i - \overline{y})}{\sqrt{\sum_{i=1}^{N}(x_i - \overline{x})^2}\sqrt{\sum_{i=1}^{N}(y_i - \overline{y})^2}} = \frac{Cov(x,y)}{\sigma_x \sigma_y} \tag{2}$$

with Cov being the covariance function, $\overline{x}$ and $\overline{y}$ the average and σ_x and σ_y the standard deviations for x and y, respectively.

In this study, the association between the GLDAS and FLDAS products in both SSM and ET variables over global arid and semi-arid regions was studied by the Pearson correlation analysis using their anomalies value X_a calculated as Equation (3):

$$X_a = X - \overline{X} \tag{3}$$

with X as the monthly value and $\overline{X}$ as the monthly mean value. The anomalies value X_a can better demonstrate the effect of the change in a variable on the corresponding variable in the correlational analysis, especially when both variables have same pattern of seasonal variations. For instance, the correlation maps between the monthly anomalies values of Chlorophyll-*a* concentrations and environmental variables provided a better representation of their connections than simply using monthly mean values in [63,64].

2.3.3. Data-Driven Modelling of Selected Variables

In this study, the monthly data from dryland area in Egypt (mainly the Nile Delta region) during the years 2010 to 2019 were sampled and selected for the SSM and ET model development and evaluation. This region has 50 randomly selected sampling locations with a 10 km scale. A total of 29 parameters, including MODIS SSM and ET (modeling objective), as well as GLDAS and FLDAS outputs (e.g., Heat flux, specific humidity, total precipitation rate, soil temperature, wind speed, etc.) were used in the modelling (see comprehensive list in Table S1). The modelling dataset (~3000 records) was randomly split into a training set and testing set, with a ratio of 8:2, respectively.

Traditionally, intensive manual experiments are needed for model selection and parameterization, because it is difficult to know which model is most suitable for the problem before researchers try all the models. Therefore, the automatic optimization methods were proposed to overcome this issue, which try to train and evaluate model groups by parameters sequentially, and then include new parameter groups according to the updated results. In this study, automated model selection and tuning techniques were performed using Bayesian parameter optimization on the OptiML platform (https://bigml.com/api/optimls), which tried different supervised models (101 in total) for the regression tasks, including one linear regression, 89 tree models, 9 ensembles (up to 256 trees), and two deep neural networks. During the process, the platform used the Monte Carlo cross-validation [65], which iteratively generated new original data sets of training and test segmentation for the upper half of the model, and discarded the remaining half of the model due to poor performance. Therefore, this approach has an advantage over k-fold cross validation for the training and validation subsets, in that it is independent of the number of iterations. For both SSM and ET, deep neural networks outperformed other candidate models with R^2 values of 0.87 and 0.81 (see comprehensive metrics in Table S2), respectively, and thus applied for the simulation.

2.3.4. Model Accuracy Assessment

Following [66], accuracy assessments of the algorithms are presented through standard statistical errors (root mean square error (RMSE), bias (ψ), the mean absolute percentage difference ($|\psi|$)), and the linear regression (i.e., slope, intercept, and coefficient of determination (R^2)). Mathematically, the value of ψ is derived from the following:

$$\psi = \frac{1}{N} \sum_{i=1}^{N} \psi_i \tag{4}$$

where N is the number of data points and ψ_i is derived from:

$$\psi_i = \left(\frac{X_{simulation}(i) - X_{obs}(i)}{X_{obs}(i)} \right) \times 100 \tag{5}$$

In addition, the value of $|\psi|$ is calculated from the following:

$$|\psi| = \frac{1}{N} \sum_{i=1}^{N} |\psi_i| \tag{6}$$

The quantity ψ determines the bias, while $|\psi|$ illustrates the scattering of data points.

3. Results

3.1. The Time Series of Multiple Hydrological Parameters

Time series boxplots of multiple variables, in the order of ET, precipitation, FPAR, LAI, NDVI and SSM (surface soil moisture), are presented in Figures 2–5 for the BSh, BSk, BWh and BWk regions, respectively. Generally, the BSh region exhibits the highest values for all variables. On the other hand, the BWh region has the lowest ET and SSM, whereas the BWk region has the lowest FPAR, LAI, NDVI and precipitation. For BSh (Figure 2), strong seasonal patterns of ET are represented by maximum and upper quartile values: (1) highest in spring (February and March); (2) second highest in early fall (August and September); (3) lowest in June, July, October and November. Such patterns can also be found by median values, yet without peaks in early fall. The years of 2011, 2014, 2017 and 2018 (highlighted in Figure 2) show higher ET compared with other years. The seasonal variation is also shown in precipitation records, while the double peaks of maximum/upper quartile values are slightly changed, occurring often in winter (December to March) and later summer. The summer precipitation peaks sometimes surpassed the spring peaks (e.g., August in 2011, 2014, 2016 and 2017) (highlighted in Figure 2). For BSk (Figure 3), ET maximum/upper quartile values show a different seasonal pattern: (1) highest in summer (June, July and August); and (2) lowest in winter (December and January), indicating temperature as the controlling factor for ET. For both BSh and BSk, the FPAR, LAI and NDVI likely have the same patterns of seasonality as shown in ET and precipitation, while their SSM has comparably irregular seasonality patterns.

For BWh (Figure 4), due to lower overall ET, the seasonal pattern of median value is less noted, showing highest ET in summer (June and July), with occasional pulses in later winter and early spring (e.g., 2011, 2014, 2017 and 2018) (highlighted in Figure 4). The precipitation maximum and upper quartile values of BWh are highest in later winter and early spring. The strong precipitation pulses in 2011 and 2017 explain the ET surges at the same time. In contrast, BWk (Figure 5) has the highest ET in spring (February and March). In addition, the SSM of BWk indicates a dry period of 3 years from 2010 to 2012. For both BWh and BWk, their FPAR, LAI and NDVI do not show such seasonality pattern as shown by ET. However, the precipitation pulses in 2011 and 2017 are not reflected in these variables.

3.2. Trends of Multiple Hydrological Parameters

Figure 6 illustrates the trends followed by multiple variables, including ET (Figure 6a), SSM (Figure 6b) and precipitation (Pp) (Figure 6c) over BSh, BSk, BWh and BWk regions calculated based on harmonic analysis during the period between 2010 and 2019, only showing regions where the linear trend was statistically significant at the 95% confidence level. Comparing these data with Figure 1b, typical vegetation types of tropical savanna and scrub (BSh regions in the south of Africa and east of Australia) faced a rapid decreasing trend for all variables, indicating serious drought conditions. SSM also dropped in the north of China. Meanwhile, some regions experienced an increase in: (1) ET in the south of Argentina and Paraguay; (2) SSM in Eastern Europe and North America; (3) precipitation in North America (middle of the United States) and Eastern Africa (Tanzania). Colossal regions of the Middle East and North Africa (MENA), as well as the western desert of China, are absent in ET maps, due to the calculated trend not reaching the statistical significance level.

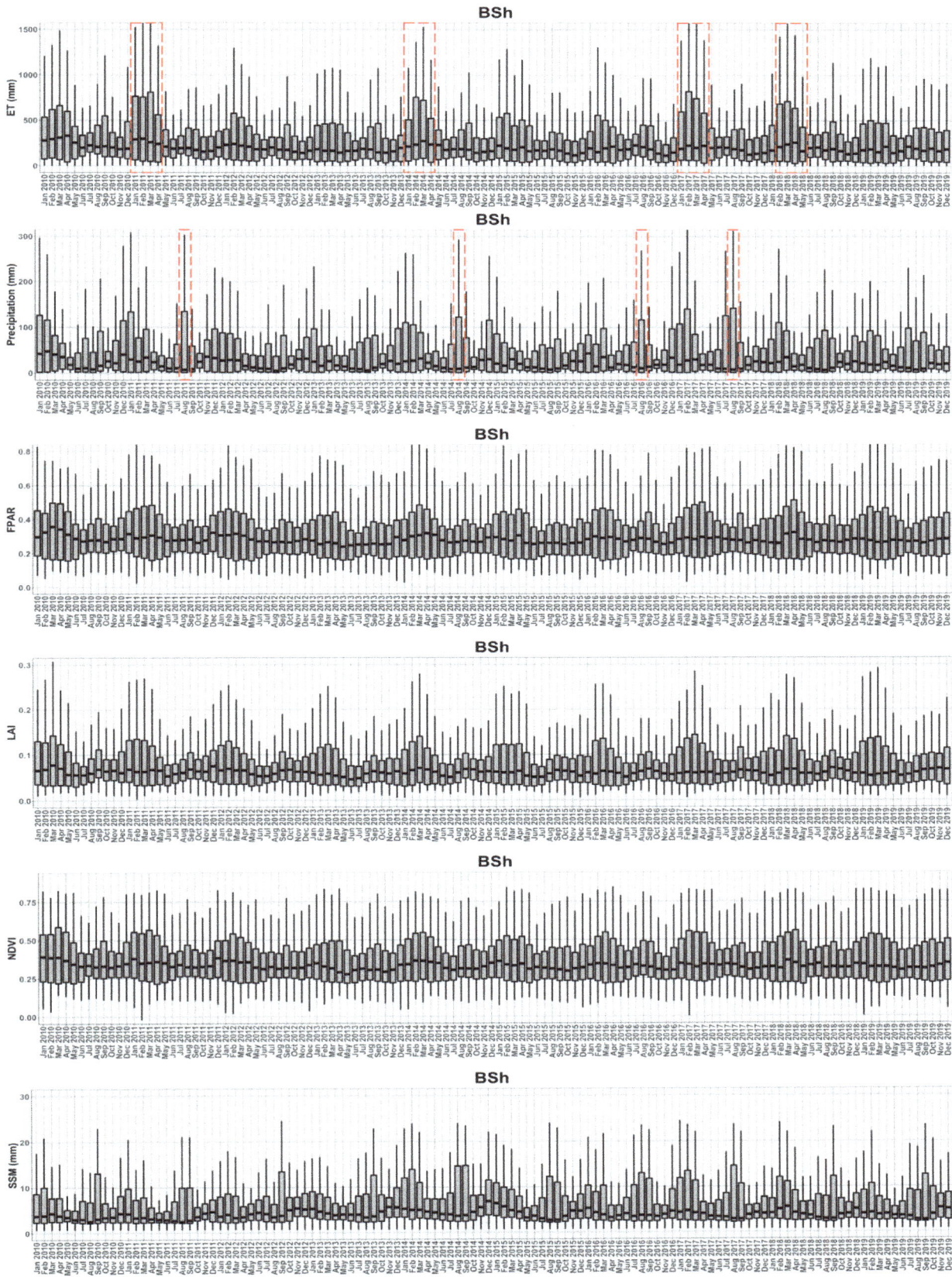

Figure 2. The time series boxplots of evapotranspiration (ET), precipitation, fraction of absorbed photosynthetically active radiation (FPAR), Leaf Area Index (LAI), Normalized Difference Vegetation Index (NDVI) and surface soil moisture (SSM) between 2010 and 2019 for BSh.

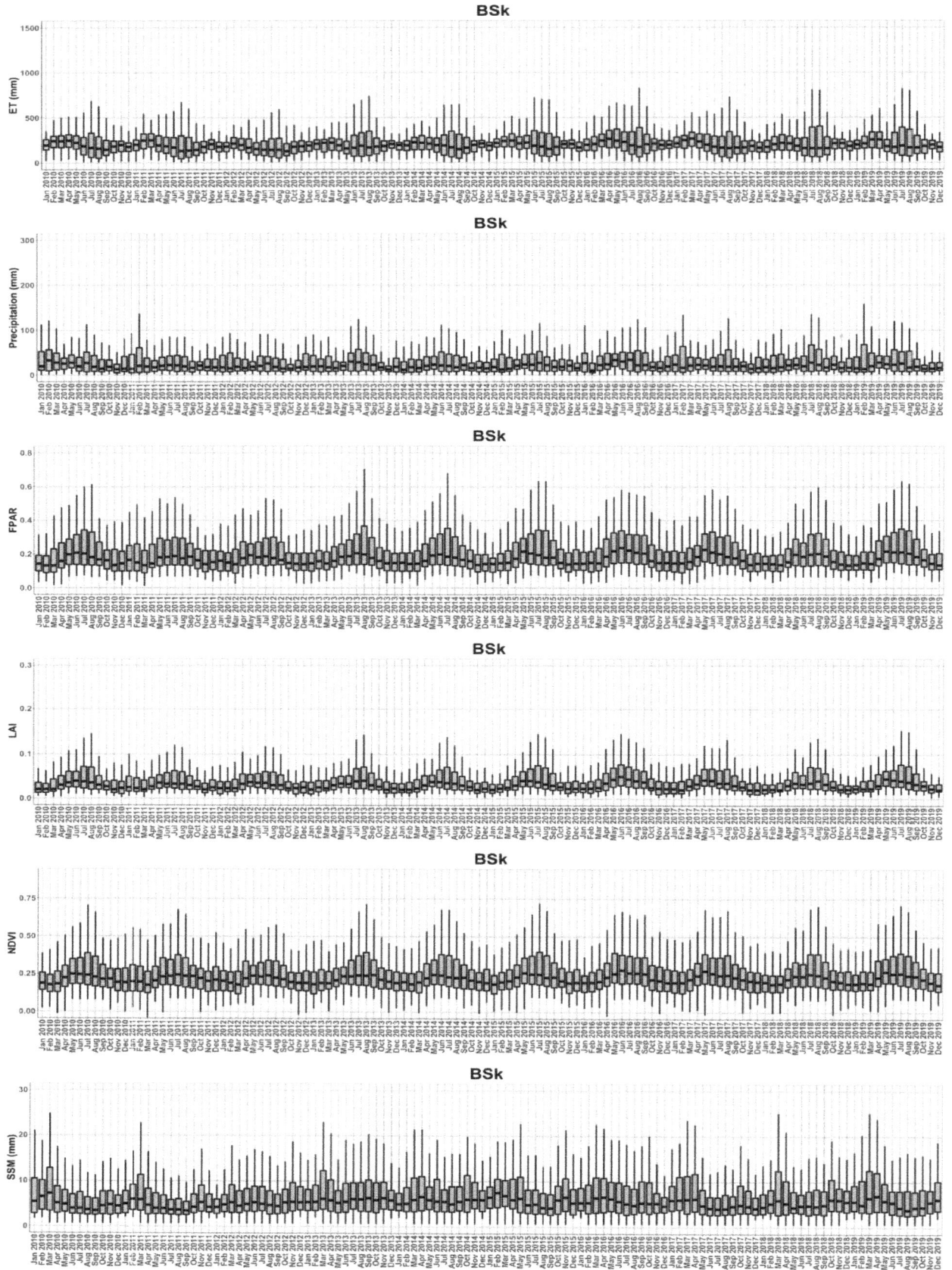

Figure 3. The time series boxplots of ET, precipitation, FPAR, LAI, NDVI and SSM between 2010 and 2019 for BSk.

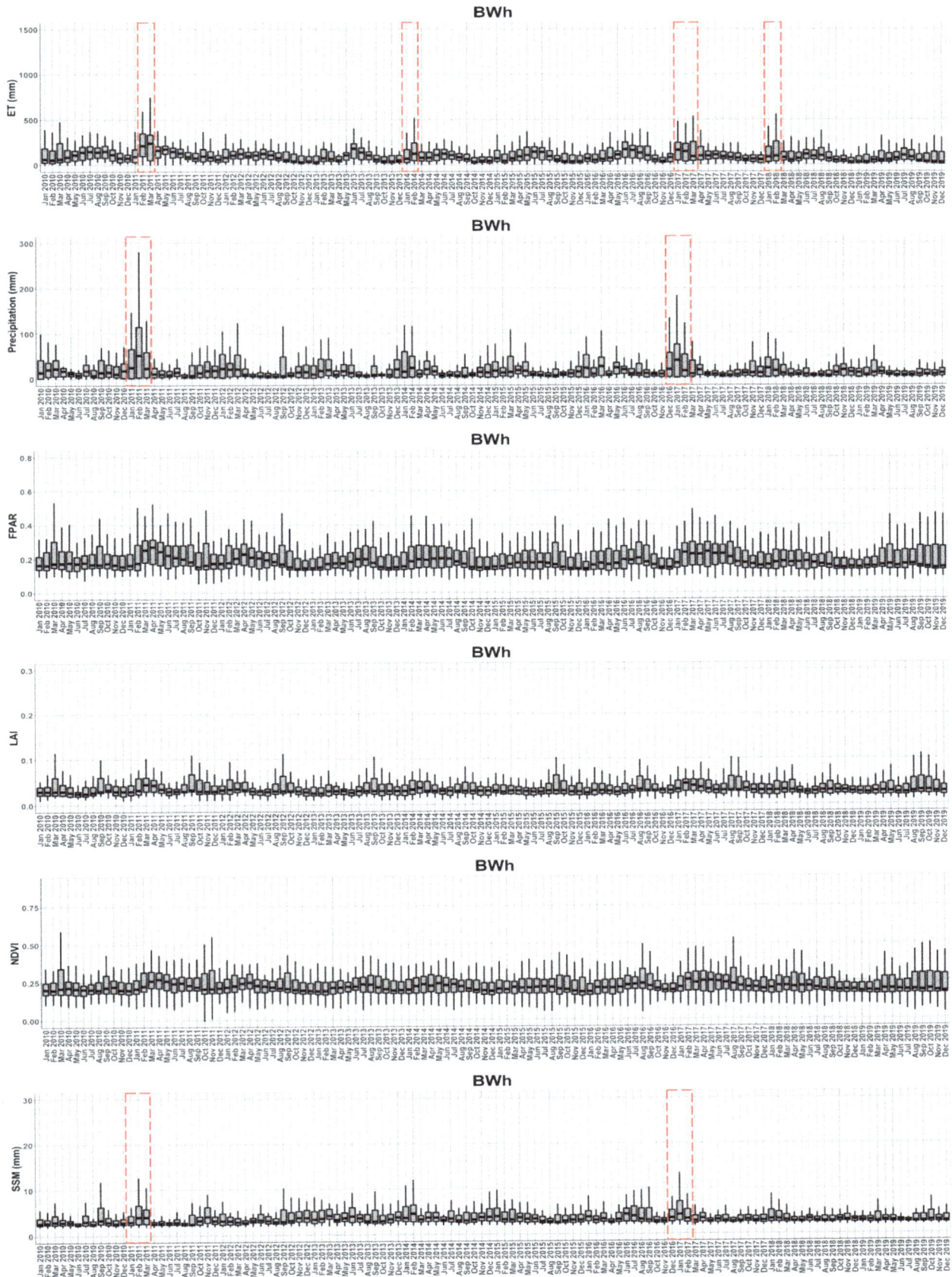

Figure 4. The time series boxplots of ET, precipitation, FPAR, LAI, NDVI and SSM between 2010 and 2019 for BWh.

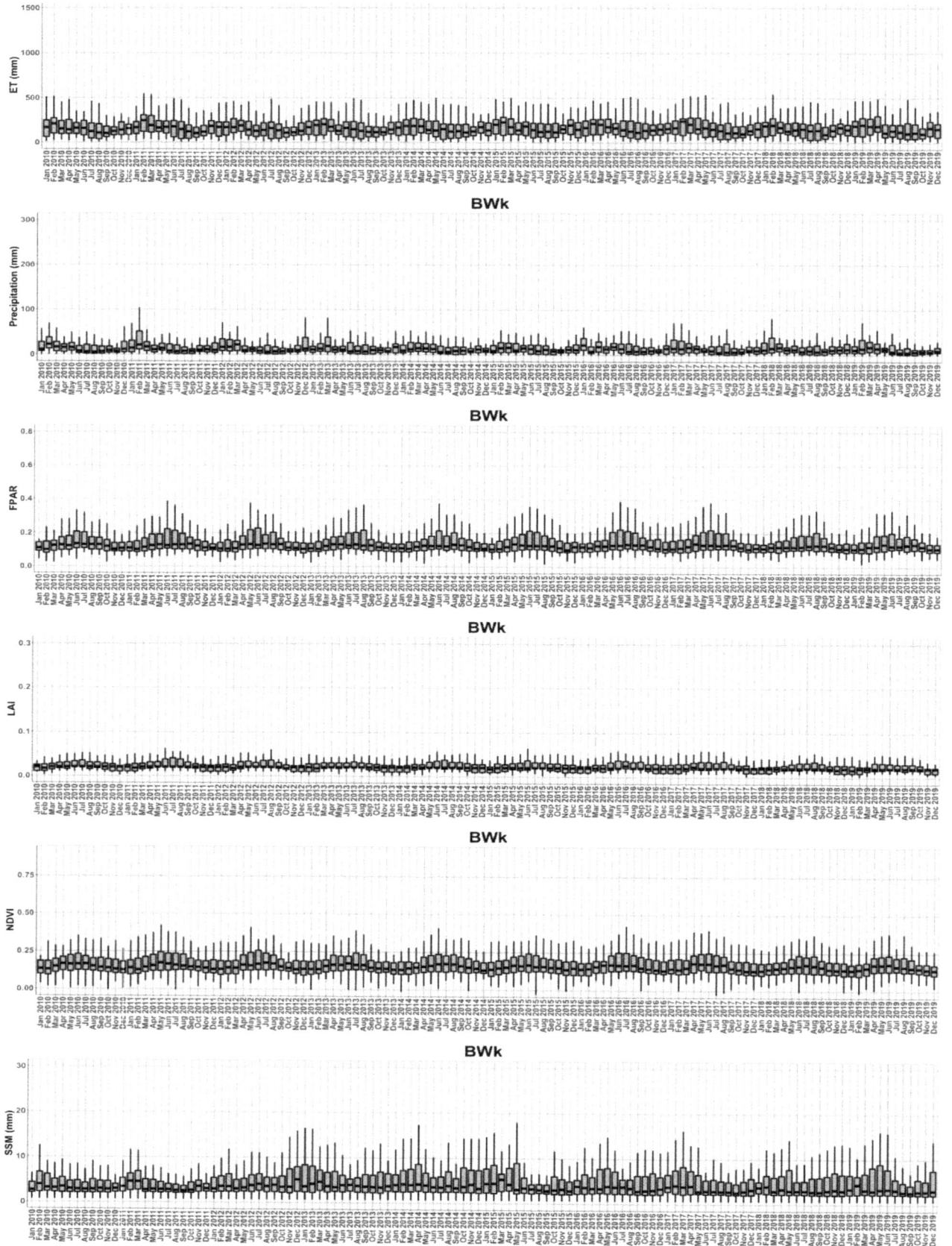

Figure 5. The time series boxplots of ET, precipitation, FPAR, LAI, NDVI and SSM between 2010 and 2019 for BWk.

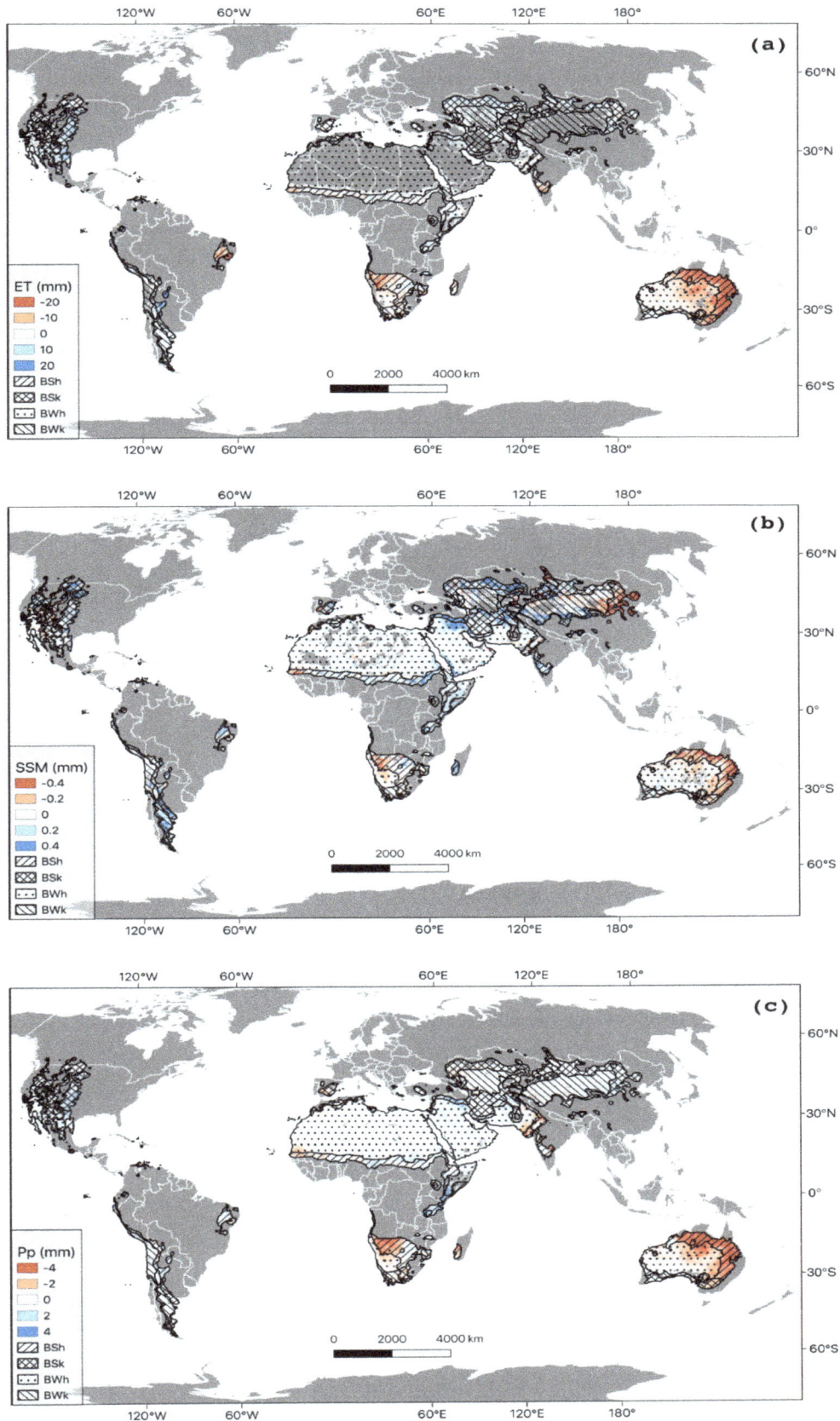

Figure 6. Yearly trends followed by multiple hydrological factors (**a**) evapotranspiration (ET), (**b**) surface soil moisture and (**c**) precipitation over the arid region are calculated based on harmonic analysis of the periods between the years 2010–2019. Only regions with a confidence level of 95% are shown.

3.3. The Landcover and Water Storage Changes

Figure 7 shows the land cover changes between 2010 and 2019 over BSh, BSk, BWh and BWk regions. Table 1 shows the pixel count of the areas that changed and the percentages for each type: BSh has the highest change of 10.57%, while BWh has the lowest change of 3.92%. It is noted that such changes are most plausible in four regions: (1) Region A: north of Syria and Iraq in the Mesopotamia area; (2) Region B: east of Ethiopia and Somalia in Eastern Africa; (3) middle and east of Australia; and (4) Region D: south of Argentina. The time series plots in Figure 8 illustrate the GRACE Water Equivalent Height values in each region with the pink areas covering the periods from 2010 to 2019, which indicate different types of water storage changes: (1) Region A had a steady decrease from 2010 to 2019 until recent recovery in 2020; (2) Region B has been increasing since 2010; (3) Region C decreased from the historical peak in 2011 and returned to the average level; (4) Region D has experienced a continuous decrease since 2002.

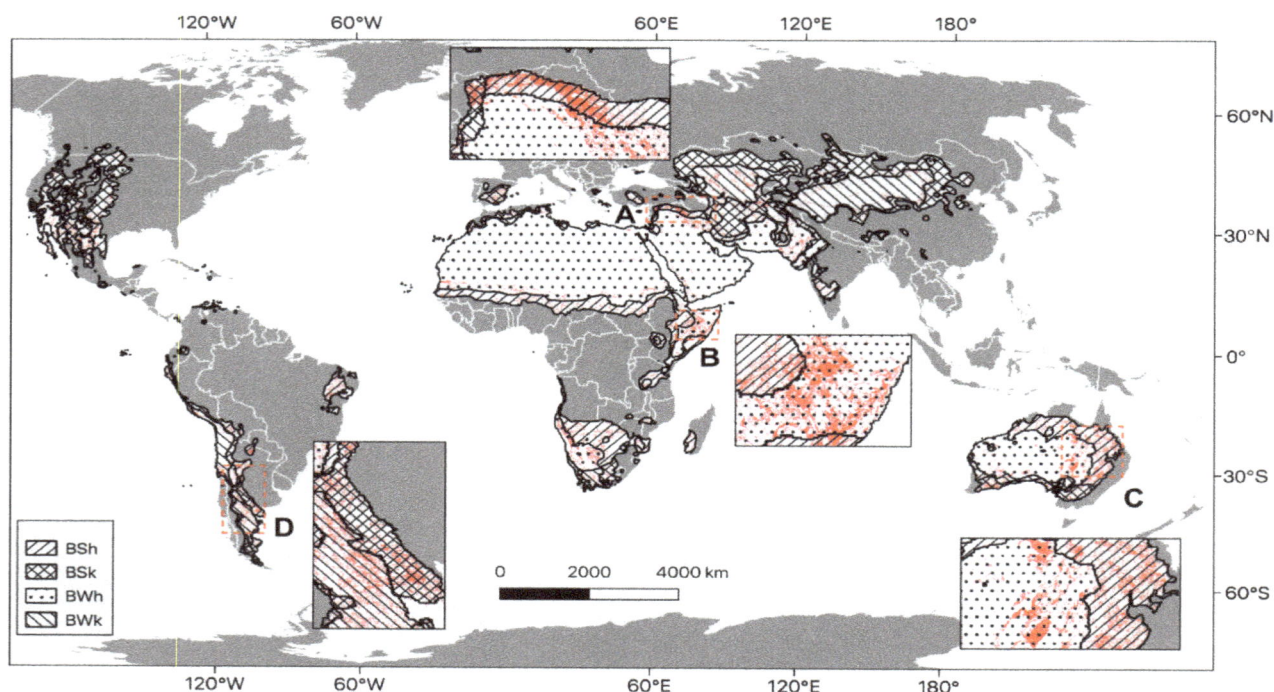

Figure 7. The land cover changes between 2010 and 2019, red color refers to the areas that has changed in terms of land cover types.

Table 1. The areas of changed landcover between 2010 and 2019 presented by pixel counts.

Climate Type	Not Changed Landcover (Pixel Count)	Changed Landcover (Pixel Count)	Changed Landcover (Percentage)
BSh	1,972,365	233,177	10.57%
BSk	2,043,856	154,848	7.04%
BWh	4,997,033	203,959	3.92%
BWk	1,392,325	89,749	6.06%
Total	10,405,579	681,733	6.15%

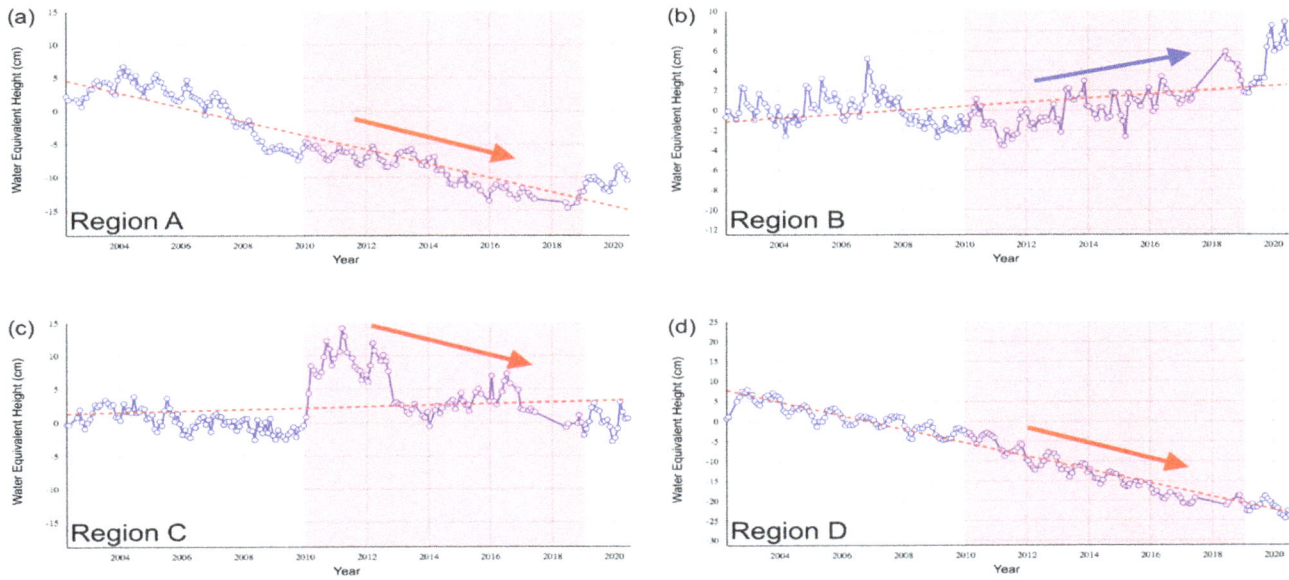

Figure 8. The GRACE Water Equivalent Height (cm) of time series for regions in Figure 7: (**a**) Region A; (**b**) Region B; (**c**) Region C and (**d**) Region D from 2002 to 2020, with pink areas covering 2010 to 2019. The blue dotted line are the observations; red dotted line is the regression line; the red/blue arrows indicate decreasing/increasing trends over the period of 2010–2019, respectively. Source: https://ccar.colorado.edu/grace/jpl.html.

3.4. The Comparison and Modeling of FLDAS and GLDAS ET and SSM Products

Figure 9 shows correlation maps between FLDAS and GLDAS ET (Figure 9a) and SSM (Figure 9b) products. It is noted that some regions (areas in China, Egypt, Saudi Arabia) are not shown due to a lower confidence value (<95%) in the statistical significance test. In general, FLDAS ET/SSM products have strong consistency with those of GLDAS. However, some BWh and BWk regions (North Africa, south of the Arabian Peninsula, west of China, west coast of South America) demonstrate less correlation, while some areas in Egypt even showing a negative correlation (small map in Figure 9b). Compared with Figure 1b, these regions are mostly covered by desert or have no biome data recorded.

Moreover, the comparisons between satellite-based products (NASA-USDA SSM and MODIS ET) and model products obtained from FLDAS and GLDAS over arid regions (BSh, BSk, BWh and BWk) are presented in Figure 10 (SSM) and Figure 11 (ET), respectively. It is noted that both models tend to overestimate SSM (linear regression coefficient <1) but underestimate ET (linear regression coefficient >1), which is indicated by the positive (SSM) and negative (ET) biases (ψ). The values of $|\psi|$ and ψ are nearly identical for all plots in Figure 10. The absolute values of ψ and $|\psi|$ are also almost identical in the cold areas (BSK and BWk), while different in the hot areas (BSh and BWh). This means model outputs of ET sometimes are estimated as being higher than ET observed from MODIS satellites in the hot areas but not in the cold areas. For SSM products (Figure 10), satellite SSM values fall into the range of around [0, 25], while ranges of [12, 48] and [0, 45] are set for FLDAS and GLDAS, respectively. For both FLDAS and GLDAS, the hot areas (BSh and BWh) have better consistency with NASA-USDA products than the cold areas (BSK and BWk). Compared with GLDAS, FLDAS achieves better consistency in the hot areas with higher R^2, yet is slightly worse in the cold areas. For ET products (Figure 11), the highest consistency is found in BSh, yet the lowest consistency in BWk. FLDAS has slightly better coherency with MODIS ET data than GLDAS in all regions.

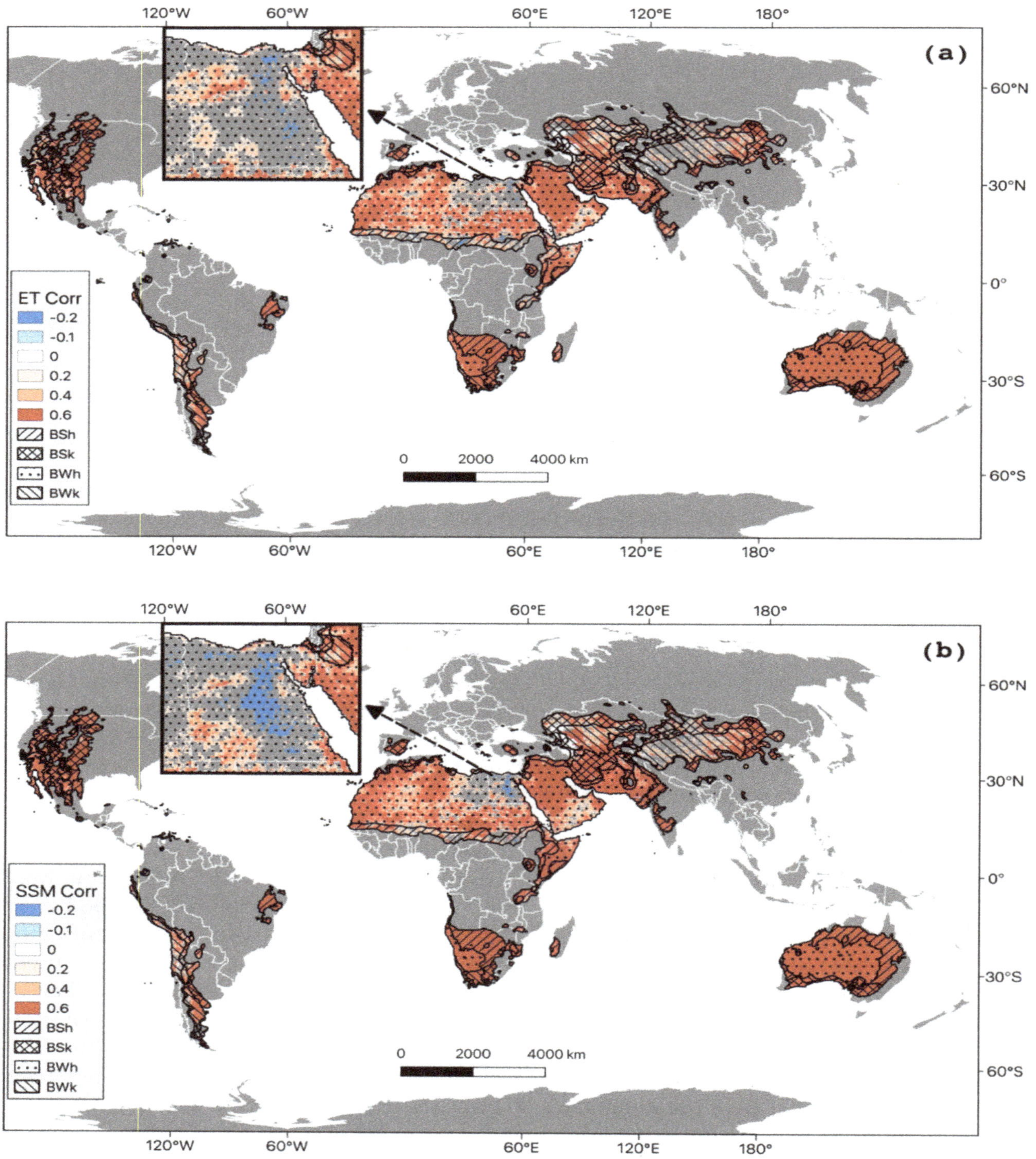

Figure 9. The correlation maps (correlation coefficient) between FLDAS and GLDAS in (**a**) ET and (**b**) SSM products calculated by monthly anomalies values during the years 2001–2019. Only regions with a confidence level of 95% are shown.

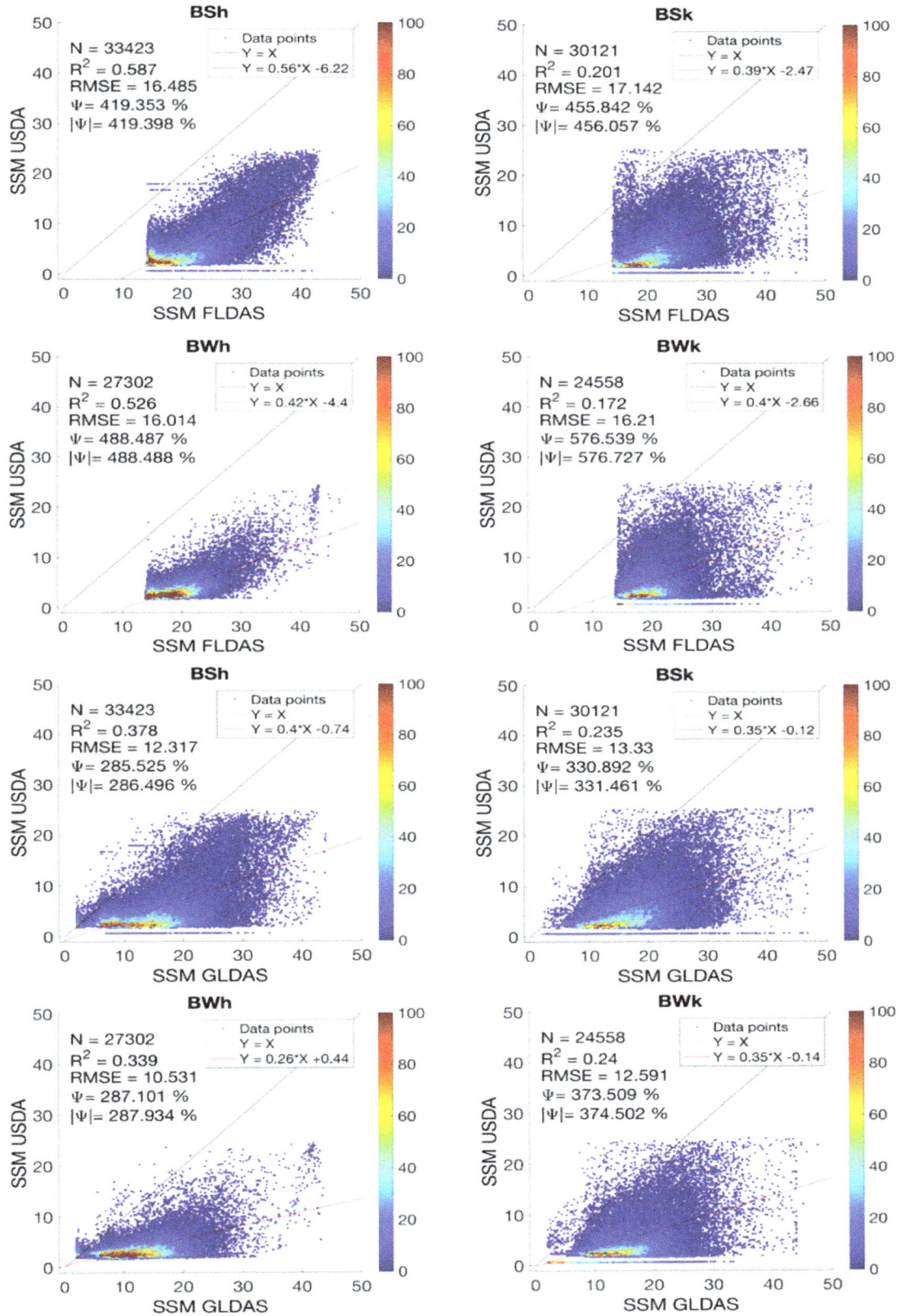

Figure 10. Comparisons of NASA-USDA SSM versus SSM values obtained from FLDAS and GLDAS for all types of arid regions. The solid black and red lines represent the one to one line and regression line, respectively.

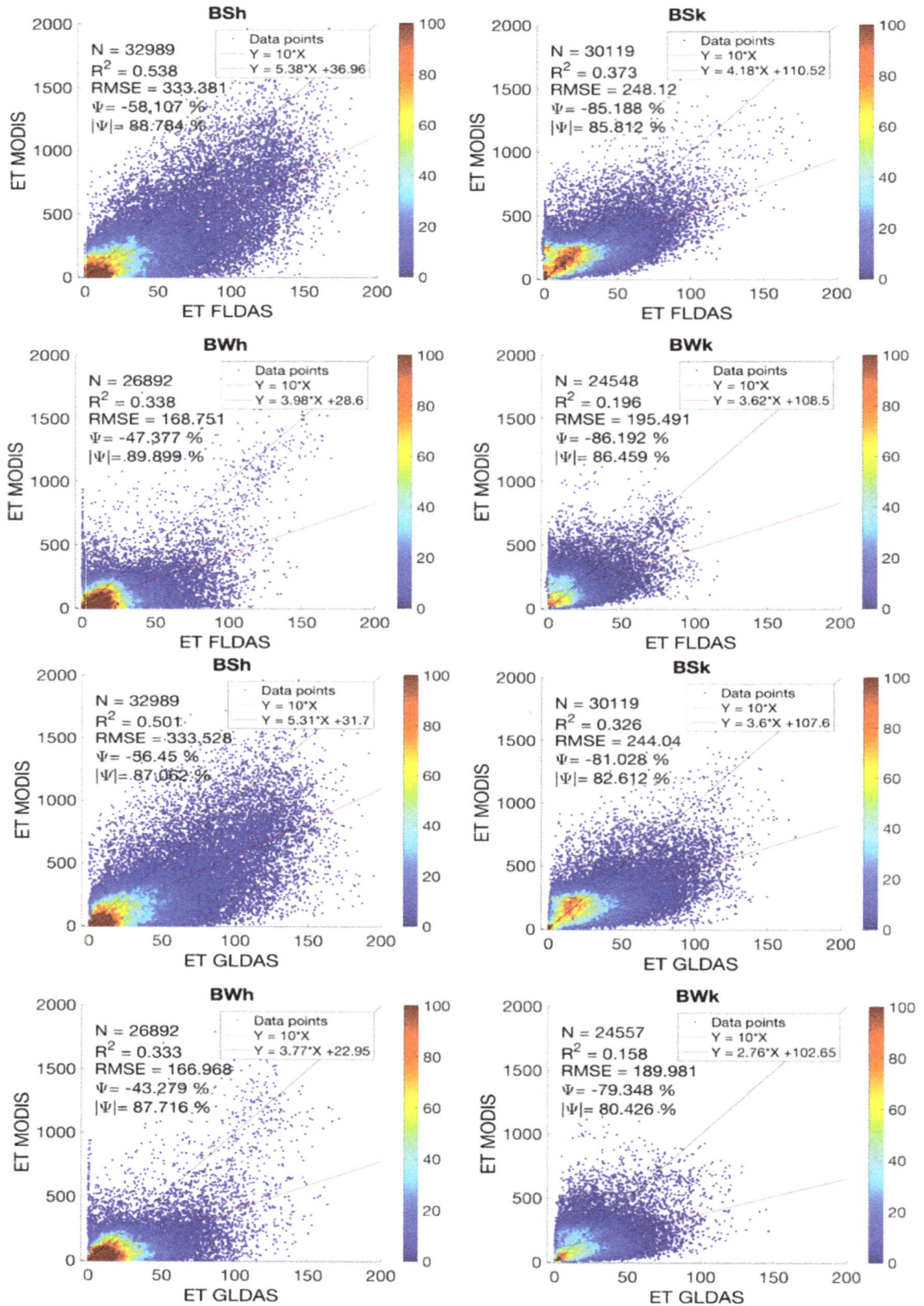

Figure 11. Comparisons of MODIS ET versus ET values obtained from FLDAS and GLDAS for all types of arid regions. The solid black and red lines represent the y = 10*x line and regression line, respectively.

As shown in Figure 9, strong disagreement between FLDAS and GLDAS products mainly exists in the Egypt territory. It is not decided which product has better performance due to a lack of ground observations. Therefore, the satellite-based dataset becomes a relatively reliable source of monitoring SSM and ET in this region. However, the satellite-based dataset is occasionally inaccessible due to atmospheric conditions (e.g., clouds) and sensor operating issues. Therefore, additional investigations are conducted in this region to compare FLDAS and GLDAS's coherency with satellite products to improve their performance through machine learning modeling. Scatterplots are applied to compare the satellite products with FLDAS, GLDAS and the proposed model. The results are presented in Figures 12 and 13 for SSM and ET, respectively. It is noted that both FLDAS and GLDAS have undesired results with near-zero R^2 and high RMSE, $|\psi|$ and ψ. The proposed models in this study remarkably improve the results with a very good correlation coefficient R^2 (>0.8) and lower RMSE and bias.

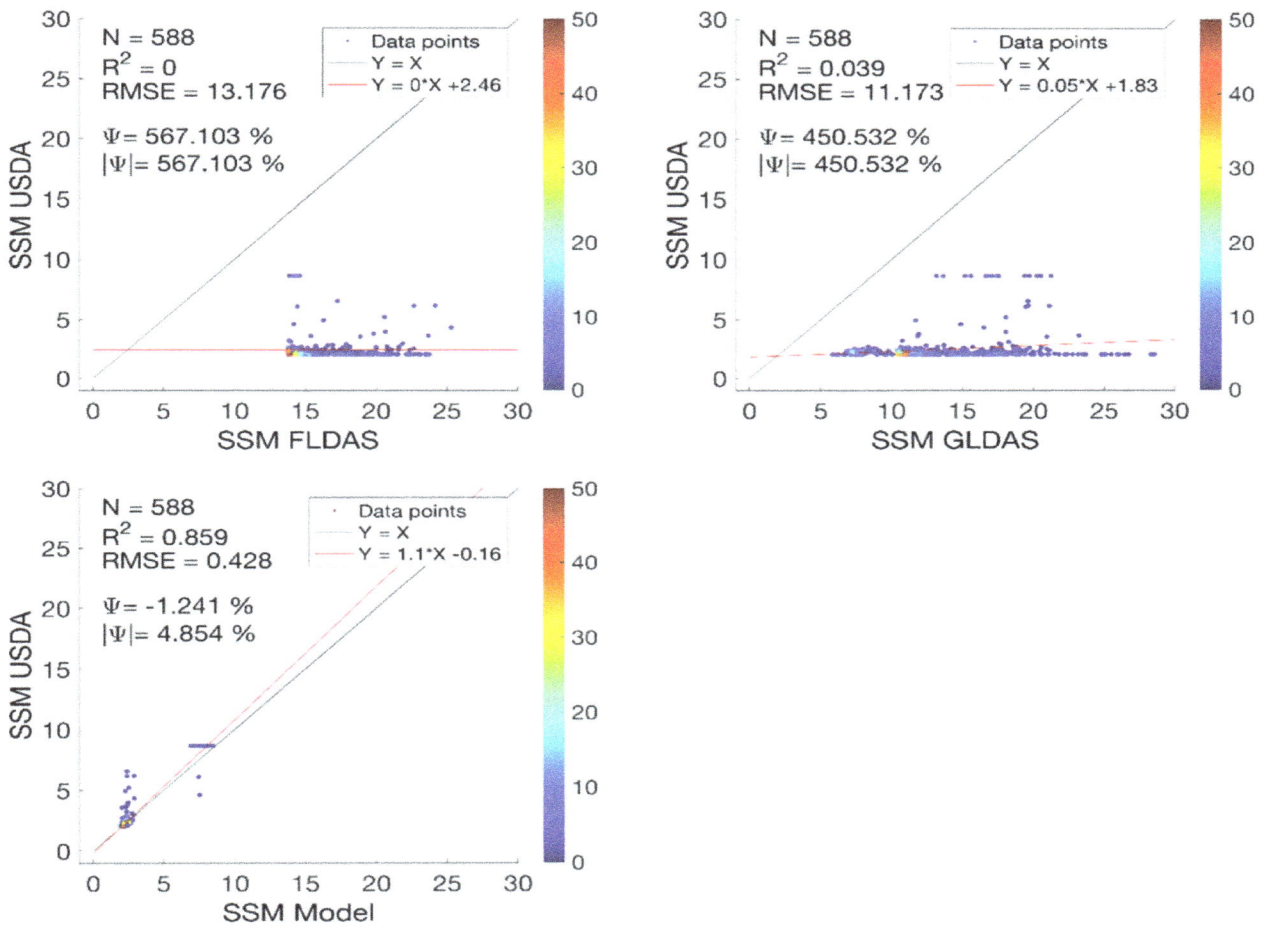

Figure 12. Comparisons of NASA-USDA SSM versus SSM values obtained from FLDAS, GLDAS and the proposed model using testing data in Egypt. The solid black and red lines represent the one to one line and regression line, respectively.

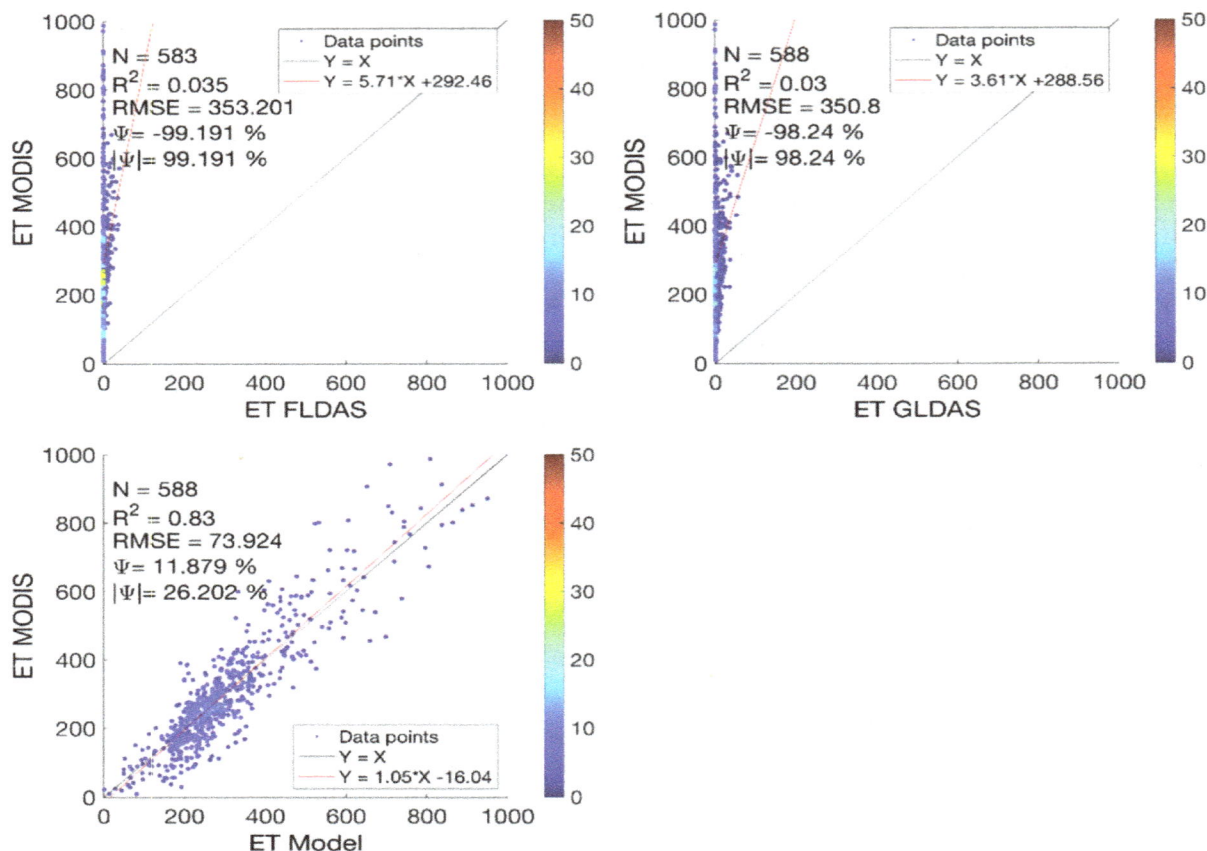

Figure 13. Comparisons of MODIS ET versus ET values obtained from FLDAS, GLDAS and the proposed model using testing data in Egypt. The solid black and red lines represent the one to one line and regression line, respectively.

4. Discussions

At local and regional scales, previous studies have been conducted on the connections between hydrological and environmental/climatic variations in arid and semi-arid regions. A study at China's Yarlung Zangbo River (YLZR) basin showed the negatively correlated relationship between the climatic indices, including Pacific decadal oscillation (PDO) and the multivariate El Niño-Southern Oscillation (ENSO) index (MEI)) and precipitation, annual average temperature, and the streamflow of selected hydrological stations [67]. Similarly, this negative relationship between precipitation and ENSO indices was also found for Jana and Karan Islands in the nearshore Saudi Arabia area [36]. The monthly precipitation datasets (the time period 1961–2007) were compared over the arid Balochistan province of Pakistan, showing an increase in drought with global warming, indicated by the time series analysis of gauge-based records and climate models [68]. The study in the Upper Omo-Ghibe river basin of Ethiopia showed the trends in precipitation, temperature, and streamflow in the period from 1981 to 2008. A detailed review of the studies in global-scale monthly hydroclimatic time series was reported [69]. Therefore, our satellite-based approach on the analysis of monthly hydroclimatic time series during the last decade (2010–2019) is an updated continuing study of the dryland hydrology at a global scale. However, one of the fundamental challenges of dryland remote sensing is the decoupling of the ecosystem function (e.g., vegetation photosynthetic ability and stomatal conductance) from optical reflectance [70]. During the drought periods of extreme moisture shortage, some vegetation types, especially drought-tolerant evergreen shrubs, can keep constant greenness even if their vegetation and soil function decrease. This situation can be found in the boxplots in Figures 2–5, where it is noted that the indices (FPAR, LAI and NDVI) do not have strong responses for

ET as compared to the precipitation pulses. Additionally, high aerosol concentrations, leaves covered by dust, marked adjacency effect, as well as surface anisotropy in sparse and heterogeneous canopies, make high-quality surface reflectance retrievals over drylands still a challenging process [71–73]. In arid and semi-arid ecosystems, the presence of soil background, standing litter and senesced vegetation, can have a significant impact on the reflectance and vegetation indices (e.g., NDVI), also adding considerable noise to LAI and the retrieval of FPAR [74–77]. The spurious changes induced by variable soil background reflectance also make it difficult to validate the actual vegetation change [76,78,79]. Therefore, the Soil-Adjusted Vegetation Index (SAVI) was introduced with a soil adjustment factor (L), typically set as L = 0.5 [75], to account for differences from the soil background in red and near-infrared reflectance. However, SAVI is limited to differentiate vegetation from soil under sparse vegetation cover conditions [79]. Prior knowledge is also required to select the optimal L value [75]. In addition, one of the major assumptions of global MODIS LAI and FPAR products is that each pixel contains only one biome type with constant within-biome soil and vegetation properties [80]. However, this simplified assumption is unlikely to hold in arid and semi-arid areas due to their heterogeneous characteristics.

Sustainable water resources management requires knowledge regarding the effects of land use and land cover change (LULCC) on groundwater recharge and surface runoff [81]. The tropical and sub-tropical semi-arid regions use groundwater to meet improving urban, industrial and agricultural water requirements [82]. Hence, a thorough understanding of these effects and the subsequent changes is required to maintain a sustainable water supply in the arid and semi-arid regions. In response to the growing food demand of an increasing human population, semi-arid regions are experiencing a major conversion of natural vegetation to agricultural land, leading to land exploitation due to agricultural practices [83]. Figure 7 shows the situation of land cover changes over drylands, where some hotspot regions were also experiencing LULCC along with groundwater variations. Some regions in Iraq depend on groundwater for sprinkler irrigation to overcome the issue of less water due to a shortage of rainfall. Before this, groundwater was one of the main sources of freshwater and for domestic uses. Because of this change, many government and private wells have been drilled to use for irrigation [84]. In Syria, during the conflict with ISIS, the dependence on groundwater for agriculture grew at an alarming rate, as ISIS controlled the Euphrates river dam [85]. In many parts of Ethiopia, such as the Shinile catchment found in the Somali regional state, the pastoral and semi pastoral communities travel a long distance in search of water and grazing land. It is largely an arid and semi-arid watershed with a large potential for groundwater which the communities are unaware of [86]. New South Wales in Australia has undergone transitions between pasture/scrubland, vineyard and built-up land [87]. These transitions due to tourism and economic development would affect the groundwater recharge in the region. In Argentina, there is evidence of an increase in groundwater recharge in areas where croplands replaced forests [88]. Some studies also show that land cover changes influence ecosystem water fluxes as well as salinity patterns [89,90]. It is noted that land cover and groundwater change during the later 2010s is possibly connected with global warming conditions between 2014 and 2019, which are among the warmest years on record, and the year 2020 is likely to become the second warmest year on record (https://www.ncdc.noaa.gov/cag/global/haywood/globe/land_ocean).

Both GLDAS and FLDAS do provide a suite of outputs over a global domain in order to provide physically consistent and spatially and temporally continuous products that are critical for global dryland studies. In particular, FLDAS has been recommended in hydroclimate studies and early warning applications over areas of Eastern, Southern, and Western Africa [57]. However, in Figure 9, the strong discrepancy between GLDAS and FLDAS for certain areas excludes North Africa from the recommendatory list. The analysis in Figures 10 and 11 shows the modeling biases in cold and dry areas from both FLDAS and GLDAS SSM/ET products, showing that they systematically overestimate the SSM and underestimate the ET, respectively. However, the higher SSM values from models may be due to their soil layer depth being configured as 0–10 cm, while the satellite generally detects the soil moisture at depths of 0-5cm. A study on the central Tibetan Plateau [91] suggested that the modeling biases were contributed by soil texture and soil organic carbon (SOC) content that alters soil thermal/hydraulic properties, which causes the significant underestimation of the surface soil moisture. The present study also suggested that products retrieved from the Advanced Microwave Scanning Radiometer-Earth Observing System (AMSR-E) evidently overestimated soil moisture, indicating that the satellite-based retrieval algorithms must be improved for the cold semi-arid regions. In arid and semi-arid ecosystems, it is still a theoretical and technical challenge to partition ecosystem scale ET fluxes between plant transpiration and soil/canopy evaporation [92]. Both numerical models and satellite products require ground-based data for calibration and validation. However, the observational data (chiefly from flux towers) are very limited, which results in inaccurate, poorly constrained estimates of energy, water, and carbon fluxes in the dryland to study its ecosystem's structural and functional dynamics.

In Figures 12 and 13, the proposed SSM/ET products by using machine-learning models demonstrate an appreciable improvement over the LDAS model. The previous studies applying similar automatic methodologies have been discussed [35,40,93], highlighting the importance of balancing the performance and explainability of the machine learning models for scientific studies. The machine learning models, along with numerical models based on physical processes (e.g., FLDAS), are the most important sources for hydrological studies in the dryland countries due to their limited scientific data and technological ability. For example, countries in the Nile River Basin (shared by Egypt, Sudan, Tanzania, Ethiopia, etc.) are undergoing accelerated population growth and urbanization [39]. The ongoing construction and filling of Ethiopia's Grand Ethiopian Renaissance Dam have brought up intensifying tensions between Ethiopia and downstream dryland countries including Egypt and Sudan. However, scientific studies are restricted to support the decision making and collaborations due to limited on-site observations, while the number of functional hydrometeorological gauging stations declined by 88% between the early 1960s and 2014 [94]. It is noted that our approach to the correction of the process-based hydrological models is also known as a process of hydrological post-processing to obtain "point" predictions. By contrast, new data-driven practices [95,96] of delivering "probabilistic" hydrological predictions (e.g., quantile regression [97]) were introduced recently with their potential in expressing complex hydrological processes considering predictive uncertainties [98].

Fortunately, new and emerging advances leverage significant opportunities for overcoming previous challenges in dryland remote sensing and thus revolutionize our ability to contextualize arid and semi-arid regions within the broader Earth system. Accordingly, the following strategies are recommended [99]: (1) exploring novel fusions of sensors and techniques across different spatiotemporal scales (e.g., solar-induced fluorescence, thermal, microwave, hyperspectral, and LiDAR) to understand dryland structural and functional dynamics; (2) capturing instant responses of dryland ecosystems to diurnal variation in water stress by utilizing near-continuous observations from geostationary satellites; (3) building up new ground observational networks for validation and calibration to better represent the heterogeneity of dryland system; (4) developing algorithms specifically for dryland ecosystems; (5) coupling physical process-based models with remote sensing observations using data assimilation to improve ecological forecasts and long-term projections.

5. Conclusions

In this paper, we investigated the hydrological conditions over global arid and semi-arid regions, including areas of BSh (Hot semi-arid climate), BSk (Cold semi-arid climate), BWh (Hot desert climate) and BWk (Cold desert climate), through synergistic approaches of remote sensing and modelling. The time series analysis during the last decade (2010–2019) of multiple hydrological variables (ET, SSM and precipitation) and vegetation indices (FPAR, LAI and NDVI) highlighted precipitation pulses that showed less impact on the vegetation indices compared to ET and SSM in some arid regions. Moreover, the trends of ET, SSM and precipitation, as well as land cover changes, that were demonstrated over the same period, emphasized drier trends in the south of Africa and the east of Australia and wetter trends in Mesopotamia and North America. Finally, the comparison of widely-used FLDAS and GLDAS models showed a general consistency, while a strong discrepancy was observed in the downstream Nile Basin, also demonstrating the fact that both LDAS models overestimate SSM and underestimate ET while model outputs of hot areas have better consistency with satellite products than cold areas.

This work further confirmed the scientific fact of decoupling between ecosystem function and observed optical characteristics (e.g., greenness indicated by vegetation indices), which calls for novel techniques and applications of remote sensing in the arid and semi-arid regions. Some areas (e.g., Mesopotamia, Eastern Africa, Australia and Argentina) experienced intense land cover changes along with noted water storage variations, indicating complex hydrological processes which need further investigations. Our data-driven modelling as a post-processing approach effectively improved the process-based models' performance and recommended that both process-based and satellite-based hydrological products have a higher potential to be improved for the arid and semi-arid regions. However, considering complex hydrological processes and predictions uncertainties, new practices of delivering data-driven probabilistic modelling are suggested for future studies. In addition, the study was fully deployed on the open-source cloud computing and analytical platforms (Google Earth Engine and BigML), which posed their great potential in other global-scale environmental studies.

Author Contributions: Conceptualization, H.E.-A. and W.L.; data analysis, W.L. and R.T.; funding acquisition, H.E.-A.; methodology, H.E.-A., W.L. and R.T.; validation, W.L. and R.T.; writing—original draft, H.E.-A., W.L.; writing—review and editing, H.E.-A., W.L., R.T., T.P., S.P.T., K.P.M. and D.S. All authors have read and agreed to the published version of the manuscript.

Acknowledgments: The first author acknowledge the support from the Earth Systems Science and Data Solutions (EssDs) Lab and Schmid College of Science and Technology, Chapman University.

References

1. Assessment, M.E. *Dryland Systems, Ecosystems and Human Well-Being: Current State and Trends*; Island Press: Washington, DC, USA, 2005.
2. Ahlstrom, A.; Raupach, M.R.; Schurgers, G.; Smith, B.; Arneth, A.; Jung, M.; Reichstein, M.; Canadell, J.G.; Friedlingstein, P.; Jain, A.K.; et al. The dominant role of semi-arid ecosystems in the trend and variability of the land CO_2 sink. *Science* **2015**, *348*, 895–899. [CrossRef]
3. Biederman, J.A.; Scott, R.L.; Bell, T.W.; Bowling, D.R.; Dore, S.; Garatuza-Payan, J.; Kolb, T.E.; Krishnan, P.; Krofcheck, D.J.; Litvak, M.E.; et al. CO_2 exchange and evapotranspiration across dryland ecosystems of southwestern North America. *Glob. Chang. Biol.* **2017**, *23*, 4204–4221. [CrossRef]
4. Humphrey, V.; Zscheischler, J.; Ciais, P.; Gudmundsson, L.; Sitch, S.; Seneviratne, S.I. Sensitivity of atmospheric CO_2 growth rate to observed changes in terrestrial water storage. *Nature* **2018**, *560*, 628–631. [CrossRef]
5. Poulter, B.; Frank, D.; Ciais, P.; Myneni, R.B.; Andela, N.; Bi, J.; Broquet, G.; Canadell, J.G.; Chevallier, F.; Liu, Y.Y.; et al. Contribution of semi-arid ecosystems to interannual variability of the global carbon cycle. *Nature* **2014**, *509*, 600–603. [CrossRef]
6. Huang, J.; Guan, X.; Ji, F. Enhanced cold-season warming in semi-arid regions. *Atmos. Chem. Phys.* **2012**, *12*, 5391–5398. [CrossRef]
7. Huang, J.; Yu, H.; Guan, X.; Wang, G.; Guo, R. Accelerated dryland expansion under climate change. *Nat. Clim. Chang.* **2016**, *6*, 166–171. [CrossRef]
8. Cayan, D.R.; Das, T.; Pierce, D.W.; Barnett, T.P.; Tyree, M.; Gershunov, A. Future dryness in the southwest US and the hydrology of the early 21st century drought. *Proc. Natl. Acad. Sci. USA* **2010**, *107*, 21271–21276. [CrossRef] [PubMed]
9. Cook, B.I.; Ault, T.R.; Smerdon, J.E. Unprecedented 21st century drought risk in the American Southwest and Central Plains. *Sci. Adv.* **2015**, *1*, e1400082. [CrossRef] [PubMed]
10. Huang, J.; Yu, H.; Dai, A.; Wei, Y.; Kang, L. Drylands face potential threat under 2 °C global warming target. *Nat. Clim. Chang.* **2017**, *7*, 417–422. [CrossRef]
11. D'Odorico, P.; Porporato, A. (Eds.) *Dryland Ecohydrology*; Kluwer Academic Publishers: Dordrecht, The Netherlands, 2006; ISBN 978-1-4020-4259-1.

12. Scott, R.L.; Huxman, T.E.; Barron-Gafford, G.A.; Darrel Jenerette, G.; Young, J.M.; Hamerlynck, E.P. When vegetation change alters ecosystem water availability. *Glob. Chang. Biol.* **2014**, *20*, 2198–2210. [CrossRef] [PubMed]

13. Noy-Meir, I. Desert Ecosystems: Environment and Producers. *Annu. Rev. Ecol. Syst.* **1973**, *4*, 25–51. [CrossRef]

14. Reynolds, J.F.; Smith, D.M.S.; Lambin, E.F.; Turner, B.L.; Mortimore, M.; Batterbury, S.P.J.; Downing, T.E.; Dowlatabadi, H.; Fernandez, R.J.; Herrick, J.E.; et al. Global Desertification: Building a Science for Dryland Development. *Science* **2007**, *316*, 847–851. [CrossRef]

15. Nagler, P.L.; Nguyen, U.; Bateman, H.L.; Jarchow, C.J.; Glenn, E.P.; Waugh, W.J.; van Riper, C. Northern tamarisk beetle (*Diorhabda carinulata*) and tamarisk (*Tamarix* spp.) interactions in the Colorado River basin: Northern tamarisk beetle and tamarisk interactions. *Restor. Ecol.* **2018**, *26*, 348–359. [CrossRef]

16. Nagler, P.L.; Jarchow, C.J.; Glenn, E.P. Remote sensing vegetation index methods to evaluate changes in greenness and evapotranspiration in riparian vegetation in response to the Minute 319 environmental pulse flow to Mexico. *Proc. IAHS* **2018**, *380*, 45–54. [CrossRef]

17. Fisher, J.B.; Melton, F.; Middleton, E.; Hain, C.; Anderson, M.; Allen, R.; McCabe, M.F.; Hook, S.; Baldocchi, D.; Townsend, P.A.; et al. The future of evapotranspiration: Global requirements for ecosystem functioning, carbon and climate feedbacks, agricultural management, and water resources: The Future of Evapotranspiration. *Water Resour. Res.* **2017**, *53*, 2618–2626. [CrossRef]

18. Baldocchi, D.; Falge, E.; Gu, L.; Olson, R.; Hollinger, D.; Running, S.; Anthoni, P.; Bernhofer, C.; Davis, K.; Evans, R.; et al. FLUXNET: A new tool to study the temporal and spatial variability of ecosystem-scale carbon dioxide, water vapor, and energy flux densities. *Bull. Am. Meteorol. Soc.* **2001**, *82*, 2415–2434. [CrossRef]

19. Novick, K.A.; Biederman, J.A.; Desai, A.R.; Litvak, M.E.; Moore, D.J.P.; Scott, R.L.; Torn, M.S. The AmeriFlux network: A coalition of the willing. *Agric. For. Meteorol.* **2018**, *249*, 444–456. [CrossRef]

20. Mirtl, M.; Borer, E.T.; Djukic, I.; Forsius, M.; Haubold, H.; Hugo, W.; Jourdan, J.; Lindenmayer, D.; McDowell, W.H.; Muraoka, H.; et al. Genesis, goals and achievements of Long-Term Ecological Research at the global scale: A critical review of ILTER and future directions. *Sci. Total Environ.* **2018**, *626*, 1439–1462. [CrossRef]

21. Schimel, D.; Hargrove, W.; Hoffman, F.; MacMahon, J. NEON: A hierarchically designed national ecological network. *Front. Ecol. Environ.* **2007**, *5*, 59. [CrossRef]

22. Kowalik, W.S.; Marsh, S.E.; Lyon, R.J.P. A relation between landsat digital numbers, surface reflectance, and the cosine of the solar zenith angle. *Remote Sens. Environ.* **1982**, *12*, 39–55. [CrossRef]

23. Marsh, S.E.; Lyon, R.J.P. Quantitative relationships of near-surface spectra to Landsat radiometric data. *Remote Sens. Environ.* **1980**, *10*, 241–261. [CrossRef]

24. Rouse, J., Jr. *Contractor Report (CR): Monitoring the Vernal Advancement and Retrogradation (Green Wave Effect) of Natural Vegetation*; NTRS—NASA Technical Reports Server: Washington, DC, USA, 1974.

25. Tucker, C.J. Red and photographic infrared linear combinations for monitoring vegetation. *Remote Sens. Environ.* **1979**, *8*, 127–150. [CrossRef]

26. Anyamba, A.; Tucker, C.J. Analysis of Sahelian vegetation dynamics using NOAA-AVHRR NDVI data from 1981–2003. *J. Arid Environ.* **2005**, *63*, 596–614. [CrossRef]

27. Donohue, R.J.; McVicar, T.R.; Roderick, M.L. Climate-related trends in Australian vegetation cover as inferred from satellite observations, 1981–2006. *Glob. Chang. Biol.* **2009**, *15*, 1025–1039. [CrossRef]

28. Fensholt, R.; Rasmussen, K. Analysis of trends in the Sahelian 'rain-use efficiency' using GIMMS NDVI, RFE and GPCP rainfall data. *Remote Sens. Environ.* **2011**, *115*, 438–451. [CrossRef]

29. Donohue, R.J.; Roderick, M.L.; McVicar, T.R.; Farquhar, G.D. Impact of CO_2 fertilization on maximum foliage cover across the globe's warm, arid environments: CO_2 Fertilization and Foliage Cover. *Geophys. Res. Lett.* **2013**, *40*, 3031–3035. [CrossRef]

30. Fensholt, R.; Langanke, T.; Rasmussen, K.; Reenberg, A.; Prince, S.D.; Tucker, C.; Scholes, R.J.; Le, Q.B.; Bondeau, A.; Eastman, R.; et al. Greenness in semi-arid areas across the globe 1981–2007—An Earth Observing Satellite based analysis of trends and drivers. *Remote Sens. Environ.* **2012**, *121*, 144–158. [CrossRef]

31. Helldén, U.; Tottrup, C. Regional desertification: A global synthesis. *Glob. Planet. Chang.* **2008**, *64*, 169–176. [CrossRef]

32. Kolby Smith, W.; Reed, S.C.; Cleveland, C.C.; Ballantyne, A.P.; Anderegg, W.R.L.; Wieder, W.R.; Liu, Y.Y.; Running, S.W. Large divergence of satellite and Earth system model estimates of global terrestrial CO_2 fertilization. *Nat. Clim. Chang.* **2016**, *6*, 306–310. [CrossRef]

33. Zhu, Z.; Piao, S.; Myneni, R.B.; Huang, M.; Zeng, Z.; Canadell, J.G.; Ciais, P.; Sitch, S.; Friedlingstein, P.; Arneth, A.; et al. Greening of the Earth and its drivers. *Nat. Clim. Chang.* **2016**, *6*, 791–795. [CrossRef]

34. Gorelick, N.; Hancher, M.; Dixon, M.; Ilyushchenko, S.; Thau, D.; Moore, R. Google Earth Engine: Planetary-scale geospatial analysis for everyone. *Remote Sens. Environ.* **2017**, *202*, 18–27. [CrossRef]

35. El-Nadry, M.; Li, W.; El-Askary, H.; Awad, M.A.; Mostafa, A.R. Urban Health Related Air Quality Indicators over the Middle East and North Africa Countries Using Multiple Satellites and AERONET Data. *Remote Sens.* **2019**, *11*, 2096. [CrossRef]

36. Maneja, R.H.; Miller, J.D.; Li, W.; El-Askary, H.; Flandez, A.V.B.; Dagoy, J.J.; Alcaria, J.F.A.; Basali, A.U.; Al-Abdulkader, K.A.; Loughland, R.A.; et al. Long-term NDVI and recent vegetation cover profiles of major offshore island nesting sites of sea turtles in Saudi waters of the northern Arabian Gulf. *Ecol. Indic.* **2020**, *117*, 106612. [CrossRef]

37. Li, W.; El-Askary, H.; Qurban, M.A.; Li, J.; ManiKandan, K.P.; Piechota, T. Using multi-indices approach to quantify mangrove changes over the Western Arabian Gulf along Saudi Arabia coast. *Ecol. Indic.* **2019**, *102*, 734–745. [CrossRef]

38. Li, W.; El-Askary, H.M.; Qurban, M.; Allali, M.; Manikandan, K. On the drying trends over the MENA countries using harmonic analysis of the enhanced vegetation index. In *Advances in Remote Sensing and Geo Informatics Applications*; El-Askary, H.M., Lee, S., Heggy, E., Pradhan, B., Eds.; Springer: Cham, Switzerland, 2019; pp. 243–245. ISBN 978-3-030-01439-1.

39. Li, W.; El-Askary, H.; Lakshmi, V.; Piechota, T.; Struppa, D. Earth Observation and Cloud Computing in Support of Two Sustainable Development Goals for the River Nile Watershed Countries. *Remote Sens.* **2020**, *12*, 1391. [CrossRef]

40. Li, W.; Ali, E.; Abou El-Magd, I.; Mourad, M.M.; El-Askary, H. Studying the Impact on Urban Health over the Greater Delta Region in Egypt Due to Aerosol Variability Using Optical Characteristics from Satellite Observations and Ground-Based aeronet Measurements. *Remote Sens.* **2019**, *11*, 1998. [CrossRef]

41. Bastin, J.-F.; Berrahmouni, N.; Grainger, A.; Maniatis, D.; Mollicone, D.; Moore, R.; Patriarca, C.; Picard, N.; Sparrow, B.; Abraham, E.M.; et al. The extent of forest in dryland biomes. *Science* **2017**, *356*, 635–638. [CrossRef]

42. Beck, H.E.; Zimmermann, N.E.; McVicar, T.R.; Vergopolan, N.; Berg, A.; Wood, E.F. Present and future Köppen-Geiger climate classification maps at 1-km resolution. *Sci. Data* **2018**, *5*, 180214. [CrossRef]

43. Hengl, T.; Walsh, M.G.; Sanderman, J.; Wheeler, I.; Harrison, S.P.; Prentice, I.C. Global mapping of potential natural vegetation: An assessment of Machine Learning algorithms for estimating land potential. *PeerJ* **2018**, *6*, e5457. [CrossRef]

44. Funk, C.; Peterson, P.; Landsfeld, M.; Pedreros, D.; Verdin, J.; Shukla, S.; Husak, G.; Rowland, J.; Harrison, L.; Hoell, A.; et al. The climate hazards infrared precipitation with stations—A new environmental record for monitoring extremes. *Sci. Data* **2015**, *2*, 150066. [CrossRef]

45. Bolten, J.D.; Crow, W.T.; Zhan, X.; Jackson, T.J.; Reynolds, C.A. Evaluating the Utility of Remotely Sensed Soil Moisture Retrievals for Operational Agricultural Drought Monitoring. *IEEE J. Sel. Top. Appl. Earth Obs. Remote Sens.* **2010**, *3*, 57–66. [CrossRef]

46. Mladenova, I.E.; Bolten, J.D.; Crow, W.T.; Anderson, M.C.; Hain, C.R.; Johnson, D.M.; Mueller, R. Intercomparison of Soil Moisture, Evaporative Stress, and Vegetation Indices for Estimating Corn and Soybean Yields Over the U.S. *IEEE J. Sel. Top. Appl. Earth Obs. Remote Sens.* **2017**, *10*, 1328–1343. [CrossRef]

47. Mohammed, I.; Bolten, J.; Srinivasan, R.; Lakshmi, V. Improved Hydrological Decision Support System for the Lower Mekong River Basin Using Satellite-Based Earth Observations. *Remote Sens.* **2018**, *10*, 885. [CrossRef] [PubMed]

48. Kerr, Y.H.; Levine, D. Foreword to the Special Issue on the Soil Moisture and Ocean Salinity (SMOS) Mission. *IEEE Trans. Geosci. Remote Sens.* **2008**, *46*, 583–585. [CrossRef]

49. Landerer, F.W.; Swenson, S.C. Accuracy of scaled GRACE terrestrial water storage estimates: Accuracy of GRACE-TWS. *Water Resour. Res.* **2012**, *48*. [CrossRef]

50. Swenson, S.; Wahr, J. Post-processing removal of correlated errors in GRACE data. *Geophys. Res. Lett.* **2006**, *33*, L08402. [CrossRef]

51. Swenson, S. *GRACE Monthly Land Water Mass Grids Netcdf Release 5.0*; PO.DAAC: Pasadena, CA, USA, 2012.

52. Mu, Q.; Zhao, M.; Running, S.W. *MODIS Global Terrestrial Evapotranspiration (ET) Product (NASA MOD16A2/A3) Collection 5*. NASA Headquarters; Report; Numerical Terradynamic Simulation Group Publications: Missoula, MT, USA, 2013.

53. Didan, K. *MOD13A2 MODIS/Terra Vegetation Indices 16-Day L3 Global 1 km SIN Grid V006*; NASA LP DAAC: Washington, DC, USA, 2015. [CrossRef]

54. Myneni, R. *MOD15A2H MODIS/Terra Leaf Area Index/FPAR 8-Day L4 Global 500 m SIN Grid V006*; NASA LP DAAC: Washington, DC, USA, 2015. [CrossRef]

55. Chen, J.M.; Black, T.A. Defining leaf area index for non-flat leaves. *Plant Cell Environ.* **1992**, *15*, 421–429. [CrossRef]

56. Fensholt, R.; Sandholt, I.; Rasmussen, M.S. Evaluation of MODIS LAI, fAPAR and the relation between fAPAR and NDVI in a semi-arid environment using in situ measurements. *Remote Sens. Environ.* **2004**, *91*, 490–507. [CrossRef]

57. McNally, A.; Arsenault, K.; Kumar, S.; Shukla, S.; Peterson, P.; Wang, S.; Funk, C.; Peters-Lidard, C.D.; Verdin, J.P. A land data assimilation system for sub-Saharan Africa food and water security applications. *Sci. Data* **2017**, *4*, 170012. [CrossRef]

58. Derber, J.C.; Parrish, D.F.; Lord, S.J. The new global operational analysis system at the National Meteorological Center. *Weather Forecast.* **1991**, *6*, 538–547. [CrossRef]

59. Adler, R.F.; Huffman, G.J.; Chang, A.; Ferraro, R.; Xie, P.-P.; Janowiak, J.; Rudolf, B.; Schneider, U.; Curtis, S.; Bolvin, D.; et al. The version-2 global precipitation climatology project (GPCP) monthly precipitation analysis (1979–present). *J. Hydrometeorol.* **2003**, *4*, 1147–1167. [CrossRef]

60. NASA GSFC Hydrological Sciences Laboratory (HSL). *FLDAS Noah Land Surface Model L4 Global Monthly 0.1 × 0.1 Degree (MERRA-2 and CHIRPS) V001*; Amy McNally NASA/GSFC/HSL: Greenbelt, MD, USA, 2018.

61. Xie, P.; Arkin, P.A. Global precipitation: A 17-year monthly analysis based on gauge observations, satellite estimates, and numerical model outputs. *Bull. Am. Meteorol. Soc.* **1997**, *78*, 2539–2558. [CrossRef]

62. Sheffield, J.; Goteti, G.; Wood, E.F. Development of a 50-Year High-Resolution Global Dataset of Meteorological Forcings for Land Surface Modeling. *J. Clim.* **2006**, *19*, 3088–3111. [CrossRef]

63. Li, W.; El-Askary, H.; Qurban, M.; Proestakis, E.; Garay, M.; Kalashnikova, O.; Amiridis, V.; Gkikas, A.; Marinou, E.; Piechota, T.; et al. An Assessment of Atmospheric and Meteorological Factors Regulating Red Sea Phytoplankton Growth. *Remote Sens.* **2018**, *10*, 673. [CrossRef]

64. Li, W.; El-Askary, H.; ManiKandan, K.; Qurban, M.; Garay, M.; Kalashnikova, O. Synergistic Use of Remote Sensing and Modeling to Assess an Anomalously High Chlorophyll-a Event during Summer 2015 in the South Central Red Sea. *Remote Sens.* **2017**, *9*, 778. [CrossRef]

65. Xu, Q.-S.; Liang, Y.-Z. Monte Carlo cross validation. *Chemom. Intell. Lab. Syst.* **2001**, *56*, 1–11. [CrossRef]

66. Tiwari, S.P.; Sarma, Y.V.B.; Kurten, B.; Ouhssain, M.; Jones, B.H. An Optical Algorithm to Estimate Downwelling Diffuse Attenuation Coefficient in the Red Sea. *IEEE Trans. Geosci. Remote Sens.* **2018**, *56*, 7174–7182. [CrossRef]

67. Zhang, R.; Xu, Z.; Zuo, D.; Ban, C. Hydro-Meteorological Trends in the Yarlung Zangbo River Basin and Possible Associations with Large-Scale Circulation. *Water* **2020**, *12*, 144. [CrossRef]

68. Ahmed, K.; Shahid, S.; Wang, X.; Nawaz, N.; Najeebullah, K. Evaluation of Gridded Precipitation Datasets over Arid Regions of Pakistan. *Water* **2019**, *11*, 210. [CrossRef]

69. Papacharalampous, G.; Tyralis, H.; Papalexiou, S.M.; Langousis, A.; Khatami, S.; Volpi, E.; Grimaldi, S. Global-scale massive feature extraction from monthly hydroclimatic time series: Statistical characterizations, spatial patterns and hydrological similarity. *arXiv* **2020**, arXiv:2010.12833.

70. Yan, D.; Scott, R.L.; Moore, D.J.P.; Biederman, J.A.; Smith, W.K. Understanding the relationship between vegetation greenness and productivity across dryland ecosystems through the integration of PhenoCam, satellite, and eddy covariance data. *Remote Sens. Environ.* **2019**, *223*, 50–62. [CrossRef]

71. Vanleeuwen, W.; Huete, A.; Duncan, J.; Franklin, J. Radiative transfer in shrub savanna sites in Niger: Preliminary results from HAPEX-Sahel. 3. Optical dynamics and vegetation index sensitivity to biomass and plant cover. *Agric. For. Meteorol.* **1994**, *69*, 267–288. [CrossRef]

72. Houborg, R.; McCabe, M.F. Adapting a regularized canopy reflectance model (REGFLEC) for the retrieval challenges of dryland agricultural systems. *Remote Sens. Environ.* **2016**, *186*, 105–120. [CrossRef]

73. Middleton, E.M.; Deering, D.W.; Ahmad, S.P. Surface anisotropy and hemispheric reflectance for a semiarid ecosystem. *Remote Sens. Environ.* **1987**, *23*, 193–212. [CrossRef]

74. van Leeuwen, W.J.D.; Huete, A.R. Effects of standing litter on the biophysical interpretation of plant canopies with spectral indices. *Remote Sens. Environ.* **1996**, *55*, 123–138. [CrossRef]

75. Huete, A.R.; Jackson, R.D. Suitability of spectral indices for evaluating vegetation characteristics on arid rangelands. *Remote Sens. Environ.* **1987**, *23*, 213-IN8. [CrossRef]

76. Huete, A.R.; Tucker, C.J. Investigation of soil influences in AVHRR red and near-infrared vegetation index imagery. *Int. J. Remote Sens.* **1991**, *12*, 1223–1242. [CrossRef]

77. Baret, F.; Guyot, G. Potentials and limits of vegetation indices for LAI and APAR assessment. *Remote Sens. Environ.* **1991**, *35*, 161–173. [CrossRef]

78. Elvidge, C.D.; Lyon, R.J.P. Influence of rock-soil spectral variation on the assessment of green biomass. *Remote Sens. Environ.* **1985**, *17*, 265–279. [CrossRef]

79. Huete, A.R. A soil-adjusted vegetation index (SAVI). *Remote Sens. Environ.* **1988**, *25*, 295–309. [CrossRef]

80. Myneni, R.B.; Hoffman, S.; Knyazikhin, Y.; Privette, J.L.; Glassy, J.; Tian, Y.; Wang, Y.; Song, X.; Zhang, Y.; Smith, G.R.; et al. Global products of vegetation leaf area and fraction absorbed PAR from year one of MODIS data. *Remote Sens. Environ.* **2002**, *83*, 214–231. [CrossRef]

81. Owuor, S.O.; Butterbach-Bahl, K.; Guzha, A.C.; Rufino, M.C.; Pelster, D.E.; Díaz-Pinés, E.; Breuer, L. Groundwater recharge rates and surface runoff response to land use and land cover changes in semi-arid environments. *Ecol. Process.* **2016**, *5*, 16. [CrossRef]

82. Siebert, S.; Burke, J.; Faures, J.-M.; Frenken, K.; Hoogeveen, J.; Döll, P.; Portmann, F.T. Groundwater use for irrigation—A global inventory. *Hydrol. Earth Syst. Sci.* **2010**, *14*, 1863–1880. [CrossRef]

83. Santoni, C.S.; Jobbágy, E.G.; Contreras, S. Vadose zone transport in dry forests of central Argentina: Role of land use. *Water Resour. Res.* **2010**, *46*. [CrossRef]

84. Jassas, H.; Kanoua, W.; Merkel, B. Actual Evapotranspiration in the Al-Khazir Gomal Basin (Northern Iraq) Using the Surface Energy Balance Algorithm for Land (SEBAL) and Water Balance. *Geosciences* **2015**, *5*, 141–159. [CrossRef]

85. Mohamed, M.A.; Anders, J.; Schneider, C. Monitoring of Changes in Land Use/Land Cover in Syria from 2010 to 2018 Using Multitemporal Landsat Imagery and GIS. *Land* **2020**, *9*, 226. [CrossRef]

86. Hassen, B.A.; Minch, A. *GIS Based Groundwater Recharge Estimation: The Case of Shinile Sub-Basin*; Arba Minch University: Arba Minch, Ethiopia, 2018.

87. Manandhar, R.; Odeh, I.O.A.; Pontius, R.G. Analysis of twenty years of categorical land transitions in the Lower Hunter of New South Wales, Australia. *Agric. Ecosyst. Environ.* **2010**, *135*, 336–346. [CrossRef]

88. Amdan, M.L.; Aragón, R.; Jobbágy, E.G.; Volante, J.N.; Paruelo, J.M. Onset of deep drainage and salt mobilization following forest clearing and cultivation in the Chaco plains (Argentina). *Water Resour. Res.* **2013**, *49*, 6601–6612. [CrossRef]

89. Nosetto, M.D.; Jobbágy, E.G.; Brizuela, A.B.; Jackson, R.B. The hydrologic consequences of land cover change in central Argentina. *Agric. Ecosyst. Environ.* **2012**, *154*, 2–11. [CrossRef]

90. Nosetto, M.D.; Acosta, A.M.; Jayawickreme, D.H.; Ballesteros, S.I.; Jackson, R.B.; Jobbágy, E.G. Land-use and topography shape soil and groundwater salinity in central Argentina. *Agric. Water Manag.* **2013**, *129*, 120–129. [CrossRef]

91. Chen, Y.; Yang, K.; Qin, J.; Zhao, L.; Tang, W.; Han, M. Evaluation of AMSR-E retrievals and GLDAS simulations against observations of a soil moisture network on the central Tibetan Plateau: Evaluate soil moisture products on tibet. *J. Geophys. Res. Atmos.* **2013**, *118*, 4466–4475. [CrossRef]

92. Wang, L.; Caylor, K.K.; Villegas, J.C.; Barron-Gafford, G.A.; Breshears, D.D.; Huxman, T.E. Partitioning evapotranspiration across gradients of woody plant cover: Assessment of a stable isotope technique: Isotopic evapotranspiration partitioning. *Geophys. Res. Lett.* **2010**, *37*. [CrossRef]

93. Li, W.; Tiwari, S.P.; ManiKandan, K.P.; El-Askary, H. Ocean colormodeling in the central red sea using oceanographical obser-vation and simulated parameters. In Proceedings of the 2020 IEEE International Geoscience and Remote Sensing Symposium (IGARSS), Waikoloa, HI, USA, 26 September–2 October 2020; pp. 5620–5623.

94. Nile Basin Initiative. *In The Nile Basin Water Resources Atlas*; Nile Basin Initiative: Entebbe, Uganda, 2016.

95. Montanari, A.; Koutsoyiannis, D. A blueprint for process-based modeling of uncertain hydrological systems: Stochastic process-based modeling. *Water Resour. Res.* **2012**, *48*. [CrossRef]

96. Todini, E. Hydrological catchment modelling: Past, present and future. *Hydrol. Earth Syst. Sci.* **2007**, *11*, 468–482. [CrossRef]

97. Papacharalampous, G.; Tyralis, H.; Langousis, A.; Jayawardena, A.W.; Sivakumar, B.; Mamassis, N.;
     Montanari, A.; Koutsoyiannis, D. Probabilistic Hydrological Post-Processing at Scale: Why and How to
     Apply Machine-Learning Quantile Regression Algorithms. *Water* **2019**, *11*, 2126. [CrossRef]
98. Montanari, A. Uncertainty of hydrological predictions. In *Treatise on Water Science 2*; Wilderer, P.A., Ed.;
     Elsevier: Amsterdam, The Netherlands, 2011; pp. 459–478.
99. Smith, W.K.; Dannenberg, M.P.; Yan, D.; Herrmann, S.; Barnes, M.L.; Barron-Gafford, G.A.; Biederman, J.A.;
     Ferrenberg, S.; Fox, A.M.; Hudson, A.; et al. Remote sensing of dryland ecosystem structure and function:
     Progress, challenges, and opportunities. *Remote Sens. Environ.* **2019**, *233*, 111401. [CrossRef]

2

Mapping Dynamic Water Fraction under the Tropical Rain Forests of the Amazonian Basin from SMOS Brightness Temperatures

Marie Parrens [1,*], Ahmad Al Bitar [1], Frédéric Frappart [2,3], Fabrice Papa [2,4], Stephane Calmant [2], Jean-François Crétaux [2], Jean-Pierre Wigneron [5] and Yann Kerr [1]

[1] Centre d'Etudes Spatiales de la BIOsphère (CESBIO—Université de Toulouse, CNES, CNRS, IRD), UMR5126, BPI 2801, 31401 Toulouse CEDEX 9, France; ahmad.albitar@cesbio.cnes.fr (A.A.B.); yann.kerr@cesbio.cnes.fr (Y.K.)

[2] Laboratoire d'Etudes en Géophysique et Océanographie Spatiales (LEGOS), UMR5566, Université de Toulouse, CNES, CNRS, IRD, Observatoire Midi-Pyrénées (OMP), 14 Avenue Edouard Belin, 31400 Toulouse, France; frederic.frappart@legos.obs-mip.fr (F.F.); fabrice.papa@ird.fr (F.P.); stephane.calmant@ird.fr (S.C.); jean-francois.cretaux@legos.obs-mip.fr (J.-F.C.)

[3] Géosciences Environnement Toulouse (GET), UMR5563, Université de Toulouse, CNES, CNRS, IRD, Observatoire Midi-Pyrénées (OMP),14 Avenue Edouard Belin, 31400 Toulouse, France

[4] Indo-French Cell for Water Sciences (IFCWS), IRD-IISc-NIO-IITM Joint International Laboratory, Bangalore 560012, India

[5] INRA, UMR 1391 ISPA, F-33140 Villenave d'Ornon, Bordeaux, France; wigneron@bordeaux.inra.fr

[*] Correspondence: marie.parrens@cesbio.cnes.fr

Academic Editor: Y. Jun Xu

Abstract: Inland surface waters in tropical environments play a major role in the water and carbon cycle. Remote sensing techniques based on passive, active microwave or optical wavelengths are commonly used to provide quantitative estimates of surface water extent from regional to global scales. However, some of these estimates are unable to detect water under dense vegetation and/or in the presence of cloud coverage. To overcome these limitations, the brightness temperature data at L-band frequency from the Soil Moisture and Ocean Salinity (SMOS) mission are used here to estimate flood extent in a contextual radiative transfer model over the Amazon Basin. At this frequency, the signal is highly sensitive to the standing water above the ground, and the signal provides information from deeper vegetation density than higher-frequencies. Three-day and (25 km × 25 km) resolution maps of water fraction extent are produced from 2010 to 2015. The dynamic water surface extent estimates are compared to altimeter data (Jason-2), land cover classification maps (IGBP, GlobeCover and ESA CCI) and the dynamic water surface product (GIEMS). The relationships between the water surfaces, precipitation and in situ discharge data are examined. The results show a high correlation between water fraction estimated by SMOS and water levels from Jason-2 (R > 0.98). Good spatial agreements for the land cover classifications and the water cycle are obtained.

Keywords: water fraction extent; L-band; Amazon Basin

1. Introduction

Terrestrial surface water covers only about 5% of the Earth's ice-free land surface [1,2], but plays a key role in global biogeochemistry, hydrology and wildlife diversity [3,4]. Consequently, it is critical to monitor the distribution of terrestrial water at large spatial and high temporal scales [5–8]. The work in [9] estimates that nearly two-thirds of all terrestrial freshwater wetlands disappeared between 1997 and 2011. A more recent study used three million Landsat images to provide a high resolution map of surface water extent [10]. The authors of this study estimated that 90,000 km^2 of permanent surface water had disappeared between 1984 and 2015.

Several different methods based on mapping water bodies from remote sensing datasets were used since the development of the Earth's space observations: (1) visible; (2) infrared; (3) active microwave; (2) passive microwave and; (4) hybrid approach (passive and active microwave). Each method offers varying degrees of success in providing quantitative estimates of wetlands and inundation extents.

Water surface can be sensed by optical remote sensing methods. These methods typically exploit the absorption of longer wavelengths of light in water, especially the near and shortwave infrared parts of the electromagnetic spectrum [11,12]. Optical remote sensing provides very accurate mapping of water bodies. For example, [13] senses lakes with a spatial resolution of 15 m, whereas [14] sensed global water bodies at 30-m resolution using the Landsat data. The majority of the studies using optical remote sensing for water bodies' detection provided only one snapshot of the hydrology stage. Due to the low revisit time of the optical sensors, few maps of a large area are available, and the minimum and/or the maximum of the flooded area are not always observed. The detection of sudden changes impacting the hydrologic cycle [10] is also not sensed with accuracy. These limitations are crucial issues for hydrology application. However, some studies [15,16] managed to follow the temporal dynamics of the water surface in specific places. The most important limitation of the optical sensors is their inability to penetrate clouds and dense vegetation cover, which is essential during tropical wet seasons over the Amazon Basin.

Active microwave (scatterometers and Synthetic Aperture Radar (SAR)) is also sensitive to the water surface and has the ability to penetrate clouds and, to a certain extent, vegetation. Open water surfaces are generally characterized by low backscattering coefficients. Contrary to passive microwave, the signal is more contaminated by the vegetation. The spatial resolution of scatterometers is about 25–50 km, whereas the SAR provides higher resolution, typically around 10–150 m. Several studies have shown the ability of active microwave to map surface water at regional scales, such as over the Amazon region [17] and over the Arctic region [18]. Satellite altimeters are radars that observe at nadir to measure surface topography. They provide accurate measurements of water heights in rivers, lakes and wetlands [19–21]. Due to their high spatial resolution, altimeters do not provide sufficient spatial coverage to analyze the water bodies' temporal dynamics, except in polar regions [22]. The future Surface Water Ocean Topography (SWOT) mission [23] intended to be launched in 2021 is expected to provide K-band SAR interferometry, enabling continental altimetry.

Passive microwaves are sensitive to the distribution of liquid water in the landscape; they can operate day and night for all weather conditions. However, they are limited by a low spatial resolution (approximately 30 km). They can sense only large wetlands or regions where the cumulative area of small wetlands comprises a significant portion of the field of view. Consequently, they provide the capability to map the temporal evolution of surface water over the land surface due to their high temporal resolution. In previous studies, passive microwave measurements have shown the capability to sense the dynamics of terrestrial surface water at coarse resolution [24–31]. The basic principle of the surface water measurements based on passive microwave is explained by the difference of the emissivity between the water and the soil. Flooding surfaces decrease the emissivity in both vertical (V) and horizontal (H) polarization and increase the difference between the two polarizations, especially at low frequencies. This approach produces ambiguous estimation of surface water over regions with mixtures of open water and other complex surfaces (topography effects). The work in [26,27] has extensively studied the inundation area over the Amazon Basin with the Scanning Multichannel Microwave Radiometer (SMMR). However, their studies focus essentially on a restricted area close to Manaus town from 1979 to 1987.

Hybrid approaches combine the strengths of different types of sensors. For example, altimetry data are characterized by a high spatial resolution and a low temporal resolution and can be combined with passive microwave data having low spatial resolution and a high temporal resolution to obtain a product with both high temporal and spatial resolution. The Global Inundation Extent from Multiple-Satellites (GIEMS) products are based on merged data from passive (Special Sensor Microwave/Imager (SMM/I)), active microwave (European Remote Sensing satellite (ERS)) and data from an optical sensor (Advanced Very High Resolution Radiometer (AVHRR)) [32].

Table 1 presents the major studies related to the observation and detection of water bodies from space by using the techniques presented above. Visible and infrared remote sensing methods were extensively used, but provided static maps of water bodies at the global scale or a dynamic map at the regional scale [15,33]. A lack of studies concerning dynamic water surface extent from 2013 to the present is clearly identified in this table.

The floodplains and wetlands of the Amazon River are important in terms of water volume and in terms of fluxes between the land and the atmosphere. Mapping water fraction under the Amazon tropical dense forest is challenging, but sensing water under dense vegetation remains a key issue in the remote sensing scientific community.

In this study, we developed a method to map the temporal evolution of the water bodies at coarse spatial resolution and weekly temporal resolution by using a microwave sensor at L-band (1.4 GHz) called Soil Moisture Ocean Salinity (SMOS) over the Amazon Basin. The SMOS satellite operates at L-band, and it was shown that this frequency is the most suitable, being less impacted by vegetation than higher frequencies [34–36]. Originally, the SMOS satellite was dedicated to sense soil moisture over land surfaces and the ocean salinity. The SMOS physical signal (brightness temperature) is highly impacted by the presence of standing water over the ground.

Our motivation is to use a contextual radiative transfer model and a single dataset to estimate the water fraction over the tropical basin. The area of study and the datasets used in this work are presented in the Section 2 and 3, respectively. Section 4 presents the algorithm permitting retrieving the water fraction extent from the SMOS data, and Section 5 contains the results and the validation. The discussion and conclusions are presented in Sections 6 and 7.

Table 1. List of selected scientific papers on the observation of the water surface over the continents from space. References, area of study and sensors are shown.

Remote Sensing Approach	Reference	Area of Study	Sensors	Frequency
Passive microwave	[27]	Amazon Basin	SSMR	Q-band
	[29]	Boreal regions	SSM/I-SSMR	K- and Ka-band
	[37]	North Eurasian	AMSR-E-QSCAT	C-band
		See [38] for a review of the SAR technique		
Active microwave	[17]	Amazon Basin	ENVISAT SAR	L-band
	[39]	High latitude regions	ENVISAT ASAR	C-band
	[40]	Mekong basin	ENVISAT ASAR	C-band
	[41]	Global scale	ENVISAT ASAR	C-band
	[22]	Boreal regions	Topex-Poseidon	C-band
Hybrid approaches	[42]	Global scale	SSM/I, ERS, AVHRR	Ka- and C-band
	[32]	Global scale	SSM/I, ERS, AVHRR	Ka- and C-band
	[43]	Global scale	SSM/I, ERS, AVHRR	Ka- and C-band
	[37]	Global scale	SSM/I, SSMI/S, ERS, QSAT, ASCAT	Ku- and C-band
Optical and infrared	[44]	Okavango Delta	AVHRR	-
	[45]	Brahmaputra	AVHRR	-
	[46]	Inner Niger Delta	MODIS	-
	[47]	China	MODIS	-
	[33]	Mekong Delta	MODIS	-
	[10]	Global scale	Landsat	-

2. Study Areas

This study focuses on the Amazon Basin, which is the largest tropical basin with an area of approximately 6,000,000 km^2 and contributes up to 15% of the global river discharge to the ocean (approximately 200,000 m^{3}s^{-1} discharge). With a sediment load of three million tons near its mouth [48] and drainage area covers about 6,200,000 km^2, almost 5% of all of the continental masses, the

Amazon Basin is one of the most impressive hydrological basins of the world. The Amazon is highly interconnected by floodplain channels, resulting in complex flow patterns. Figure 1 presents the Amazon Basin with the main rivers and floodplains. Covering more than 300,000 km^2, the Amazon extensive floodplains play a crucial role for global climate and biodiversity, but they are still poorly monitored at a large scale, limiting our understanding of their role in flood hazard, carbon production, sediment transport, nutriment exchange and air-land interactions. Surface water stored in floodplains represents about half of the terrestrial water storage and 15–20% of the water that flowed out of the Amazon floodplains [49–53]. Because it extends over two hemispheres, the Amazon region is characterized by several rainfall regimes. Rainfall shows opposing phases between the Northern and the Southern Hemisphere with a rainy season in austral winter in the Northern Hemisphere and summer in the Southern Hemisphere. The rainfall shows a gradient from northwest to southeast with decreasing rainfall amount and increasing length in the dry season. For the eastern part of the basin, the rainy season occurs from March–May, and the dry season prevails from September–November. For the northern regions, low rainfall seasonality is observed with wet conditions throughout the year. For more information on the Amazon hydrological regime, see [54–56].

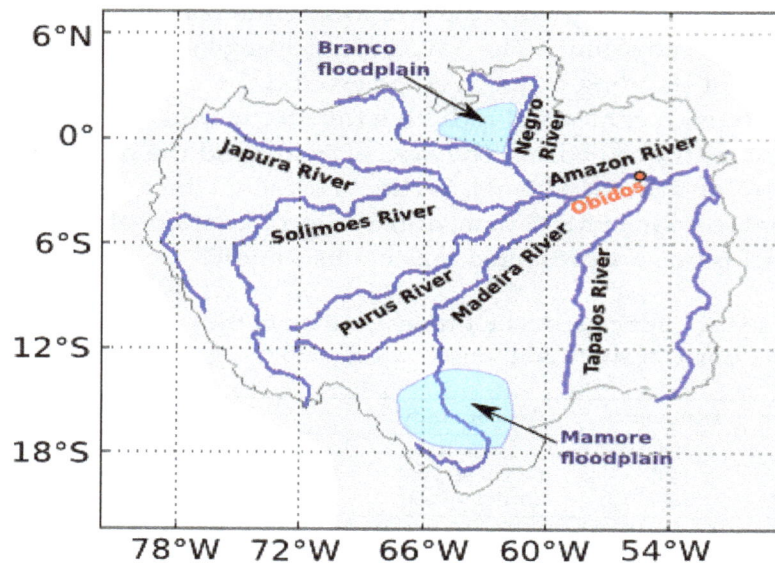

Figure 1. Map of the Amazon Basin with the main rivers and floodplains.

3. Data

This section describes the data used to compute the water fraction extent from passive microwave at L-band (SMOS data, topography data and skin temperature) and the data used to compare and validate this product (precipitation data, static land cover maps, other dynamic water fraction products, water level from altimetry and in situ river discharge data).

3.1. L-Band Brightness Temperatures from SMOS

The SMOS mission is a joint program of the European Space Agency (ESA), the Centre National d'Etudes Spatiales (CNES) and the Centro para el Desarrollo Teccnologico Industrial (CDTI) in the framework of the Earth Explorer Opportunity Mission initiative. It is the first satellite specifically dedicated to soil moisture retrievals with a passive microwave radiometer at 1.4 GHz (L-band). The physical signal of SMOS is the Brightness Temperature (TB). This signal is highly sensitive to the water under the ground [34]. Clouds and rain have a negligible effect [57], and the atmospheric contribution is limited and known [34]. The microwave signal is to a lesser extent sensitive to the vegetation, but actually, the L-band signal is less impacted by the vegetation than higher frequencies.

SMOS has a Sun-synchronous orbit at a 757-km altitude with a 06:00 LST ascending Equator crossing time and an 18:00 LST descending Equator crossing time. The globe is fully imaged twice every three days. The main innovative feature of SMOS is the capability for multi-incidence-angle observations at full polarization across a 900-km swath. In this study, the SMOS Level (L) 3 TB (RE04v300) products [58] produced by the Centre Aval de Traitement des Données SMOS (CATDS) are used. These data are projected on the Equal-Area Scalable Earth (EASE) Grid 2 [59] with a spatial resolution of 25 km × 25 km. The main differences between the SMOS L3 TB and the other lower levels of data are: (i) the L3 TB products are expressed at the top of the atmosphere over the terrestrial reference frame (H and V); and (ii) they are bin averaged every $5°$ from $2.5°$–$62.5°$. In the present study, SMOS L3 TB were used from 2010–2015 over the Amazon Basin. Angles of $32 ± 5°$, $37 ± 5°$, $42 ± 5°$ and $47 ± 5°$ in both H and V polarization were considered to retrieve the water fraction over the tropical basin. The SMOS data were downloaded from the CATDS servers (www.catds.fr).

3.2. Topography

The digital elevation model obtained by the Shuttle Radar Topography Mission (SRTM) [60] with a spatial resolution of 30 arc sec(approximately 1 km) was used over the Amazon Basin. These data result from the Global 30 Arc-Second Elevation (GTOPO30) computed at the U.S. Geological Survey's EROS Data Center (USGS) and were available at https://Ita.cr.usgs.gov/GOTO30. The elevation and topographic index maps over the Amazon watersheds were computed by averaging all of the SRTM elevation values present in an SMOS pixel (Figure 2). These data are used to flag areas with an elevation higher than 500 and/or a topographic index indicated as moderate in the SMOS flag.

3.3. Skin Temperature

The surface skin temperature produced by the European Centre for Medium-range Weather Forecasting (ECMWF) was used in this study. This product was obtained by the SMOS L3 preprocessor, which computed the spatiotemporal average of the ECMWF reanalysis products on the EASE 2.0 grid.

3.4. Precipitation Data

Precipitation measured by the Tropical Rainfall Measuring Mission (TRMM) were used over the entire Amazon Basin from 2010–2015. TRMM is a joint mission between NASA and the Japan Aerospace Exploration (JAXA) Agency and provides rainfall estimates at $0.25° × 0.25°$ spatial resolution. The TRMM-3B42 product [61] uses microwave data to calibrate the infrared-derived estimates and creates estimates that contain microwave-derived rainfall estimates when and where microwave data are available and the calibrated infrared estimates where microwave data are not available.

3.5. Static Land Cover Maps

The International Geosphere-Biosphere Programme (IGBP), the GlobeCover land cover classification and the ESA CCI maps were used. The IGBP land cover map was obtained using images from the Moderate Resolution Imaging Spectroradiometer (MODIS) with a spatial resolution of $0.005°$ during 2001–2012 [62]. The GlobeCover land surface map is based on data from the Advanced Very High Resolution Radiometer (AVHRR) data and is at a 1-km spatial resolution. Data were acquired during 1992–1993 [63]. Recently, a new release of the ESA Climate Change Initiative (CCI) Land Cover map was made available [64]. The new water/no water global mask at 150 m was built on previous achievements using SAR systems and further improved thanks to a combination with recent Landsat-derived products. This dataset was based on acquisitions from the years 200–2012. Data can be downloaded at: http://www.esa-landcover-cci.org/?q=node/162. For the three products, the water classes were aggregated and re-sampled on the EASE v2.0 grid to obtain the water fraction (%) present in each cell of the EASE v2.0 grid. The three products are static and are presented in Figure 2.

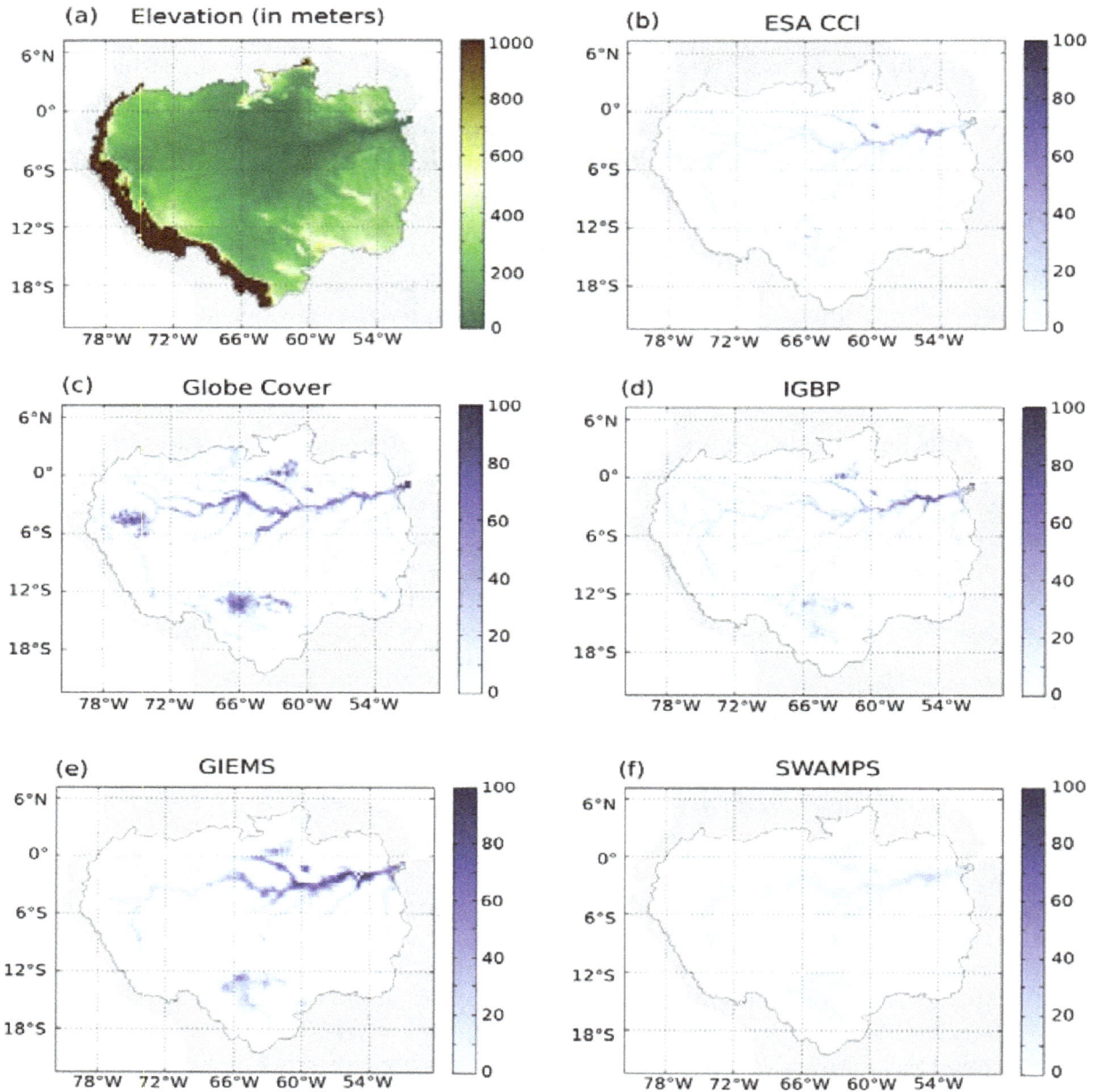

Figure 2. (**a**) Elevation (in meters) of the Amazon Basin from the SRTM data rescaled in the EASE v2.0 grid; spatial distribution of the water surface from: (**b**) ESA CCI; (**c**) IGBP; (**d**) Globe Cover; (**e**) average inundation extent from GIEMS from 1993–2007 over the Amazon Basin; and (**f**) average inundation extent from Surface Water Microwave Product Series (SWAMPS) from 2010–20over the Amazon Basin.

3.6. Dynamics and Climatoloy of Water Fraction Data

The GIEMS products provided a long-term global map of inundation at coarse resolution by merging passive and active data (SSM/I, ERS, AVHRR) from 1993–2007 at monthly time steps. The GIEMS product is gridded on an equal area grid of $0.25° \times 0.25°$ at the Equator. Over highly vegetated areas, the GIEMS product has some limitations. For example, over the Amazon Basin, the GIEMS product has a tendency to overestimate higher inundation fractions [32]. This product is fully described in [42]. These data were subsequently employed in estimating surface water storage variations in large river basins [50,65,66]. To be compared with our data, the GIEMS product was averaged during the full period and considered as a static climatological product. Figure 2 shows the temporal GIEMS average over the Amazon Basin from 1993–2007.

The recent Surface Water Microwave Product Series (SWAMPS) data provided daily surface water globally at 25-km resolution from 1992–2013. This product was based on the combination of passive and active microwave sensors (SSM/I, SSMIS, ERS, QuikSCAT, ASCAT) and visible sensors (MODIS). The data are described in detail and validated in [37]. In our study, only data from 2010–2013, coinciding with SMOS data availability, over the Amazon Basin were considered. Figure 2 shows the temporal SWAMPS average over the Amazon Basin from 2010–2013.

3.7. Water Level from Satellite Altimetry

In the Amazon Basin, water level (in meters) time series for virtual stations calculated from the Jason-2 altimeter satellite over the period 2008–2012 were downloaded from the Hydroweb database (http://hydroweb.theia-land.fr/). The Jason-2 satellite was launched on 20 June 2008 in the follow-on mission to the Jason-1 satellite (2002–2008, CNES/NASA). It operates at Ku-band (13.575 GHz) and C-band (5.3 GHz) and has a time period of ∼9.9 days [67]. The water level computation method and the location of the virtual stations are presented in [68]. In the present study, water level time series over 83 virtual stations from 2010–2012 inclusive were used.

3.8. In Situ River Discharge Data

Monthly in situ discharge (m^3/s) observations for the Amazon River were obtained from the Obidos gauge station (1°00′ S, 55°00′ W) for the period of January 2010–March 2014. Data are available on Hidroweb (http://www.hidroweb.ana.gov.fr) from the Brazilian water agency.

4. Methods

This section describes the approach used to derive surface water extent (expressed in %) contained in an SMOS pixel from the SMOS L3 TB. The contextual radiative transfer model and the selection of two reference points are described below, followed by the statistical method used to compare and validate the SWAF product.

4.1. Contextual Radiative Transfer Model to Retrieve the Water Fraction under Dense Forests

4.1.1. Description of the SWAF Algorithm

It is well known that the microwave emission is highly impacted by the dielectric constant [57]. In the present algorithm, we assumed that over the tropical basins, the pixels are only composed of two contributions: the water and the forest, such as represented in Figure 3.

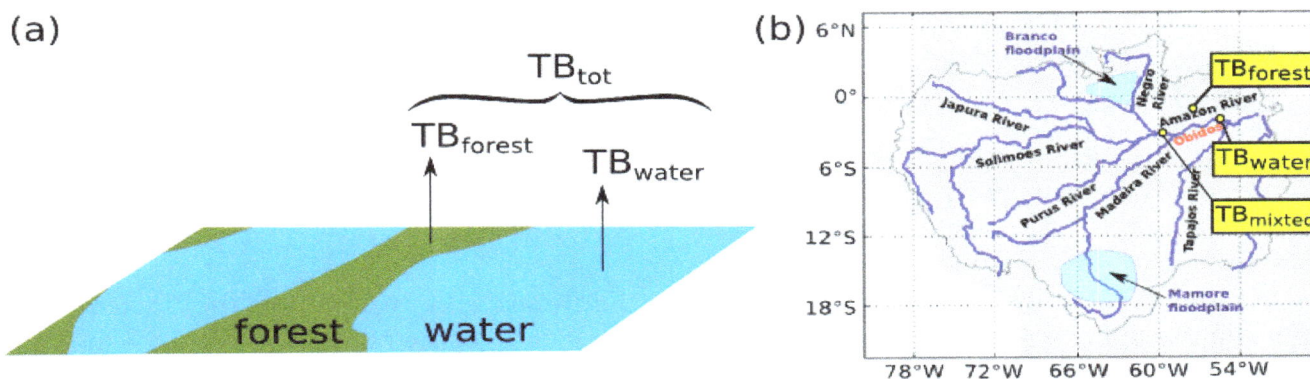

Figure 3. (a) Pixel representation with the two contributions: forest and water; (b) location of the "water", the "forest" and the mixed pixels.

Therefore, for a given pixel, the TB for the entire pixel is the sum of the water contribution and the forest contribution. The water contribution is the TB of the water weighted by the fraction of the pixel flooded, called the Surface WAterFraction (SWAF). In the same way, the contribution of the forest is the TB of the forest weighted by the part of the pixel not flooded (1-SWAF). The TB of the total pixel can be expressed as:

$$TB(\theta, p)_{tot} = SWAF(\theta, p)TB(\theta, p)_{water} + (1 - SWAF(\theta, p))TB(\theta, p)_{forest} \qquad (1)$$

where θ is the incidence angle, p the polarization (p = H or V), $TB(\theta, p)_{tot}$ is the TB of the total pixel, $TB(\theta, p)_{water}$ the TB of the water and $TB(\theta, p)_{forest}$ the TB of the forest. The fraction of the water present in an SMOS pixel depends on both the incidence angle and the polarization and can be expressed as:

$$SWAF(\theta, p) = \frac{TB(\theta, p)_{tot} - TB(\theta, p)_{forest}}{TB(\theta, p)_{water} - TB(\theta, p)_{forest}} \qquad (2)$$

$TB(\theta, p)_{tot}$ is the brightness temperature observed by the SMOS satellite. This observation is done over each pixel over the Amazon Basin. However, the $TB(\theta, p)_{forest}$ values are extracted over a selected pixel located at 2.137° S, 60.803° W (Figure 3) and composed exclusively of forest. The $TB(\theta, p)_{forest}$ are interpolated in time from the ascending SMOS overpass. In the same way, the $TB(\theta, p)_{water}$ time series are computed over a selected pixel located close to Obidos (2.142° S, 55.449° W) and composed of more than 80% water (Figure 3). Indeed, an SMOS pixel composed only of freshwater cannot been found (water from the ocean is salty, which modifies the emissivity). Therefore, the $TB(\theta, p)_{water}$ has been computed as the product of the emissivity and the skin temperature provided from the ECMWF. The Klein and Swift model [69] has been used to calculate the water emissivity. An average of the $TB(\theta, p)_{water}$ over the full period is computed to add more stability to the model. The $TB(\theta, p)_{water}$ is the only contribution that is constant in time in the model. Details about the value and the time series of both the $TB(\theta, p)_{forest}$ and $TB(\theta, p)_{water}$ are presented in the next section.

At this stage, an illustration is needed for a better comprehension of the algorithm. Figure 4 presents the time series of the $TB(\theta, p)_{water}$, $TB(\theta, p)_{forest}$ and a pixel annually flooded considered as "mixed" i.e., composed of both forest and water ($TB(\theta, p)_{mixed}$) located in Figure 3.

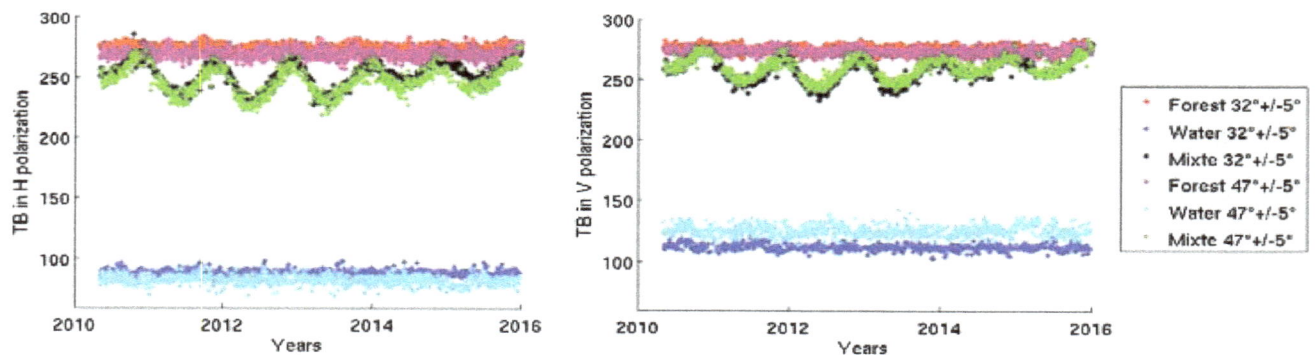

Figure 4. Time series of TB at H polarization (left) and V polarization (right) at two incidence angles (32° ± 5° and 47° ± 5°): the "water", "forest" and "mixed" pixels.

As shown in Figure 4, the $TB(\theta, p)$ values over the "forest", the "water" and the "mixed" pixels differ slightly following the angles and the polarization. For all of the polarization and the incidence angles, the $TB(\theta, p)$ values of the "water" pixel are the lowest, whereas the values of $TB(\theta, p)$ over the "forest" pixel are highest. The values of $TB(\theta, p)$ over the "mixed" pixel are included between the two contributions. Over the six years, the values of $TB(\theta, p)_{water}$ and $TB(\theta, p)_{forest}$ are very constant with time. This is not the case for the time series of $TB(\theta, p)_{mixed}$, which shows annual cycles due to the annual inundation of the Amazon River. For the "mixed" pixel, at each date, Equation (2) is computed to obtain the SWAF value. This method is generalized over all of the pixels of the Amazon Basin (the localization of the "mixed" pixel moved, but the water and forest reference pixels are always the same). The $SWAF(\theta, p)$ data are computed each day with the SMOS ascending overpass and smoothed with a sliding window of 17 days.

The passive interferometric technique has some limitations over the areas with complex topography [70]. To avoid artifacts, SWAF data are not computed over areas with moderate topography according to the SMOS flag. Moreover, pixels with a topography index estimated as moderate or higher in the SMOS flag over the Amazon Basin were not considered.

4.1.2. Forest and Water TB Reference

As explained in the previous section, the time series of TB values over the "water" pixel are averaged in time to be sure of the $TB(\theta, p)_{water}$ stability. Results for each angle and polarization used in the algorithm are presented in Table 2. Lower values of TB are obtained in H polarization than in V polarization. In H polarization, the mean TB over the "water" pixel decreases with the increase of the incidence angles. The reverse is observed in V polarization. The signal over the "water" pixel is really stable during the full period as shown by the standard deviation, which does not exceed 1 K.

To compare the value obtained over the "water" pixel, the TB over the "forest" pixel are also averaged. However, note that these values are not used in the SWAF algorithm. For the "forest" pixel, the mean TB values increase for decreasing incidence angles, in both H and V polarization. This behavior is more marked in H polarization. The standard deviation of the TB over the "forest" pixel is higher than that observed over the "water" pixel, but does not exceed 4 K. Lower standard deviations are obtained in V polarization than in H polarization, except at $32° \pm 5°$.

Table 2. Average and standard deviation (σ) of the TB over the "water" and "forest" pixel in both H and V polarization and at four incidence angles ($32° \pm 5°$, $37° \pm 5°$, $42° \pm 5°$, $47° \pm 5°$).

| Incidence Angle | "Water" Pixel | | | | "Forest" Pixel | | | |
| | H-pol | | V-pol | | H-pol | | V-pol | |
	Mean (K)	σ (K)	Mean (K)	σ (K)	Mean (K)	σ (K)	Mean (K)	σ (K)
$32° \pm 5°$	94.52	0.51	122.58	0.64	274.43	2.71	276.61	2.72
$37° \pm 5°$	89.96	0.49	128.25	0.67	272.44	2.94	276.12	2.61
$42° \pm 5°$	84.72	0.46	135.27	0.70	271.88	3.57	275.72	2.71
$47° \pm 5°$	78.78	0.43	143.93	0.74	269.71	3.22	274.26	2.51

4.2. Statistic Scores' Computation

In this study, we use a common set of skill scores: (i) the coefficient of correlation (r); (ii) the p-value; (iii) the cross-correlation; (iv) the bias; and (v) the Root Mean Square Error (RMSE) value. The Pearson correlation coefficient (r) is used to compare the dynamic behavior of the SWAF data (x) with the dynamic evolution of other variables (y):

$$r = \frac{\sum_{i=1}^{n}(x_i - \overline{x})(y_i - \overline{y})}{\sqrt{(\sum_{i=1}^{n}(x_i - \overline{x})^2}\sqrt{(\sum_{i=1}^{n}(y_i - \overline{y})^2}} \quad \text{with} \quad \overline{x} = \frac{1}{n}\sum_{i=1}^{n}x_i \quad \text{and} \quad \overline{y} = \frac{1}{n}\sum_{i=1}^{n}y_i \quad (3)$$

with n the number of elements in the x and y series. Associated with the r, the p-value is also computed for the null hypothesis. The authors consider that for a p-value higher than 0.05, correlation values are not significant.

The cross-correlation measures the similarity of two time series (x and y) as a function of the displacement of one relative to the other. In this study, the displacement corresponds to the time (in months). Therefore, the cross-correlation value is the higher correlation value obtained if the x time series is moved by n months with respect to the y time series.

The bias between two series (x and y) is defined as:

$$bias = \sum_{i=1}^{n} \frac{(x_i - y_i)}{n} \tag{4}$$

and the RMSE value is usually used to define the accuracy of the data. It is computed as:

$$RMSE = \sqrt{\frac{\sum_{i=1}^{n}(x_i - y_i)^2}{n}} \tag{5}$$

5. Results over the Amazon Basin

This section provides the SMOS water fraction results and analysis with a focus on the comparison and the validation of the product using a set of multi-source datasets described in Section 3.

5.1. Spatial Patterns and Temporal Dynamics of the SWAF Maps

This section described the spatial and temporal behavior of the water surface extent estimated by SMOS for the four angles and the two polarizations presented in Section 3.

Figure 5 shows the SMOS water fraction (SWAF) averaged over the 2010–2015 period for the entire Amazon Basin. Results are presented for four angle bins: ($32° \pm 5°$, $37° \pm 5°$, $42° \pm 5°$, $47° \pm 5°$) and the two polarizations (H and V pol). From Figure 5, it can be seen that independent of the incidence angles and polarizations, the major spatial patterns of the Amazon Basin are observed: the Amazon River and its tributaries, the Mamore floodplain in the south of the basin, the Branco floodplain in the north of the basin and the Balbina lake located in the north of Manaus. For a given angle, the spatial distribution of the SWAF is close in H and V polarization. However, the percentage of water fraction estimate is slightly higher at H polarization than V polarization. The major difference concerns the spatial distribution of the SWAF sensed at low incidence angle ($32° \pm 5°$) and at high incidence angle ($47° \pm 5°$). Low incidence angles reveal small patterns of SWAF and, in particular, the smaller Amazon River west affluent. Conversely, SWAF sensed with a higher incidence angle shows only the major structure of the flooded areas (Amazon River, Rio Negro River, Mamore plain, etc.).

Figure 6 shows the temporal dynamic of the SWAF for the full basin for the eight SMOS configurations. For all of the angles and the polarizations, water surface extent exhibits a clear seasonal cycle. The minimum of the inundation is observed in March, whereas the maximum of the flooding is reached during October. This observation is valid for all of the angles and polarization. Both in H and V polarization, for all of the incidence angles, the temporal dynamics of the SWAF are in good agreement, except for the incidence angle of $47° \pm 5°$. SWAF sensed with the highest incidence angle underestimates the water fraction in the Amazon Basin with respect to lower angles. In V polarization, on average, the Amazon Basin is less flooded than in H polarization. These results are in accordance with Figure 5. During the wet season, almost 1% of the Amazon Basin is flooded, whereas during the dry season, only 0.2% of the basin is flooded.

Figure 7 shows the monthly spatial variability of the SWAF product over the Amazon Basin. The monthly average has been computed during the full period (2010–2015) at V-polarization and at $32° \pm 5°$ of incidence angle. The Amazon River, its main tributaries and the major floodplains are well represented for all of the months. The maximum of the inundation of the Amazon River is observed between March and July. The spatial and temporal variation of the Mamore floodplain is well described by the SWAF data. This floodplain is inundated from January–June, and the least flooding is observed during September and October. The major Amazon tributaries are more flooded between January and May. Similar results are observed at H polarization and higher incidence angles (not shown).

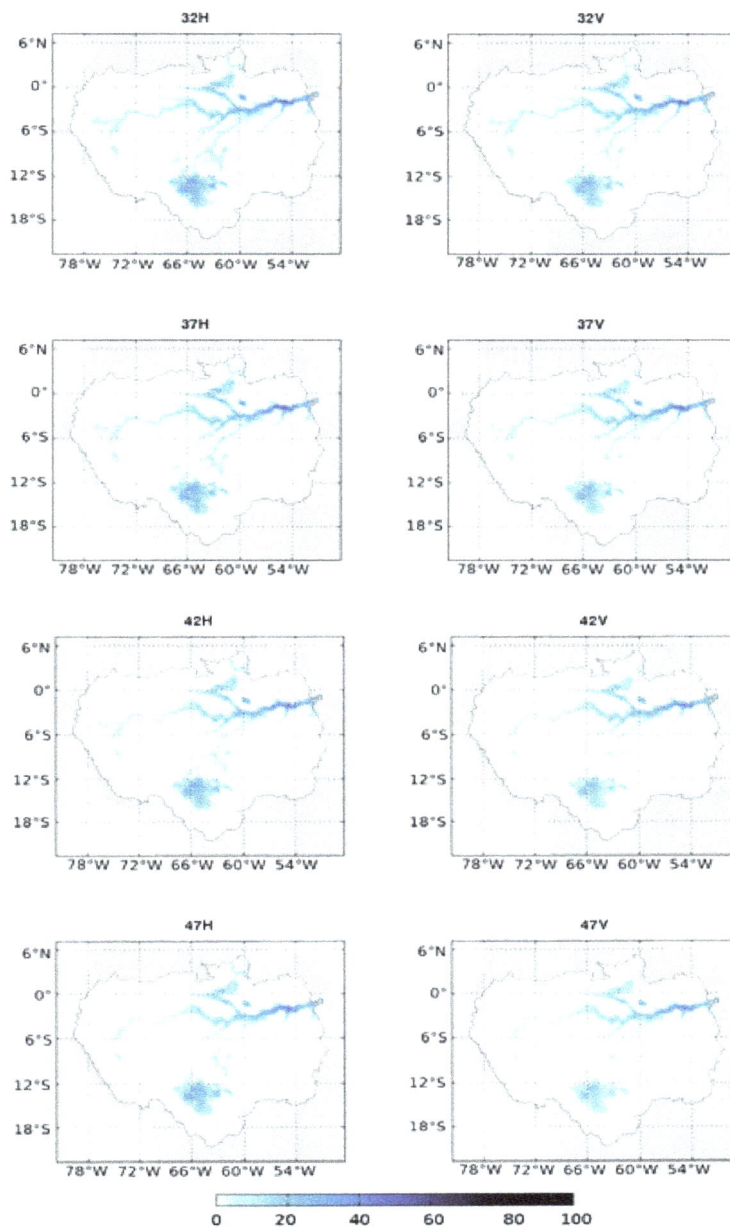

Figure 5. Average in time of the SMOS water fraction during 2010–2015 over the full Amazon Basin. Both H and V polarization and the four incidence angles (32°± 5°, 37°± 5°, 42°± 5°, 47°± 5°) are considered.

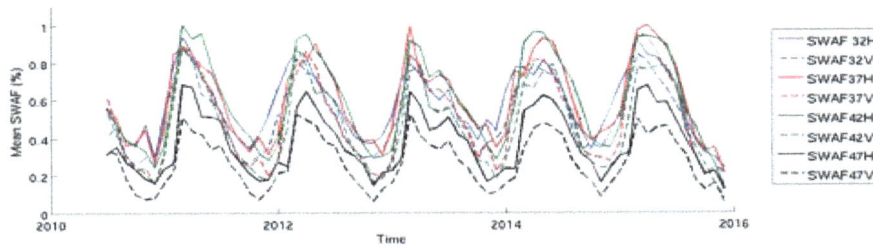

Figure 6. Spatial average of the SWAF over the full Amazon Basin in H polarization (lines) and V polarization (dashed lines) for the four incidence angles: 32°± 5° (blue), 37°± 5° (red), 42°± 5° (green) and 47°± 5° (black).

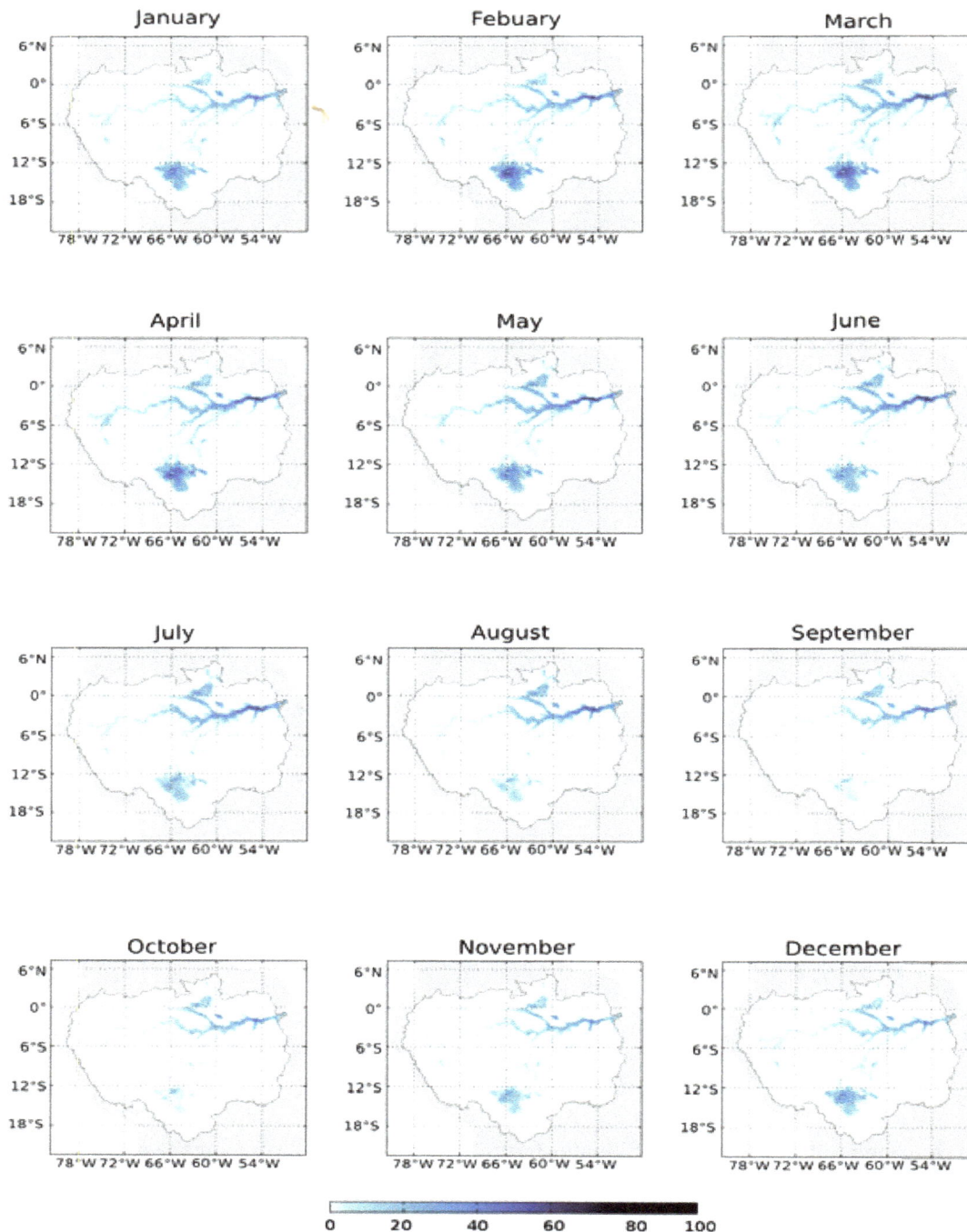

Figure 7. Monthly average of the SWAF product from 2010–2015 at V-polarization and at $32° \pm 5°$ incidence angle.

5.2. Comparison and Validation

In the following section, the SWAF product is compared to other data sources and variables: static and dynamic water extent maps, water level measured by altimetry satellite, in situ river discharge at the outlet of the basin and precipitation data. The static land surface maps obtained by optical sensors and the GIEMS product were also used to analyze the spatial patterns. The SWAF data were also compared to the dynamic water fraction product available over the Amazon Basin, the SWAMPS product. Note that all of these datasets are completely independent from the SMOS water surface extent maps.

5.2.1. Comparison to Static and Climatological Water Extent Maps

The SWAF temporal average over the entire Amazon Basin had been compared to three static land cover maps (IGBP [62], GlobeCover [63] and ESA CCI [64]) and the mean of the GIEMS water fraction maps. These maps are detailed in Section 3.5 and presented in Figure 2. Main similar patterns can be observed from the three different static land cover maps. The number of pixels partially flooded over the Amazon Basin is higher for the GlobeCover map than for both the IGBP and ESA CCI maps. Conversely, the ESA CCI map produces less pixels totally flooded over the Amazon Basin than the others maps. IGBP, GlobeCover and ESA CCI maps present inundation surfaces of $220,000 \, \text{km}^2$, $360,000 \, \text{km}^2$ and $210,000 \, \text{km}^2$, respectively. Note that the static maps are based on data from different time periods (see Section 3.5), and the inland water occupation can change with time and anthropogenic activities.

The average GIEMS product over the 1993–2007 period is presented in Figure 2. The Amazon River, its major tributaries, southern and northern floodplains are well depicted. On average, over the 1993–2007 period, $440,000 \, \text{km}^2$ of the surface were flooded.

Figure 8 presents the mean distribution (2010–2015) of the water fraction extent for the IGBP map, GlobeCover map, ESA CCI map, GIEMS averaging and the temporal average of all of the SWAF products for the eight SMOS configurations over the Amazon Basin. For all of the products, the majority of the pixels are partly flooded, and a few of them are totally flooded. The distributions of both IGBP and ESA CCI are very close. The GlobeCover map and the mean GIEMS product (1993–2007) provide larger estimates of pixels flooded or partially flooded with respect to the other products, in particular for water fraction higher than 0.4. For all of the SMOS configurations, the distributions of the SWAF products are comprised between both the GlobeCover and GIEMS distributions and both the IGBP and ESA CCI distributions. Between the eight SMOS configurations, only a few differences in their spatial distributions can be noticed. Both in H and V polarizations, a decrease of the incidence angle leads to a decrease of the detection of pixels partially flooded. In V polarization, fewer pixels are flooded than in H polarization. This behavior is particularly marked for water fraction ranges between 10% and 40%. This trend makes the SWAF computed in V polarization closer to the IGBP and ESA CCI maps than the SWAF calculated in H polarization for the moderately flooded pixels. Table 3 presents an average of the number of square kilometers flooded in the Amazon Basin. Figures 5 and 8 and Table 3 confirm that, on average, the number of square kilometers flooded decreases for increasing incidence angles, and more flooded areas are detected in H polarization than in V polarization. Independent of the selected SMOS configuration, the number of square kilometers flooded is in the range between the IGBP and GlobeCover estimates.

Figure 9 presents the bias (reference static maps, SWAF configurations) and the RMSE between water surface extent for each reference maps used in Figure 8 and the water surface extent for the eight SMOS configurations. For all of the SWAF configurations, lower bias is obtained by comparing the SWAF data with IGBP (mean bias = 0.6%) and ESA CCI (mean bias = −0.4%). Higher bias is obtained by comparing SWAF data with the GlobeCover map (mean bias = 5.9%). Bias between the static maps and the SWAF data is always higher at V-polarization than at H-polarization for all of the static maps. Moreover, the bias values increase with the growth of the incidence angles. A lower RMSE value is obtained by comparing the SWAF data with the ESA CCI map (mean RMSE = 5.8%), and a higher RMSE value is obtained with the GlobeCover map (mean RMSE = 17.8%). Following the static map considered, the behavior of the SWAF data with respect to angles and polarization differs. For IGBP and ESA CCI maps, lower values of RMSE are obtained at V-polarization than at H-polarization, and a slight decrease of the RMSE values is observed with the increase of the incidence angles. The contrary is noticed for the comparison with the GIEMS data. No trend concerning the RMSE behavior is observed for the comparison between the SWAF data and the GlobeCover data.

Table 3. Average of the number of square kilometers flooded in the Amazon Basin for the eight SMOS configurations.

Incidence Angle (°)	H-pol	V-pol
32° ± 5°	290,000	270,000
37° ± 5°	280,000	260,000
42° ± 5°	280,000	250,000
47° ± 5°	280,000	250,000

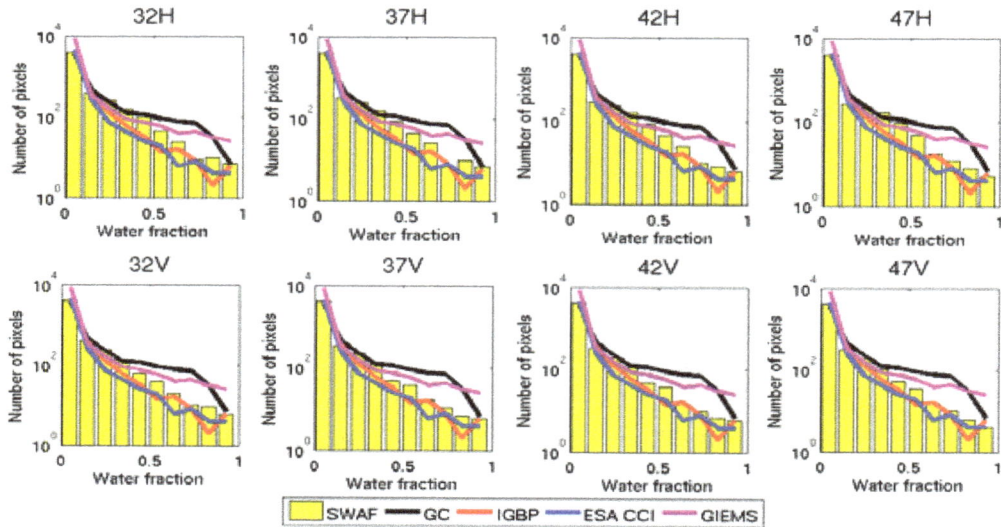

Figure 8. Histogram of water fraction for the IGBP map (red), GlobeCover map (black), the ESA CCI (blue) and SWAF (yellow columns) for eight SMOS TB configurations (32–47 angle bins and H/V configurations).

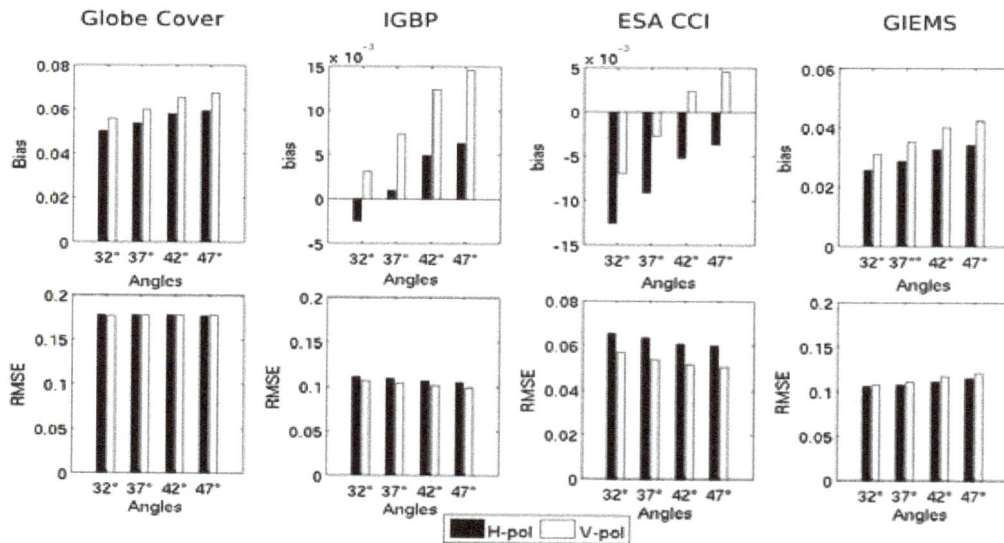

Figure 9. Bias (reference static maps, SWAF) and RMSE values computed between each reference maps (GlobeCover, IGBP, ESA CCI, GIEMS) and SWAF for the eight configurations.

5.2.2. Comparison with Water Height Measured by Altimetry

For low topography slopes, the water surface extent can be related to the water height. Other studies [32,66] had already shown that the seasonal and inter-annual variation patterns of the surface water extent and the water level agree well. In this present study, the water levels measured by the Jason-2 satellite were compared with the SWAF product over 83 virtual stations during the 2010–2012 periods. Results are presented in Figure 10. Only stations with significant results (p-value < 0.01) are represented. The color dot indicates the correlation value for each virtual station. The gray color dots show the virtual station with non-significant results. For all of the SMOS configurations, the correlation value between the water fraction and the water level measured by Jason-2 is very high ($r > 0.8$) throughout the Amazon River and the major tributaries. The lower correlation values are located over areas where the relation between the water surface's extent and the water level is not direct. For a given angle, slight differences in terms of correlation values are observed with respect to the polarization choice. The number of not significant stations varies following angles and polarization. At H polarization, the number of no significant stations is equal to 36, 33, 33 and 44 for angles from $32° \pm 5°–47° \pm 5°$, respectively. At V-polarization, the number of no significant stations is slightly lower and equal to: 35, 30, 31 and 44, respectively.

To formalize this information, the sum of the correlation values for each SMOS configuration is presented in Figure 11. Only stations with significant correlations for the eight SMOS configuration are considered to compute this figure. The sum of the correlation values between the SWAF and the water level estimated by altimetry is always higher at V-polarization than at H-polarization, except at $32° \pm 5°$. The higher sum of correlation is obtained at V polarization and at $47° \pm 5°$ incidence angle. At high incidence angles, higher correlation values are obtained, but the number of significant stations is lower. The contrary is observed at low incidence angles.

5.2.3. Comparison with SWAMPS Dynamic Surface Extent

The recent SWAMPS product provides a daily estimation of the surface water extent. Note that the SWAMPS products are obtained by more complex algorithm merging active and passive microwaves than the SWAF data. Figure 2 shows the average of the SWAMPS water fraction over the Amazon Basin from January 2010–March 2013. Spatially, the mean SWAF and SWAMPS products are in good agreement. Both in the SWAF and SWAMPS products, the Amazon River and its tributaries are well represented, and the major floodplains are present. In the SWAF data, no data are provided over the southeast part of the Amazon Basin due to high topographic index where as the SWAMPS product shows some patterns of water surface over the same region. An important difference between the two products is the spread of the rivers. In the SWAMPS product, all of the rivers have a larger floodplain area than in the SWAF data. Over the two major floodplains of the Amazon Basin, different patterns are observed in the SWAF data. This behavior is not observed for the SWAMPS data. The water surface extents estimated by the SWAMPS data are lower than those estimated by the SWAF data.

Figure 12 shows the temporal correlation values between the SWAMPS and SWAF data from January 2010–March 2013 for all of the SMOS configurations. The correlation value is computed only over pixels where both SWAMPS and SWAF water fraction are present. For all of the SMOS configurations, a good agreement ($r > 0.8$) between the SWAF and SWAMPS products is observed over the Amazon River and the two largest floodplains of the basin. Over these locations, the temporal dynamics of the surface water are well described by the two products. Concerning the Amazon tributaries, results are more contrasting, and the SWAF and the SWAMPS seem to have a different temporal dynamics. In terms of SMOS configurations, results are very similar whatever the incidence angle and polarization chosen. For example, the number of pixels that obtained a high correlation value ($r > 0.8$) between the SWAF and the SWAMPS products ranges between 147 (for $32° \pm 5°$ in H-pol) and 172 (for $47° \pm 5°$ in H-pol). No trend is noticed between high and low angles or V and H polarization. For the accepted correlation value ($r > 0.6$), the number of pixels that satisfied these criteria ranges between 1097 (for $47° \pm 5°$ in V-pol) and 1247 (for $37° \pm 5°$ in V-pol). In this case, a clear trend is observed: increasing the incidence angle leads to decreasing the number of pixels with an accepted correlation.

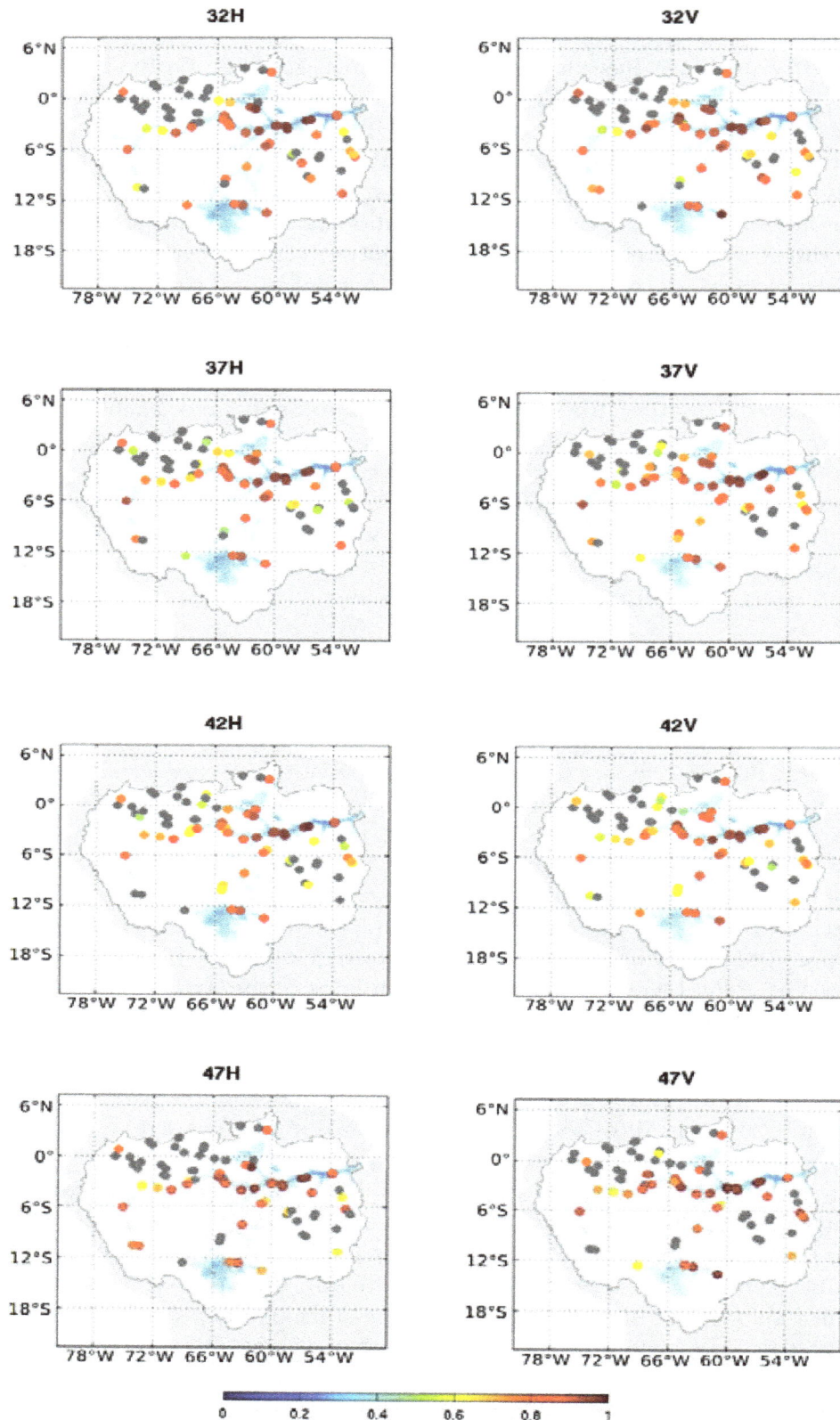

Figure 10. For each SMOS configuration, correlation values against the SWAF water surface extent and the water level measured by Jason-2 during 2010–2012. The color dot represents the correlation value. Gray color dots show no significant results (p-value > 0.05).

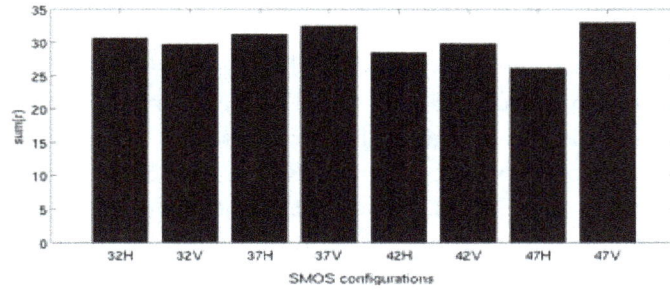

Figure 11. For each SMOS configuration, the sum of the correlation value (r) obtained in Figure 8. Only significant stations for all of the SMOS configuration are used for the computation.

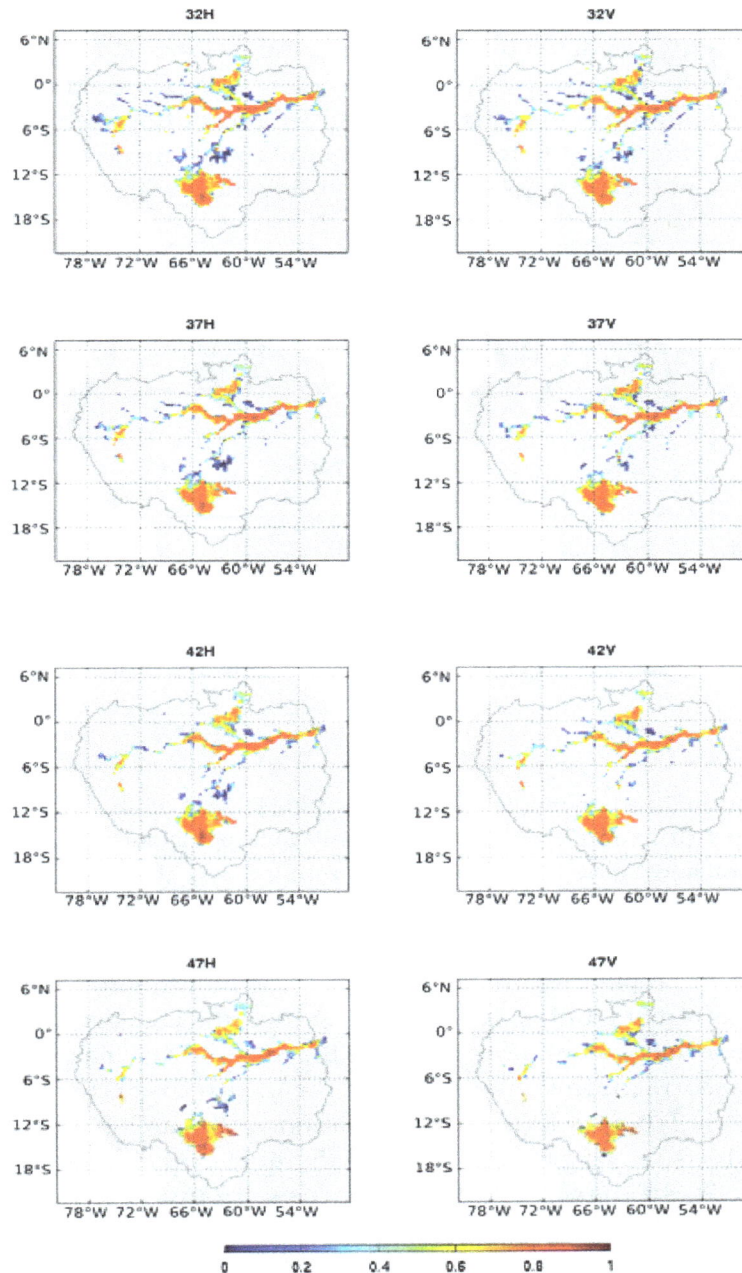

Figure 12. Temporal correlation values between the SWAMPS and SWAF products from January 2010–March 2013 for each SMOS configuration.

5.2.4. Link between SWAF and the Hydrological Components

Strong seasonal and interannual variations can be observed in both precipitation and surface water extent in the Amazon Basin. Figure 13 presents the standardized anomalies (i.e., the time series of a hydrological parameter minus the average of the time series divided by its standard deviation of the observation period) of both precipitation from TRMM and SWAF products for the eight SMOS configurations over the entire Amazon Basin. Note that inundation is highly linked to the precipitation events, but can also occur in response to snow melt or heavy precipitation at upstream locations. In this case, flooded areas and precipitation are separated in both time and space. A good agreement between all of the SWAF products whatever the polarization or the incidence angles can be noticed, except at $47° \pm 5°$ in H polarization. For the entire Amazon Basin, the cross-correlation values between the TRMM precipitation and the SWAF products are shown in Table 4. The best correlation values are obtained at $42° \pm 5°$ and $32° \pm 5°$. In H and V polarization, similar correlation values were obtained. The correlation value is highly impacted by the choice of the incidence angle, whereas the polarization plays a negligible role. Note that using the angle $47° \pm 5°$ strongly degrades the correlation value between the precipitation and the water surface extent. By computing the time lag correlation values, a time lag of two months was found between the precipitations and the water surface for all of the SMOS configurations, except for the angle of $47° \pm 5°$ (four months).

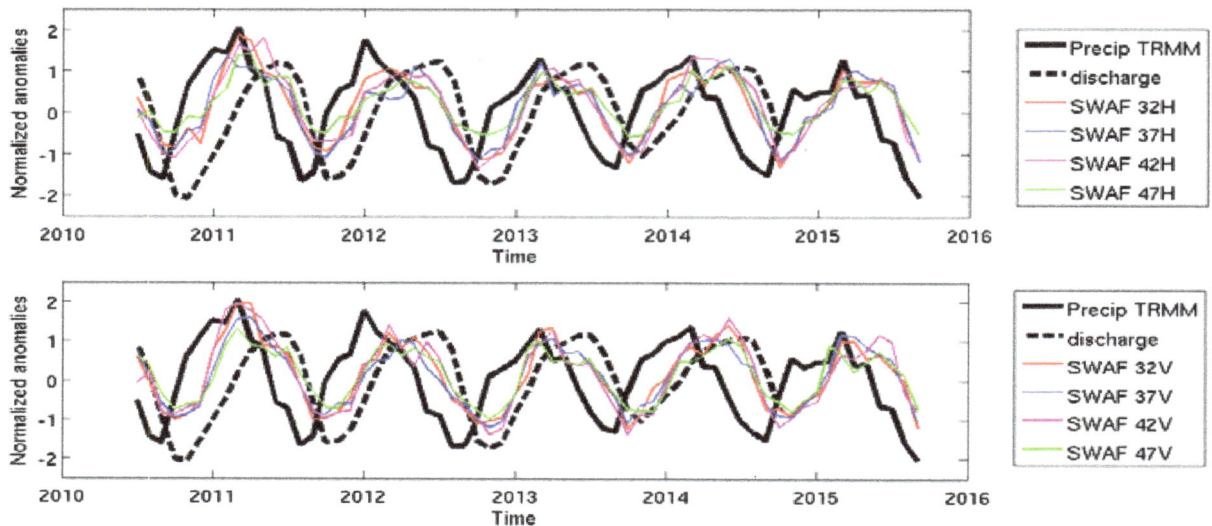

Figure 13. Monthly normalized anomalies of precipitation (TRMM data), in situ discharge at Obidos and the SWAF at H-pol (top) and V-pol (bottom) with the four incidence angles considered in this study. The precipitation and SWAF anomalies were computed over the entire Amazon Basin.

The time series of the Amazon River discharge is closely linked to the total amount of the surface water extent in the whole basin [43]. Figure 13 also shows the monthly normalized anomalies of the in situ river discharge measured at Obidos. By computing the time lag correlation values, it was found that the maximum water surface extent often precedes the maximum Amazon discharge. For the entire Amazon Basin, the cross-correlation values between the in situ discharge at Obidos and the SWAF products are reported in the Table 4. High cross-correlation values are obtained between the discharge and all of the SWAF products varying from 0.78 ($47° \pm 5°$ at H-pol) to 0.88 ($42° \pm 5°$ at H-pol). For all of the SWAF configurations, the maximum water surface extent precedes the discharge by one month. The normalized anomalies of SWAF whatever the SMOS configuration are better correlated with the normalized anomalies of discharge than those of precipitation. The cross-correlation value between the normalized anomalies of precipitation and discharge is equal to 0.84 with a time lag of four months for the whole Amazon Basin.

Table 4. Cross-correlation (r) values between TRMM precipitation, discharge at the outlet and the SWAF products for the eight configurations over the 2010–2015 period.

Incidence Angle (°)	Precipitation		Discharge	
	H-pol	V-pol	H-pol	V-pol
$32° \pm 5°$	0.91	0.90	0.83	0.84
$37° \pm 5°$	0.88	0.89	0.86	0.85
$42° \pm 5°$	0.92	0.88	0.88	0.82
$47° \pm 5°$	0.83	0.85	0.78	0.83

6. Discussion

6.1. Water Surface Validation

The SWAF products and the aggregated high spatial resolution of the IGBP and ESA CCI maps over the Amazon Basin showed a good agreement (see Figures 8 and 9). The mean bias and RMSE between the SWAF for all of the configurations and the IGBP map is equal to 0.6% and 10%, respectively. Better results are obtained for the comparison of the SWAF and the ESA CCI maps (mean bias = −0.4% and mean RMSE = 5.8%). The SWAF sensitivity to seasonal and annual water fraction extent was also demonstrated in the comparison against precipitation and discharge dynamics. The comparison between the anomaly of water surface dynamics estimated from the SWAF products over the entire Amazon Basin and the anomaly of discharge at the mouth of the Amazon showed a shift of one month. This result is in good agreement with previous research papers. For example, results presenting the same time-lag for surface water extent [43], surface water storage [49] and terrestrial water storage [71]. Our findings showed that precipitation often preceded the water surface dynamic by two months over the Amazon Basin. These results also agree well with previous studies [43,71].

6.2. Impact of the Angles and Polarization on the Water Surface Retrievals

The SMOS mission provides data in multi-angular and full polarization modes. A specific analysis is presented to determine the best acquisition configuration to retrieve the water fraction from SMOS. The use of combined polarization and angles was used in this study as the single angle and channel gave satisfactory results. As shown in the previous section, the spatio-temporal evolution of the SWAF estimated using different angles and polarizations was very close in the Amazon Basin. Differences are mainly present in areas far from the main Amazon river stream characterized by high vegetation density. Table 5 summaries the best SMOS configurations (low or high incidence angles and H or V polarizations) in order to obtain the higher agreement between the SWAF product and each variable used for validation (land cover classification, water level, dynamic water fraction, precipitation and discharge). "NS" stands for Non-Significant, and it is used when no trend in the results (see previous section) was observed.

Table 5. Summary of the best SMOS configurations permitting a good agreement between SWAF and the other variables. Green color means that good agreement was found, whereas the red color means the opposite. See Section 5 for the results. NS, Non-Significant.

Variables	H-pol	V-pol	Low angles	High angles
Land cover classification			NS	NS
Water level			NS	NS
Dynamic water fraction	NS	NS	NS	NS
Precipitation	NS	NS	NS	NS
Discharge	NS	NS	NS	NS

Table 5 shows that the low incidence angles were more suitable to detect the dynamic water surface extent over the tropical regions. This result was expected based on the microwave signal theory [72]. The H-polarization tends to overestimate the water fraction extent for a low fraction

(<50%). Over the Amazon Basin, low incidence angles at V polarization are the best configuration to compute the water fraction extent with the SMOS data. For future studies, the authors advise to use the SWAF products at low incidence angles and at V polarization over tropical areas. In the near future, SWAF estimated in both H and V polarization at low incidence angle will be combined to extend the domain of applicability of the algorithm to other environments to increase the water sensibility to the SWAF product.

6.3. Impact of Vegetation Cover

Estimating water surface extent under dense vegetation with passive microwave is challenging. The work in [26,27] provided estimates of the water surface extent over the Amazon Basin by using the passive microwave SMMR sensor at 37 GHz (Q-band). They found good agreement in the seasonal changes in inundation area and good correlations values with the water level in Manaus. However, they noticed the effects of the vegetation on their results, in particular for small patches of open water intermixed with vegetation canopies or from an attenuating effect of homogeneous canopies overlying water surfaces. To overcome the vegetation attenuation, the GIEMS product mixed passive and active microwave products at coarse spatial resolution with the optical dataset at finer resolution, which enhances the capability to detect the small water fraction. However, this capability could be hampering over dense vegetated area and frequently cloud-covered regions, such as the tropical ones, due to the limitation of the optical sensors. The frequencies of the passive microwave data from SSM/I are at 19 GHz (K band) and 85 GHz (E band), and the active one is at 5.25 GHz (C band). The work in [34] showed that the low frequencies were less sensitive to the vegetation effects than the higher frequency. The results presented in this study are the first to demonstrate the potential of L-band (1.4 GHz) brightness temperature to estimate surface water extent under dense vegetation.

6.4. Limitations and Prospects of the SWAF Dataset

The most important limitation of the SWAF product is its coarse resolution inherent to the passive microwave sensor. This limitation was already noticed for the GIEMS [42] products using microwave sensors. This limitation implies that the water surface lower than 4% for all of the angles and polarization could not be mapped by the SWAF products.

The SMOS sensor is based on dual polarizations and multi-angular measurements. Mountainous areas modify local incidence angles and multi-scattering, which impacts the TB values [70] and, consequently, the water surface estimation. The effects led to an overestimation of the water fraction. To overcome this effect, areas with moderate to strong topographic slopes were not considered in this study.

The snow is also an important component of the hydrological cycle. Due to the impact of the topography slopes on the TB and the presence of the snow only over the mountainous areas in the tropical basins, the temporal evolution of the snow coverage from the Andes Mountains has not been investigated in this study.

The method developed in this study is based on the impact of surface water at the L-band signal and based on the stability over time of the TB measured over dense forests. Some additional computation was performed to measure the sensitivity of the SWAF product to the "forest" reference point. It was found that instead of choosing only one reference "forest" pixel, but a set of pixels composed only of forest, the mean TB increases by 3 K for all of the incidence angles and polarization. This value is included in the incertitude range. However, this growth on the TB "forest" reference value tends to increase the SWAF by 8%. Concerning the "water" reference point, no change was observed by choosing a set of pure water pixels. Moreover, the method applied in this study could not be applied in areas where the TB over vegetation is not stable over time or not dense enough. Future work will concentrate on the extension of the current algorithm to other environments by using multi-angular information. By solving this limitation, the water fraction would be estimated at the global scale by using only one dataset and a simple approach.

7. Conclusions and Prospect

This study presents the validation and the link to other hydrological components of regional (Amazon Basin) daily and multi-year (2010–2015) water surface extent maps from the SMOS mission at coarse resolution (25 km × 25 km). The SWAF product is based on L-band acquisitions. At such a frequency, the signal is highly sensitive to the standing water above the ground, and it is expected to penetrate deeper in the vegetation than at higher frequencies, such as visible and infrared or microwave at higher frequencies. As the L-band signal is more sensitive to open water under dense vegetation, the SWAF product provides surface water extent estimates (percentage of inundation in a pixel of 25 × 25 km) with a high temporal resolution (<3 days) based on the accumulation of daily surface water extent in the Amazon Basin between 2010 and 2015. The SWAF product is computed from the L-band, and it can be computed easily and quickly without any ancillary data. Over this basin, the water surface extent showed a strong seasonal and interannual variability with two marked droughts in 2010 and 2015.

The SWAF data were compared to three sets of static land cover maps provided from visible sensors (IGBP, GlobeCover and ESA CCI) and the average inundation extent from GIEMS over 1993–2007. It was found that the SWAF products are close to the IGBP and ESA CCI maps. On average and during the 2010–2015 period, 270,000 km^2 were inundated over the Amazon Basin. A slight overestimation of the flooded areas could be noticed. Over the Amazon Basin, the SWAF products were highly correlated with water levels measured by Jason-2 ($r > 0.8$) for the significant stations. The temporal dynamics of the SWAF products were also validated against precipitation (TRMM data) and in situ discharge at the mouth of each river. It was found that over the Amazon Basin, the precipitations often precede the inundation by three months, and the water surface extent impacts the discharge at the mouth of the Amazon after one month. As expected by the microwave theory, the mall water fraction could not be detected by the large footprint of SMOS. This implied that low water fraction extent (<4%) could not be mapped by the SWAF products. The mountainous areas were also a limitation of the SWAF products. The topography-modified local incidence angles implied significant impact on the microwave signal and, consequently, on the water surface estimation. The effects led to overestimation of the water fraction. To avoid this effect, the areas with high topography slopes were flagged in the SWAF products.

Based on the SMOS product, the SWAF products declined with several incidence angles at two polarizations (H and V). It was clear that high incidence angles ($>47° \pm 5°$) were not suitable to sense the water surface from the L-band microwave signal. The H-polarization tended to increase the lower value of the water fraction extent with respect to the V-polarization. The SWAF products computed with different angles and polarizations led to similar results with very slight differences over the Amazon Basin. For future use, the authors advise the use of SWAF computed with low incidence angles ($32° \pm 5°$ and/or $37° \pm 5°$) at V for the Amazon Basin.

The methodology permitting retrieval of the water fraction applied in this study does not require much computation time and can be easily be applied to another L- band microwave dataset, such as the new Soil Moisture Active and Passive (SMAP) data or an older dataset (SSM/I...). The method had been validated over the Amazon Basin by taking advantage of the numerous data and research performed over this area.

In the near future, this recent water surface fraction product can be easily extended with the future SMOS data and the Soil Moisture Active and Passive (SMAP) data to obtain a long record of inundation products under dense vegetation. These data will be useful to better understand the water, carbon and methane cycles over the tropical areas. By adding a third component (saturated soil) on the first-order radiative transfer, this method is likely to be applied in other regions in the world.

Acknowledgments: This work was funded by the program Terre Océan Surfaces Continentales et Atmosphère (TOSCA, France) and the CNES under the project TOSCA-SOLE and Marie Parrens was funded by the CNES PostDoc program.

Author Contributions: Ahmad Al Bitar and Marie Parrens conceived of and designed the algorithms. Marie Parrens performed the analysis. Frédéric Frappart has contributed to the evaluation of the products. Frédéric Frappart, Fabrice Papa, Jean-François Crétaux and Stephane Calmant, Jean-Pierre Wigneron and Yann Kerr provided scientific expertise, datasets and corrections to the manuscript. Marie Parrens and Ahmad Al Bitar wrote the manuscript with contributions from all of the co-authors.

Abbreviations

The following abbreviations are used in this manuscript:

SAR	Synthetic Aperture Radar
SWOT	Surface Water Ocean Topography
V	Vertical
H	Horizontal
SMMR	Scanning Multichannel Microwave Radiometer
GIEMS	Global Inundation Extent from Multi-Satellites
SSMI/I	Special Sensor Microwave/Imager
ERS	European Remote Sensing
QSAT	QuickSCAT
ASCAT	Advanced Scatterometer
SSM/S	Special Sensor Microwave/Sounder
SMOS	Soil Moisture and Ocean Salinity
SMAP	Soil Moisture Active and Passive
ESA	European Space Agency
CNES	Centre National d'Etude Spatiale
CDTI	Centro para el Desarrollo Teccnologico Industrial
L	Level
EASE	Equal-Area Scalable Earth
SRTM	Shuttle Radar Topography Mission
USGS	U.S. Geological Survey
IGBP	International Geosphere Biosphere Programme
AVHRR	Advanced Very High Resolution Radiometer
MODIS	Moderate Resolution Imaging Spectroradiometer
CCI	Climate Change Initiative
NASA	North America Space Agency
TRMM	Tropical Rainfall Measuring Mission
ECMWF	European Center for Medium range Weather Forecasting
SWAF	SMOS WAter Fraction
ENVISAT	ENVironment SATellite
TB	Brightness Temperature

References

1. Matthews, E.; Fung, I. Methane emission from natural wetlands: Global distribution, area, and environmental characteristics of sources. *Glob. Biogeochem. Cycles* **1987**, *1*, 61–86.
2. Mitsch, W.; Gosselink, J. *Wetlands*, 3rd ed.; John Wiley & Sons, Inc.: Hoboken, NJ, USA, 2000.
3. Cole, J.J.; Prairie, Y.T.; Caraco, N.F.; McDowell, W.H.; Tranvik, L.J.; Striegl, R.G.; Duarte, C.M.; Kortelainen, P.; Downing, J.A.; Middelburg, J.J.; et al. Plumbing the global carbon cycle: Integrating inland waters into the terrestrial carbon budget. *Ecosystems* **2007**, *10*, 172–185.
4. Sjögersten, S.; Black, C.R.; Evers, S.; Hoyos-Santillan, J.; Wright, E.L.; Turner, B.L. Tropical wetlands: A missing link in the global carbon cycle? *Glob. Biogeochem. Cycles* **2014**, *28*, 1371–1386.
5. Alsdorf, D.E.; Rodriguez, E.; Lettenmaier, D.P. Measuring surface water from space. *Rev. Geophys.* **2007**, *45*, doi:10.1029/2006RG000197.
6. Bakker, K. Water security: Research challenges and opportunities. *Science* **2012**, *337*, 914–915.

7. Finlayson, C.; Davidson, N. *Global Review of Wetland Resources and Priorities for Wetland Inventory*; Preface iv Summary Report; CM Finlayson & AG Spiers: Canberra, CT, USA, 1999; p. 15.

8. Vörösmarty, C.; Hoekstra, A.; Bunn, S.; Conway, D.; Gupta, J. Fresh water goes global. *Science* **2015**, *349*, 478–479.

9. Costanza, R.; de Groot, R.; Sutton, P.; van der Ploeg, S.; Anderson, S.J.; Kubiszewski, I.; Farber, S.; Turner, R.K. Changes in the global value of ecosystem services. *Glob. Environ. Chang.* **2014**, *26*, 152–158.

10. Pekel, J.F.; Cottam, A.; Gorelick, N.; Belward, A.S. High-resolution mapping of global surface water and its long-term changes. *Nature* **2016**, *540*, 418–422.

11. Smith, L.C. Satellite remote sensing of river inundation area, stage, and discharge: A review. *Hydrol. Process.* **1997**, *11*, 1427–1439.

12. Frazier, P.S.; Page, K.J. Water body detection and delineation with Landsat TM data. *Photogramm. Eng. Remote Sens.* **2000**, *66*, 1461–1468.

13. Verpoorter, C.; Kutser, T.; Seekell, D.A.; Tranvik, L.J. A global inventory of lakes based on high-resolution satellite imagery. *Geophys. Res. Lett.* **2014**, *41*, 6396–6402.

14. Feng, M.; Sexton, J.O.; Channan, S.; Townshend, J.R. A global, high-resolution (30-m) inland water body dataset for 2000: First results of a topographic–spectral classification algorithm. *Int. J. Digit. Earth* **2016**, *9*, 113–133.

15. Crétaux, J.F.; Jelinski, W.; Calmant, S.; Kouraev, A.; Vuglinski, V.; Bergé-Nguyen, M.; Gennero, M.C.; Nino, F.; Del Rio, R.A.; Cazenave, A.; et al. SOLS: A lake database to monitor in the Near Real Time water level and storage variations from remote sensing data. *Adv. Space Res.* **2011**, *47*, 1497–1507.

16. Frappart, F.; Do Minh, K.; L'Hermitte, J.; Cazenave, A.; Ramillien, G.; Le Toan, T.; Mognard-Campbell, N. Water volume change in the lower Mekong from satellite altimetry and imagery data. *Geophys. J. Int.* **2006**, *167*, 570–584.

17. Hess, L.L.; Melack, J.M.; Novo, E.M.; Barbosa, C.C.; Gastil, M. Dual-season mapping of wetland inundation and vegetation for the central Amazon Basin. *Remote Sens. Environ.* **2003**, *87*, 404–428.

18. Morrissey, L.A.; Durden, S.L.; Livingston, G.P.; Steam, J.A.; Guild, L.S. Differentiating methane source areas in arctic environments with multitemporal ERS-1 SAR data. *IEEE Trans. Geosci. Remote Sens.* **1996**, *34*, 667–673.

19. Birkett, C.M. Contribution of the TOPEX NASA radar altimeter to the global monitoring of large rivers and wetlands. *Water Resour. Res.* **1998**, *34*, 1223–1239.

20. Fung, L.; Cazenave, A. *Satellite Altimetry and Earth Science*. A Handbook of Techniques and Applications, International Geophysical Series; Academic Press: San Diego, CA, USA, 2001; Volume 69.

21. Ričko, M.; Birkett, C.M.; Carton, J.A.; Crétaux, J.F. Intercomparison and validation of continental water level products derived from satellite radar altimetry. *J. Appl. Remote Sens.* **2012**, *6*, 061710.

22. Papa, F.; Prigent, C.; Rossow, W.; Legresy, B.; Remy, F. Inundated wetland dynamics over boreal regions from remote sensing: The use of Topex-Poseidon dual-frequency radar altimeter observations. *Int. J. Remote Sens.* **2006**, *27*, 4847–4866.

23. Biancamaria, S.; Lettenmaier, D.P.; Pavelsky, T.M. The SWOT mission and its capabilities for land hydrology. *Surv. Geophys.* **2016**, *37*, 307–337.

24. Giddings, L.; Choudhury, B. Observation of hydrological features with Nimbus-7 37 GHz data, applied to South America. *Int. J. Remote Sens.* **1989**, *10*, 1673–1686.

25. Choudhury, B.J. Passive microwave remote sensing contribution to hydrological variables. In *Land Surface—Atmosphere Interactions for Climate Modeling*; Springer: Berlin, Germany, 1991; pp. 63–84.

26. Sippel, S.; Hamilton, S.; Melack, J.; Choudhury, B. Passive microwave satellite observations of seasonal variations of inundation area in the Amazon River floodplain. Brazil. *Remote Sens. Environ.* **1994**, *4*, 70–76.

27. Sippel, S.; Hamilton, S.; Melack, J.; Novo, E. Passive microwave observations of inundation area and the area/stage relation in the Amazon River floodplain. *Int. J. Remote Sens.* **1998**, *19*, 3055–3074.

28. Fily, M.; Royer, A.; Goïta, K.; Prigent, C. A simple retrieval method for land surface temperature and fraction of water surface determination from satellite microwave brightness temperatures in sub-arctic areas. *Remote Sens. Environ.* **2003**, *85*, 328–338.

29. Mialon, A.; Royer, A.; Fily, M. Wetland seasonal dynamics and interannual variability over northern high latitudes, derived from microwave satellite data. *J. Geophys. Res. Atmos.* **2005**, *110*, doi:10.1029/2004JD005697.

30. Temimi, M.; Leconte, R.; Brissette, F.; Chaouch, N. Flood monitoring over the Mackenzie River Basin using passive microwave data. *Remote Sens. Environ.* **2005**, *98*, 344–355.

31. Grippa, M.; Mognard, N.; Le Toan, T.; Biancamaria, S. Observations of changes in surface water over the western Siberia lowland. *Geophys. Res. Lett.* **2007**, *34*, doi:10.1029/2007GL030165.

32. Prigent, C.; Papa, F.; Aires, F.; Rossow, W.; Matthews, E. Global inundation dynamics inferred from multiple satellite observations, 1993–2000. *J. Geophys. Res. Atmos.* **2007**, *112*, doi:10.1029/2006JD007847.

33. Sakamoto, T.; Van Nguyen, N.; Kotera, A.; Ohno, H.; Ishitsuka, N.; Yokozawa, M. Detecting temporal changes in the extent of annual flooding within the Cambodia and the Vietnamese Mekong Delta from MODIS time-series imagery. *Remote Sens. Environ.* **2007**, *109*, 295–313.

34. Kerr, Y.; Waldteufel, P.; Wigneron, J.; Martinuzzi, J.; Font, J.; Berger, M. Soil moisture retrieval from space: The Soil Moisture and Ocean Salinity (SMOS) mission. *IEEE Trans. Geosci. Remote Sens.* **2001**, *39*, 1729–1735.

35. Parrens, M.; Wigneron, J.P.; Richaume, P.; Al Bitar, A.; Mialon, A.; Wang, S.; Fernandez-Moran, R.; Al-Yaari, A.; O'Neill, P.; Kerr, Y. Considering Combined or Separated Roughness and Vegetation Effects in Soil Moisture Retrievals. *Int. J. Appl. Earth Obs. Geoinform.* **2017**, *55*, 73–86.

36. Parrens, M.; Wigneron, J.P.; Richaume, P.; Mialon, A.; Al Bitar, A.; Fernandez-Moran, R.; Al-Yaari, A.; Kerr, Y.H. Global-scale surface roughness effects at L-band as estimated from SMOS observations. *Remote Sens. Environ.* **2016**, *181*, 122–136.

37. Schroeder, R.; McDonald, K.C.; Chapman, B.D.; Jensen, K.; Podest, E.; Tessler, Z.D.; Bohn, T.J.; Zimmermann, R. Development and Evaluation of a Multi-Year Fractional Surface Water Data Set Derived from Active/Passive Microwave Remote Sensing Data. *Remote Sens.* **2015**, *7*, 16688–16732.

38. Henderson, F.M.; Lewis, A.J. Radar detection of wetland ecosystems: A review. *Int. J. Remote Sens.* **2008**, *29*, 5809–5835.

39. Bartsch, A.; Trofaier, A.; Hayman, G.; Sabel, D.; Schlaffer, S.; Clark, D.; Blyth, E. Detection of open water dynamics with ENVISAT ASAR in support of land surface modelling at high latitudes. *Biogeosciences* **2012**, *9*, 703–714.

40. Kuenzer, C.; Guo, H.; Huth, J.; Leinenkugel, P.; Li, X.; Dech, S. Flood mapping and flood dynamics of the Mekong Delta: ENVISAT-ASAR-WSM based time series analyses. *Remote Sens.* **2013**, *5*, 687–715.

41. Santoro, M.; Wegmüller, U. Multi-temporal synthetic aperture radar metrics applied to map open water bodies. *IEEE J. Sel. Top. Appl. Earth Obs. Remote Sens.* **2014**, *7*, 3225–3238.

42. Prigent, C.; Matthews, E.; Aires, F.; Rossow, W.B. Remote sensing of global wetland dynamics with multiple satellite data sets. *Geophys. Res. Lett.* **2001**, *28*, 4631–4634.

43. Papa, F.; Prigent, C.; Aires, F.; Jimenez, C.; Rossow, W.; Matthews, E. Interannual variability of surface water extent at the global scale, 1993–2004. *J. Geophys. Res. Atmos.* **2010**, *115*, doi:10.1029/2009JD012674.

44. McCarthy, J.; Gumbricht, T.; McCarthy, T. Ecoregion classification in the Okavango Delta, Botswana from multitemporal remote sensing. *Int. J. Remote Sens.* **2005**, *26*, 4339–4357.

45. Jain, S.K.; Saraf, A.K.; Goswami, A.; Ahmad, T. Flood inundation mapping using NOAA AVHRR data. *Water Resour. Manag.* **2006**, *20*, 949–959.

46. Bergé-Nguyen, M.; Crétaux, J.F. Inundations in the inner Niger delta: monitoring and analysis using MODIS and global precipitation datasets. *Remote Sens.* **2015**, *7*, 2127–2151.

47. Xiao, X.; Boles, S.; Liu, J.; Zhuang, D.; Frolking, S.; Li, C.; Salas, W.; Moore, B. Mapping paddy rice agriculture in southern China using multi-temporal MODIS images. *Remote Sens. Environ.* **2005**, *95*, 480–492.

48. Molinier, M.; Guyot, J.L.; De Oliveira, E.; Guimarães, V. *Les Regimes Hydroiogiques de L'Amazone et de Ses Affluents*; IAHS Publication: Paris, France, 1996; pp. 209–222.

49. Frappart, F.; Papa, F.; da Silva, J.S.; Ramillien, G.; Prigent, C.; Seyler, F.; Calmant, S. Surface freshwater storage and dynamics in the Amazon Basin during the 2005 exceptional drought. *Environ. Res. Lett.* **2012**, *7*, 044010.

50. Papa, F.; Frappart, F.; Güntner, A.; Prigent, C.; Aires, F.; Getirana, A.C.; Maurer, R. Surface freshwater storage and variability in the Amazon Basin from multi-satellite observations, 1993–2007. *J. Geophys. Res. Atmos.* **2013**, *118*, doi:10.1002/2013JD020500.

51. Diegues, A.C.S. *An Inventory of Brazilian Wetlands*; Number 15; IUCN: Gland, Switzerland, 1994.

52. Frappart, F.; Seyler, F.; Martinez, J.M.; Leon, J.G.; Cazenave, A. Floodplain water storage in the Negro River basin estimated from microwave remote sensing of inundation area and water levels. *Remote Sens. Environ.* **2005**, *99*, 387–399.

53. Junk, W.J. General aspects of floodplain ecology with special reference to Amazonian floodplains. In *The Central Amazon Floodplain*; Springer: Berlin, Germany, 1997; pp. 3–20.

54. Salati, E.; Marques, J. Climatology of the Amazon region. In *The Amazon*; Springer: Berlin, Germany, 1984; pp. 85–126.

55. Marengo, J.A. Interannual variability of surface climate in the Amazon Basin. *Int. J. Climatol.* **1992**, *12*, 853–863.

56. Espinoza Villar, J.C.; Ronchail, J.; Guyot, J.L.; Cochonneau, G.; Naziano, F.; Lavado, W.; De Oliveira, E.; Pombosa, R.; Vauchel, P. Spatio-temporal rainfall variability in the Amazon Basin countries (Brazil, Peru, Bolivia, Colombia, and Ecuador). *Int. J. Climatol.* **2009**, *29*, 1574–1594.

57. Ulaby, F.T. *Microwave Remote Sensing: Active and Passive, Radar Remote Sensing and Surface Scattering and Emission Theory*; Addison-Wesley: Reading, MA, USA, 1982; Volume 2.

58. Bitar, A.; Mialon, A.; Kerr, Y.; Cabot, F.; Richaume, P.; Jacquette, E.; Quesney, A.; Mahmoodi, A.; Tarrot, S.; Parrens, M.; et al. The global SMOS Level 3 daily soil moisture and brightness temperature maps. *ESSD* **2017**, doi:10.5194/essd-2017-1.

59. Brodzik, M.J.; Billingsley, B.; Haran, T.; Raup, B.; Savoie, M.H. EASE-grid 2.0: Incremental but significant improvements for Earth-gridded data sets. *ISPRS Int. J. Geo-Inform.* **2012**, *1*, 32–45.

60. Jarvis, A.; Reuter, H.I.; Nelson, A.; Guevara, E. the CGIAR-CSI SRTM 90 m Database. Hole-Filled SRTM for the Globe Version 4. 2008. Available online: http://srtm.csi.cgiar.org (accessed on 16 May 2017).

61. Huffman, G.J.; Bolvin, D.T.; Nelkin, E.J.; Wolff, D.B.; Adler, R.F.; Gu, G.; Hong, Y.; Bowman, K.P.; Stocker, E.F. The TRMM multisatellite precipitation analysis (TMPA): Quasi-global, multiyear, combined-sensor precipitation estimates at fine scales. *J. Hydrometeorol.* **2007**, *8*, 38–55.

62. Friedl, M.A.; Sulla-Menashe, D.; Tan, B.; Schneider, A.; Ramankutty, N.; Sibley, A.; Huang, X. MODIS Collection 5 global land cover: Algorithm refinements and characterization of new datasets. *Remote Sens. Environ.* **2010**, *114*, 168–182.

63. Loveland, T.; Reed, B.; Brown, J.; Ohlen, D.; Zhu, Z.; Yang, L.; Merchant, J. Development of a global land cover characteristics database and IGBP DISCover from 1 km AVHRR data. *Int. J. Remote Sens.* **2000**, *21*, 1303–1330.

64. Bontemps, S.; Defourny, P.; Radoux, J.; Van Bogaert, E.; Lamarche, C.; Achard, F.; Mayaux, P.; Boettcher, M.; Brockmann, C.; Kirches, G.; et al. Consistent global land cover maps for climate modelling communities: current achievements of the ESA's land cover CCI. In Proceedings of the ESA Living Planet Symposium, Edimburgh, UK, 9–13 September 2013; pp. 9–13.

65. Papa, F.; Güntner, A.; Frappart, F.; Prigent, C.; Rossow, W.B. Variations of surface water extent and water storage in large river basins: A comparison of different global data sources. *Geophys. Res. Lett.* **2008**, *35*, doi:10.1029/2008GL033857.

66. Frappart, F.; Papa, F.; Famiglietti, J.S.; Prigent, C.; Rossow, W.B.; Seyler, F. Interannual variations of river water storage from a multiple satellite approach: A case study for the Rio Negro River basin. *J. Geophys. Res. Atmos.* **2008**, *113*, doi:10.1029/2007JD009438.

67. Birkett, C.M.; Beckley, B. Investigating the performance of the Jason-2/OSTM radar altimeter over lakes and reservoirs. *Mar. Geod.* **2010**, *33*, 204–238.

68. Da Silva, J.S.; Calmant, S.; Seyler, F.; Rotunno Filho, O.C.; Cochonneau, G.; Mansur, W.J. Water levels in the Amazon Basin derived from the ERS 2 and ENVISAT radar altimetry missions. *Remote Sens. Environ.* **2010**, *114*, 2160–2181.

69. Klein, L.; Swift, C. An improved model for the dielectric constant of sea water at microwave frequencies. *IEEE Trans. Antennas Propag.* **1977**, *25*, 104–111.

70. Mialon, A.; Coret, L.; Kerr, Y.H.; Sécherre, F.; Wigneron, J.P. Flagging the topographic impact on the SMOS signal. *IEEE Trans. Geosci. Remote Sens.* **2008**, *46*, 689–694.

71. Frappart, F.; Ramillien, G.; Ronchail, J. Changes in terrestrial water storage versus rainfall and discharges in the Amazon Basin. *Int. J. Climatol.* **2013**, *33*, 3029–3046.

72. Ulaby, F.; Moore, R.; Fung, A. *Microwave Remote Sensing: Active and Passive*; Advanced Systems and Applications Inc.: Dedham, MA, USA, 1986; pp. 1797–1848.

3

Fifteen Years (1993–2007) of Surface Freshwater Storage Variability in the Ganges-Brahmaputra River Basin using Multi-Satellite Observations

Edward Salameh [1,2], Frédéric Frappart [1,3], Fabrice Papa [1,4,*], Andreas Güntner [5], Vuruputur Venugopal [6], Augusto Getirana [7,8], Catherine Prigent [9], Filipe Aires [9], David Labat [3] and Benoît Laignel [2]

[1] Laboratoire d'Etudes en Géophysique et Océanographie Spatiales (LEGOS), Université de Toulouse, IRD, CNES, CNRS, UPS, Toulouse 31400, France; edward.salameh@legos.obs-mip.fr (E.S.); frederic.frappart@legos.obs-mip.fr (F.F.)

[2] Université de Rouen, UMR CNRS 6143, Mont-Saint-Aignan 76821, France; benoit.laignel@univ-rouen.fr

[3] Géosciences Environnement Toulouse (GET), Université de Toulouse, IRD, CNES, CNRS, UPS, Toulouse 31400, France; david.labat@get.omp.eu

[4] Indo-French Cell for Water Sciences (IFCWS), IRD-IISc-NIO-IITM Joint International Laboratory, Bangalore 560012, India

[5] GFZ German Research Centre for Geosciences, Potsdam 14473, Germany; andreas.guentner@gfz-potsdam.de

[6] Centre for Atmospheric and Oceanic Sciences (CAOS), Indian Institute of Science, IISc, Bangalore 560012, India; venu@caos.iisc.ernet.in

[7] Hydrological Sciences Laboratory, NASA Goddard Space Flight Center, Greenbelt, MD 20771, USA; augusto.getirana@nasa.gov

[8] Earth System Science Interdisciplinary Center, University of Maryland, College Park, College Park, MD 20742, USA

[9] Laboratoire d'Etudes du Rayonnement et de la Matière en Astrophysique (LERMA), CNRS, Observatoire de Paris, Paris 75014, France; Catherine.Prigent@obspm.fr (C.P.); filipe.aires@obspm.fr (F.A.)

* Correspondence: fabrice.papa@ird.fr

Academic Editor: Karl-Erich Lindenschmidt

Abstract: Surface water storage is a key component of the terrestrial hydrological and biogeochemical cycles that also plays a major role in water resources management. In this study, surface water storage (SWS) variations are estimated at monthly time-scale over 15 years (1993–2007) using a hypsographic approach based on the combination of topographic information from Advance Spaceborne Thermal Emission and Reflection Radiometer (ASTER) and Hydrological Modeling and Analysis Platform (HyMAP)-based Global Digital Elevation Models (GDEM) and the Global Inundation Extent Multi-Satellite (GIEMS) product in the Ganges-Brahmaputra basin. The monthly variations of the surface water storage are in good accordance with precipitation from Global Precipitation Climatology Project (GPCP), river discharges at the outlet of the Ganges and the Brahmaputra, and terrestrial water storage (TWS) from the Gravity Recovery And Climate Experiment (GRACE), with correlations higher than 0.85. Surface water storage presents a strong seasonal signal (~496 km^3 estimated by GIEMS/ASTER and ~378 km^3 by GIEMS/HyMAPs), representing ~51% and ~41% respectively of the total water storage signal and it exhibits a large inter-annual variability with strong negative anomalies during the drought-like conditions of 1994 or strong positive anomalies such as in 1998. This new dataset of SWS is a new, highly valuable source of information for hydrological and climate modeling studies of the Ganges-Brahmaputra river basin.

Keywords: Ganges-Brahmaputra; surface water storage; multi-satellite; floodplains

1. Introduction

Continental freshwater is crucial for all forms of life. Despite their minor quantitative contribution to the total water storage on Earth (<1%), surface water stored in rivers, lakes, wetlands, floodplains and even man-made reservoirs plays a major role in climate variability, also affecting biogeochemical and trace gas cycles. As a part of the hydrologic cycle, investigating the spatio-temporal variation of the surface water storage (SWS) is fundamental to the study of the global water cycle while providing a critical parameter for water resources management [1].

Until recently, our knowledge of surface water dynamics relied on sparse in situ observations and hydrological models. Traditional in situ gauge measurements quantify the water discharge in river channels, but no information is provided by these measurements regarding the diffusive flow over floodplains associated to rivers or wetlands. Furthermore, the number of gauging networks is, in general, limited especially in remote areas with difficult access such as tropical regions. When available, discharge data and hydrological observations are often classified by governments due to transboundary issues and their access is restricted for scientific usage [2,3].

The ability to estimate SWS variability at large scales is becoming increasingly important because of the need to predict the availability of freshwater resources and also to link this variability to climate change and extreme events such as droughts and floods [4]. Recent advancements in remote sensing made the study of surface water dynamics possible at regional to global scale [3]. Multi-satellite remote sensing techniques now offer important information on land surface waters, such as the variations of surface water extents at the global scale [2,5,6] provided by the Global Inundation Extent from Multi-Satellites (GIEMS). This information is complementary to radar altimetry observations that systematically monitor the water levels in lakes, large rivers, wetlands and floodplains [7,8].

Recent efforts have been undertaken to quantify the surface freshwater storage and its variations at seasonal to inter-annual time scales using satellite observations. A technique developed by [9] to estimate SWS variations combines surface water extent observations with altimeter-derived height variations in rivers, wetlands, and inundations [10]. This technique was firstly developed over the Rio Negro, a sub-basin of the Amazon [9], and it was tested over the Ob River basin [11] and the Orinoco [12]. Over the Amazon [4], SWS variations over the period 2003–2007 helped to quantify and characterize the extreme drought of 2005. Over the Ganges-Brahmaputra River system [13], SWS obtained from a combination of GIEMS and ENVISAT observations helped to map sub-surface water variations by decomposing the total terrestrial water storage (TWS) variations measured by the Gravity Recovery And Climate Experiment (GRACE).

Another technique to estimate SWS was proposed by [14], and combines surface water extents from GIEMS with topographic data derived from Global Digital Elevation Model (GDEM), using a hypsographic curve approach. The latter technique was firstly developed and assessed over the Amazon watershed [14] and helped to characterize the SWS anomaly during the 1997 and 2005 extreme droughts.

In the present study, we propose to estimate and analyze SWS variations over the Ganges-Brahmaputra (GB) system using the hypsographic curve technique proposed by [14]. The GB system drains a large part of the Indian sub-continent and hosts more than 700 million people. It is the third largest freshwater outlet to the world's oceans, being exceeded only by the Amazon and the Congo drainage basins [15]. The basin is facing strong climate variability with alternate periods of floods and droughts. Due to the population growth, the excessive use of water for industrial and agricultural purposes, many water management challenges are emerging in this region [16]. The Indian sub-continent is facing acute shortages of drinking and agricultural water supply, aggravated by geogenic arsenic contamination of groundwater reservoirs, especially in Bangladesh [17].

Sections 2 and 3 present respectively the study domain and the datasets used in this study. In Section 4, we briefly describe the SWS estimation technique [14]. The results are presented and discussed in Section 5; an evaluation is performed by comparing the new estimates with SWS estimates provided by the GIEMS-ENVISAT combination technique [13], as well as other external datasets such as GRACE-derived TWS variations, satellite altimetry-derived river discharge observations and precipitation. Conclusions and perspectives are provided in Section 6.

2. Study Area

The Ganges-Brahmaputra-Meghna (hereafter referred to as the Ganges-Brahmaputra basin or GB) is a transboundary river system draining a large area of ~1.7 million km^2 and crossing India, China, Nepal, Bhutan, and Bangladesh (Figure 1).

The headwaters of the Ganges (G) and Brahmaputra (B) Rivers originate in the Himalayan range in China. The Ganges heads in the Gangotri glacier whilst the Brahmaputra River head is located in the southern slopes of Kailash Mountain in the Trans-Himalaya [15]. After flowing southwest into India and then turning southeast, the Ganges River converges with Brahmaputra in Bangladesh and flow into the Bay of Bengal where the GB delta is formed. Before merging with the Ganges, the Brahmaputra River flows east through the southern area of China, then flows south into eastern India and turns southwest crossing Bangladesh borders. The Ganges watercourse is classed as a meandering channel while Brahmaputra is a braided one [18,19].

The GB River basin is unique in the world in terms of its climate and great availability of freshwater that is highly seasonal and driven primarily by monsoonal rainfall that dominates discharge, with a lesser contribution from snowmelt [15].

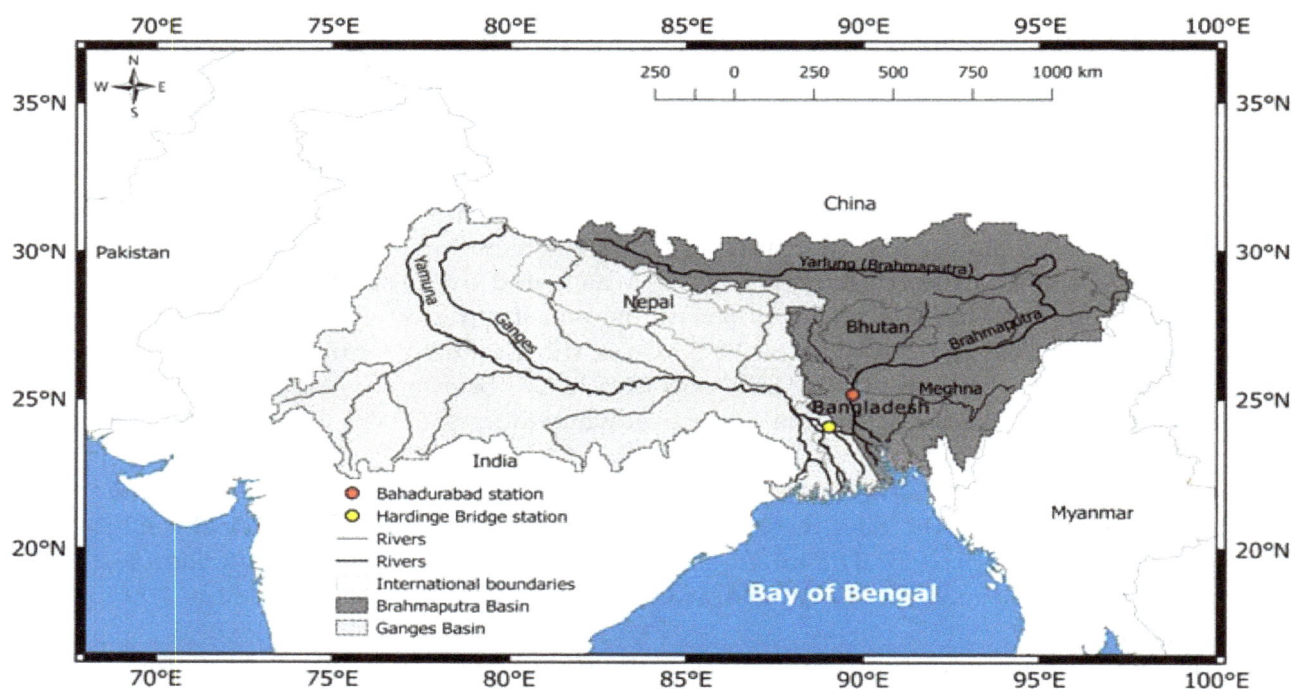

Figure 1. Ganges and Brahmaputra River basin, with the respective catchment areas shown in light gray (Ganges) and dark gray (Brahmaputra). The black lines show the main rivers (thick line) and associated tributaries (thin line) hydrography. Political borders are shown in a gray line. The red and the yellow circles correspond to the locations of Bahadurabad and Hardinge Bridge in situ gauging stations respectively where altimeter-derived river discharges are estimated.

3. Datasets

3.1. Global Inundation Extent from Multi-Satellites (GIEMS)

The complete methodology that captures the extent of episodic and seasonal inundation, wetlands, rivers, lakes and irrigated agriculture, at the global scale, is described in detail in [2,5,6,20,21].

The technique uses a complementary suite of satellite observations covering a large wavelength range: (1) passive microwave emissivities between 19 and 85 GHz. These are estimated from the Special Sensor Microwave/Imager (SSM/I) observations by removing the contributions of the atmosphere (water vapor, clouds, rain) and the modulation by the surface temperature [22,23]. The technique uses ancillary data from the International Satellite Cloud Climatology Project (ISCCP) [24] and the National Centers for Environment Prediction (NCEP) reanalysis [25]; (2) Advanced Very High Resolution Radiometer (AVHRR) visible (0.58–0.68 μm) and near-infrared (0.73–1.1 μm) reflectances and the derived Normalized Difference Vegetation Index (NDVI); (3) backscatter at 5.25 GHz from the European Remote Sensing (ERS) satellite scatterometer.

Observations are averaged over each month and mapped to an equal area grid of 0.25° resolution at the equator (each pixel covers 773 km^2) [2,6]. An unsupervised classification of the three sources of satellite data is performed and the pixels with satellite signatures likely related to inundation are retained. For each inundated pixel, the monthly fractional coverage by open water is obtained using the passive microwave signal and a linear mixture model with end-members calibrated with scatterometer observations to account for the effects of vegetation cover [6,20]. As the microwave measurements are also sensitive to the snow cover, snow and ice masks are used to filter the results and avoid any confusion with snow-covered pixels [2]. Because the ERS scatterometer encountered serious technical problems after 2000, the processing scheme had to be adapted to extend the dataset and monthly mean climatology of ERS and NDVI-AVHRR observations are used [2,5]. Fifteen years of global monthly water surfaces extent for the period 1993–2007 are available [5]. This dataset has been extensively evaluated at the global scale [2,6,26] and for a wide range of environments [9,21,27,28]. It has also been used for climatic and hydrological analyses, such as the evaluation of methane surface emissions models [29,30] and the validation of the river flooding schemes coupled with land surface models [31–35].

The spatial distribution of GIEMS was evaluated against high-resolution (100 m) SAR images in [6] and in [36] over the Amazon basin leading to an overall GIEMS uncertainty of ~10% for GIEMS. Over the Indian Sub-Continent (and especially GB), the spatial distribution of GIEMS was evaluated against static surface water dataset (Global Lake and Wetland Dataset, GLWD-3, [37]) and other related hydrological variables (precipitation, altimeter-derived river heights, river discharge) in [5,21], as well as using other regional surveys representing various components of wetland and open-water distributions [38].

Figure 2 shows GIEMS characteristics over the GB basin. Figure 2a,c show the annual mean and annual maximum extent of surface water respectively, averaged over 15 years (180 months). They exhibit very realistic distributions of major rivers (Ganges-Brahmaputra-Meghna River systems) and their tributaries and distributaries. Associated inundated areas, wetlands and the region of the Bengal delta are well delineated even in the presence of complex areas characterized by extensive flooding. The associated standard deviations (Figure 2b) show relatively high values (<50%), illustrating the high seasonal and inter-annual variability of hydrologic processes within the GB system. This is also shown in Figure 2d where the mean amplitude of the water extent (difference between the mean maximum and the mean minimum over the record for each pixel) exhibits very high values. It should be noted here that despite the fact that GIEMS is able to capture the distributions and variations of surface freshwater in the GB basin, some very high maximum values of surface water extents (Figure 2c) could be related to the fact that the method encounters difficulties in some regions to discriminate between very saturated/moist soil and standing open water, especially in the delta region. This can lead to an overestimation of the actual surface water extents, especially for pixels with high flood coverage (see the histograms in Figure 2d of [2]).

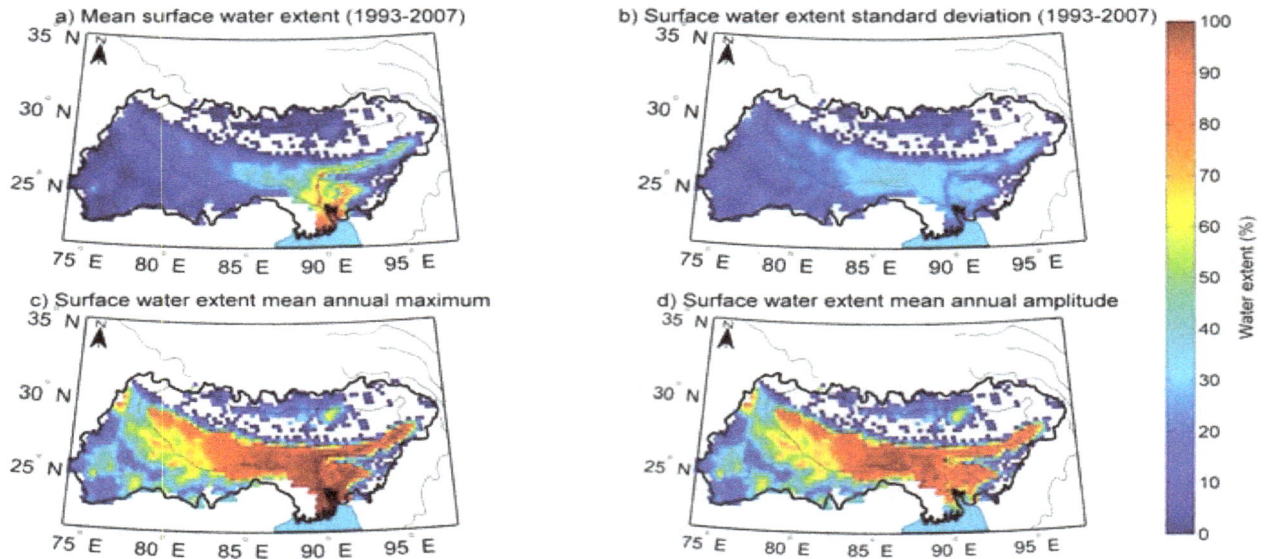

Figure 2. Main characteristics of the spatial distribution of the surface water extents provided by Global Inundation Extent Multi-Satellite (GIEMS) for the Ganges-Brahmaputra (GB) basin (all values as areal fractions of 773 km² GIEMS cells): (**a**) Mean surface water extent for the 1993–2007 period; (**b**) Associated standard deviation; (**c**) Mean annual maximum; (**d**) Mean annual amplitude.

3.2. ASTER-GDEM to Derive Hypsographic Curves

Advanced Spaceborne Thermal Emission and Reflection Radiometer (ASTER) GDEM was developed jointly by the National Aeronautic and Space Administration (NASA) and the Ministry of Economy, Trade and Industry (METI) of Japan. The ASTER instrument, launched onboard NASA's Terra spacecraft in December 1999, has an along-track stereoscopic capability using its near infrared spectral band and its nadir-viewing and backward-viewing telescopes to acquire stereo image data with a base-to-height ratio of 0.6. The basic characteristics of stereoscopy and its application to the ASTER system for GDEM generation are explained in detail in [39]. The horizontal spatial resolution is 15 m and one nadir-looking ASTER visible and near-infrared (VNIR) scene corresponds to 60 km². The methodology used to produce the ASTER GDEM involves automated processing of the entire 1.5-million ASTER scenes archived from the start of observation until August 2008 [39,40]. The processing includes stereo-correlation to produce ~1,264,000 individual scene-based ASTER DEMs, cloud masking to remove cloudy pixels, stacking all cloud-screened DEMs removing residual bad values and outliers, averaging selected data to create final pixel values, and then correcting residual anomalies. The ASTER-GDEM covers land surfaces between 83° N and 83° S and is partitioned into 22,600 tiles of 1° × 1° (containing at least 0.01% of land area). ASTER-GDEM has a 1″ (30 m) spatial horizontal resolution and is referenced with respect to the WGS84/EGM96 geoid. Several studies have dealt with the evaluation of ASTER-GDEM at local to regional scales [41–44]. Pre-production accuracies for the global product were estimated at ~20 m vertically and ~30 m horizontally. In this study, we use the ASTER-GDEM Version 2 released in October 2011 [45].

3.3. SRTM-GDEM to Derive Hypsographic Curves as Used in CaMa-Flood and HyMAP Models

The SRTM (Shuttle Radar Topography Mission) [46] mission is a joint effort between the NASA, the National Geospatial Intelligence Agency (NGA), the German (Deutsches Zentrum für Luft-und Raumfahrt) and the Italian (Agenzia Spaziale Italiana) spatial agencies. The instruments of SRTM mission embarked on Endeavour in February 2000 and acquired radar data during its 11-day mission which allows the construction of a GDEM of all land surfaces between 60° N and 56° S [46]. In this study, we use SRTM30 DEM modulated (error correction) as in [47] (CaMa-Flood) and [33]

(HyMAP). See [48] for a detailed description of error corrections and the construction of SRTM-derived hypsographic curves.

3.4. Complementary Datasets Used for Validation

3.4.1. Multi-Satellite Surface Water Storage from GIEMS and ENVISAT Radar Altimeter

Maps of water levels over the floodplains of the Ganges-Brahmaputra basin were obtained by combining observations from GIEMS and altimetry-based water levels at monthly time-scale over the 2003–2007 period where all the datasets overlap [13]. Water levels for 58 ENVISAT RA-2 altimetry stations were interpolated with respect to the inverse of the distance from the gridpoint over inundated surfaces from GIEMS [2,5]. Each monthly map of surface water levels has a spatial resolution of 0.25° and is referenced to the EGM2008 geoid [49]. The error on these estimates is lower than 10% [13]. A map of minimum water levels was estimated for the entire observation period using a hypsometric approach to take into account the difference of altitude between the river and the floodplain (see [12,13] for more details). This dataset is made available by the Centre de Topographie des Océans et de l'Hydrosphère (CTOH) [50].

3.4.2. GRACE Level-2 Monthly Solutions

The Gravity Recovery And Climate Experiment (GRACE) mission, launched in March 2002, provides measurements of the spatio-temporal changes in Earth's gravity field. At basin scale, GRACE data can be used to derive the monthly changes of the total land water storage (TWS) [51,52] with an accuracy of ~1.5 cm of equivalent water thickness when averaged over surfaces of a few 100 km^2 [53]. Three processing centers, including the Center for Space Research (CSR), Austin, TX, USA, the German Research Centre for Geosciences (GFZ), Potsdam, Germany, the Jet Propulsion Laboratory (JPL), Pasadena, CA, USA, and the Science Data Center (SDC) are in charge of the processing of the GRACE data and the production of Level-1 and Level-2 solutions. Level-2 solutions consist of time series of monthly averages of Stokes coefficients (i.e., dimensionless spherical harmonics coefficients of geopotential) developed up to a degree between 90 and 150 that are adjusted from along-track inter-satellite range GRACE measurements. These coefficients are mostly related to water storage variations on land. In this study, we use the Level-2 Release 05 solutions. The presence of an unrealistic high frequency noise corresponding to north-south striping is caused by orbit resonance during the Stokes coefficients determination and aliasing of poorly modeled short-term phenomena. To attenuate the noise in the Level-2 GRACE solutions, we applied an Independent Component Analysis (ICA) approach to the combination of GFZ/UTCSR/JPL solutions of the same monthly period to isolate statistically independent components of the observed gravity field (i.e., the continental water storage contribution from the high frequency noise) [54,55].

3.4.3. GPCP Monthly Rainfall Product

In order to further evaluate our various estimates of the satellite-derived surface water storage, we will compare them with precipitation over the Ganges and Brahmaputra watersheds estimated by the Global Precipitation Climatology Project (GPCP). GPCP, established in 1986 by the World Climate Research Program, quantifies the distribution of precipitation over the globe [56]. We use the Satellite-Gauge Combined Precipitation Data product of GPCP Version 2.1 data (monthly means from 1993 to 2008) with a spatial resolution of 2.5° in latitude and longitude. Over land surfaces, the uncertainty in the rate estimates from GPCP is generally lower than over the oceans due to the in situ gauge input (in addition to satellite) from the GPCC (Global Precipitation Climatology Center). Over land, validation experiments have been conducted in a variety of locations worldwide and the results suggest that, while there are known problems in regions of persistent convective precipitation, non-precipitating cirrus or regions of complex terrain, the estimated uncertainties range between 10% and 30% [56].

3.4.4. River Discharges

In situ river water level and discharge are infrequently recorded by the Bangladesh Water Development Board (BWDB) [57] at the two basin outlet stations before the confluence of the two rivers, at the Hardinge Bridge station (24.07° N; 89.03° E) for the Ganges and the Bahadurabad station (25.15° N; 89.70° E) for the Brahmaputra (Figure 1). Here, we will use the discharge for each river derived from satellite altimeter as in [58] for the 1993–2007 period.

4. Methods

4.1. GIEMS Surface Water Extent Thresholding

Previous analysis [2,21,38] suggested that GIEMS overestimates the actual surface water extents in regions of very saturated soils. To overcome this issue, we use external information on flood coverage from the Dartmouth Flood Observatory (DFO) database that provides surface water extent for the period 2002–2015 [59]. It comprises the Surface Water Record (SWR), a comprehensive record of satellite-observed changes in the Earth's inland surface waters, compiled from the flooding history over the period 2002–2015. Extent of surface water is mostly derived from NASA MODerate-resolution Imaging Spectroradiometer (MODIS) Terra and Aqua sensors with, in some cases, additional information from Radarsat, ASTER, or other higher spatial resolution data [60]. Water areas are accumulated over 10 days to minimize the effect of cloud cover. Inundation maps are made available at a spatial resolution of 250 m on $10° \times 10°$ tiles. A color code indicates maximum flood extent each year. This dataset is commonly used for estimating flood extent limits when processing other remotely sensed observations (e.g., [13,61,62]). It was resampled on the GIEMS low resolution grid providing a percentage of inundation for each grid point equal to the inundation extent given by the SWR divided by the area of the GIEMS grid element (773 km^2). The resulting MODIS inundation mask over the Ganges-Brahmaputra is presented in Figure 3. One can see that the main river channels, along with the major floodplains and wetlands are well depicted over the basin. However, compared to GIEMS estimates (Figure 2), many pixels present lower maximum extents, especially in the upstream regions.

In the following, the DFO MODIS-derived inundation map is used to create an inundation mask in order to limit the surface water extent given by GIEMS over the Ganges-Brahmaputra. For each pixel of GIEMS, monthly surface water extent $S_{GIEMS}(\alpha, t)$ is modulated by multiplying it by the ratio of maximum inundation extent of MODIS $S_{MODIS}(\alpha)$ and the maximum monthly extent value of GIEMS over the record such as:

$$S_{GIEMS}(\alpha, t) = S_{GIEMS}(\alpha, t) \times \frac{S_{MODIS}(\alpha)}{\max(S_{GIEMS}(t_{i=1,...,180}))} \tag{1}$$

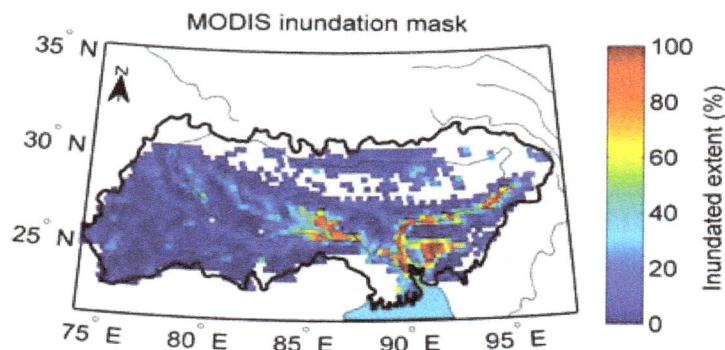

Figure 3. MODerate-resolution Imaging Spectroradiometer (MODIS)-derived surface water extent (MODIS inundation extent mask in the followings) over the GB system given in percentage of the pixel area.

4.2. The Hypsographic Curve Approach

The method to estimate surface freshwater storage consists in the combination of the surface water extent from GIEMS product with a global digital elevation model (GDEM), using a hypsographic curve approach that relates the flooded area to the elevation. We will derive here two estimates using two global datasets of hypsographic curves derived from ASTER-GDEM and SRTM30 DEM (simply named as HyMAP in the following) as processed for CaMa-Flood and HyMAP models. The three-step methodology to construct the hypsographic curve and estimate SWS is described in details in [14], and is briefly summarized below:

1. The first step is to construct the hypsographic curve for each pixel of the surface water extent dataset (GIEMS). The corresponding GDEM elevation points for each pixel of GIEMS product (equal-grid of 773 km^2) are selected. The elevation distribution function is then created and converted (by integration) to a curve of cumulative frequencies. The latter function presents the so-called hypsographic curve that consists of an area-elevation relationship, constructed for each pixel of the GIEMS data set.

2. In the second step, a translation is applied to set to zero the lowest elevation of the hypsographic curve by subtracting the lowest value from all other elevations. The hypsographic curve is then converted into an area-surface water volume relationship by estimating the surface water volume associated with an increase of the pixel fractional open water coverage by filling the hypsographic curve from its base level to an upward level, following [13]:

$$V(\alpha) = \sum_{i=1}^{\alpha} (h(i) - h(i-1)) \times \frac{S}{100} \times i \qquad (2)$$

where V is the surface water volume (in km^3) for a percentage of flood/inundation α (an increment i of 1% in percentage of inundation is chosen), S is the area of a GIEMS pixel (773 km^2), and h the elevation (in km) for a percentage of flood/inundation α given by the hypsographic curve.

3. In the last step, the surface water storage of each pixel is estimated for each month by combining the hypsographic curve with the monthly variations of surface water extent from GIEMS using Equation (2). The estimated surface water storages are not absolute. They correspond to the water volume present over a reference surface that is the topography or the elevation of the surface corresponding to the minimum water levels during the observation period. Thus, the estimated water storage represents the increment above the minimum storage.

Examples of hypsographic curves at several locations in the GB basin are shown in Figure 4.

One can see that, in most cases, the hypsographic curves from ASTER-GDEM and STRM30-HyMAP are very similar (Figure 4c,h,i,l,o) or showing small differences of less than few km^3 (Figure 4a,b,j,n,p). In some cases, the differences are large (Figure 4e,g). Those differences can be attributed to the raw DEM product specificities (mode of acquisition, resolution, uncertainties, and errors). Indeed, both DEMs have been estimated with extremely different techniques (SRTM-30 is based on radar observations while ASTER-GDEM is made using near infrared spectral band). Moreover, one of the major limitations of satellite-derived DEMs is that they are not always representing bare earth but can include vegetation and man-made structures. ASTER-GDEM can also be affected by cloud cover, such as very low but dense boundary layer clouds in tropical regions. All those effects are difficult to filter in the raw data products despite large processing and can result in erroneous high elevation topographic data, inducing further large errors in the hypsographic curves.

In order to prevent overestimation of surface water volumes due to high elevation values at the upper edge of the hypsographic curve for some pixels, we proposed a correction method which is thoroughly explained in Section 3.2 of [14] over the Amazon basin. This method mainly consists in calculating for each pixel the standard deviations (STD) of the water volume derived from ASTER-GDEM and STRM30-HyMAP hypsographic curves over 5% flood coverage windows. Standard deviation values are then used as proxies of realistic magnitude of surface water volume changes. Unrealistic variations expressed by high standard deviations (higher than the threshold fixed at 0.4 km^3) are carefully replaced by a fitter value of the surface water storage based on a simple linear regression analysis using the 10 previous water volume values of the hypsographic curve.

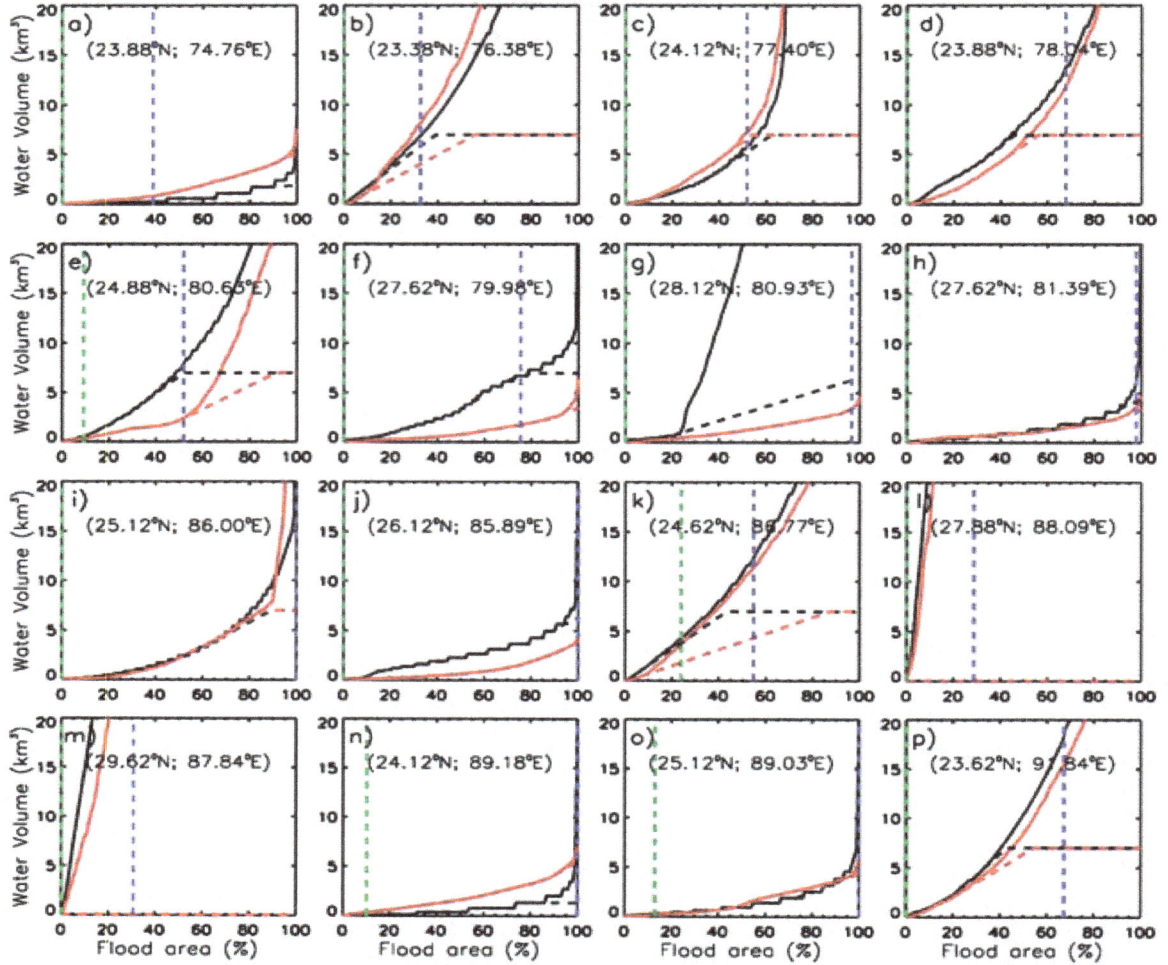

Figure 4. Surface volume profile (equivalent to the hypsographic curve), i.e., the relationship between the surface water storage within each grid cell and the inundated area of a 773 km^2 pixel (in percent) for several locations over the GB basin. Red curves are from Advanced Spaceborne Thermal Emission and Reflection Radiometer (ASTER)-Global Digital Elevation Model (GDEM) and black curves from Hydrological Modeling and Analysis Platform (HyMAP)-GDEM. The blue dashed line is the maximum coverage of surface water and the green dashed line is the minimum coverage of surface water observed by GIEMS during the period 1993–2007. The dashed red curves and the dashed black curves represent the hypsographic curves from ASTER-GDEM and HyMAP-GDEM respectively, after corrections are applied (see [14] for details.). (**a–p**) correspond to 16 different locations in the Ganges-Brahmaputra basin.

4.3. Time Series of Basin Scale Total Water Storage (TWS)

The time variations of volume of TWS anomalies from Level-2 GRACE solutions are computed following [51]:

$$\Delta V_{TWS}(t) = R_e^2 \sum_{j \in S} \Delta h_{tot}(\lambda_j, \varphi_j, t) \cos(\varphi_j) \Delta\lambda\Delta\varphi \qquad (3)$$

where $\Delta h_{tot}(\lambda_j, \varphi_j, t)$ is the anomaly of TWS at time t of the pixel of coordinates (λ_j, φ_j) and R_e is the radius of the Earth (6378 km).

5. Results and Discussion

Combining the corrected ASTER-GDEM (and HyMAP) hypsographic curves and the GIEMS satellite-derived observations corrected from MODIS, we can now estimate, for the first time, the long-term Ganges-Brahmaputra SWS and spatio-temporal variations for the period 1993–2007.

Figure 5 shows the spatial distribution of SWS characteristics (annual mean, standard deviation, mean annual maximum, and mean annual amplitude estimated for the study period) for the entire GB basin. Realistic spatial patterns are observed with the upstream pixels characterized by smaller water volumes in contrast with the downstream region. Major river channels are well delineated from the head to the outlet (or the confluence in case of a tributary), as well as the extensive floodplains present along their stream. Following the spatial distribution observed in GIEMS products (Figure 2), both floodplains associated with the river channels in the GB basin and delta plains in southern Bangladesh are well represented. SWS standard deviation (Figure 5b) and mean annual amplitude (Figure 5d) highlight the regions with strong variability such as the Meghna floodplains (between longitudes 90° E and 94° E and latitudes 22° N and 26° N) and the river confluences in Bangladesh that form the Bengal delta.

Figure 6 presents the monthly variations of basin-scale SWS for the 1993–2007 period for the Ganges (a), Brahmaputra (b) and the entire GB system (c). It corresponds to surface water volumes estimated before (green for ASTER and black for HyMAP) and after (blue for ASTER and red for HyMAP) the use of the MODIS inundation mask.

In order to evaluate the present method, our results were compared to SWS variation estimates from [13]. The technique used in the latter study is based on the combination of water extents given by GIEMS and altimetry-based water levels. It should be noted that the technique used by [13] also applies a MODIS mask to the GIEMS dataset. The SWS time series obtained by the present study (ASTER in blue and HyMAP in red) and by [13] (green) are presented in Figure 7 along with GRACE-derived TWS variations (black) over the same period and same geographical locations mentioned above.

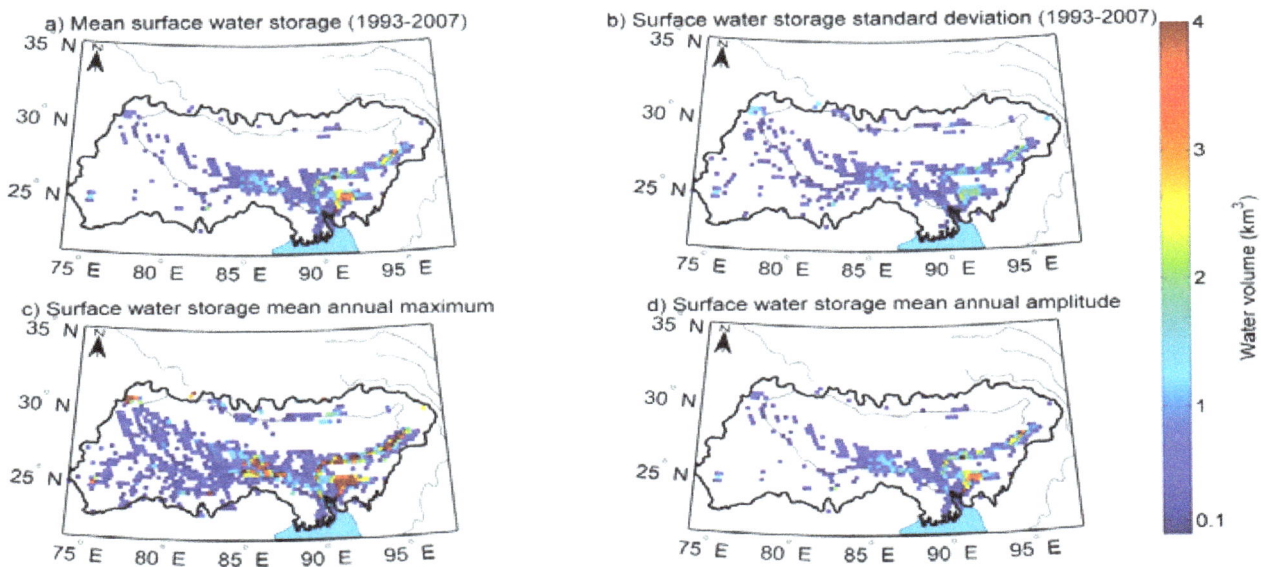

Figure 5. Main characteristics of the surface water storage (SWS) spatial distribution provided by the hypsographic curve approach (GIEMS/ASTER): (**a**) Mean surface water storage over the 1993–2007 period; (**b**) Associated standard deviation; (**c**) Mean annual maximum of SWS; (**d**) Mean annual amplitudes of SWS. It should be noted that water volumes below 0.1 km^3 are not shown in this figure.

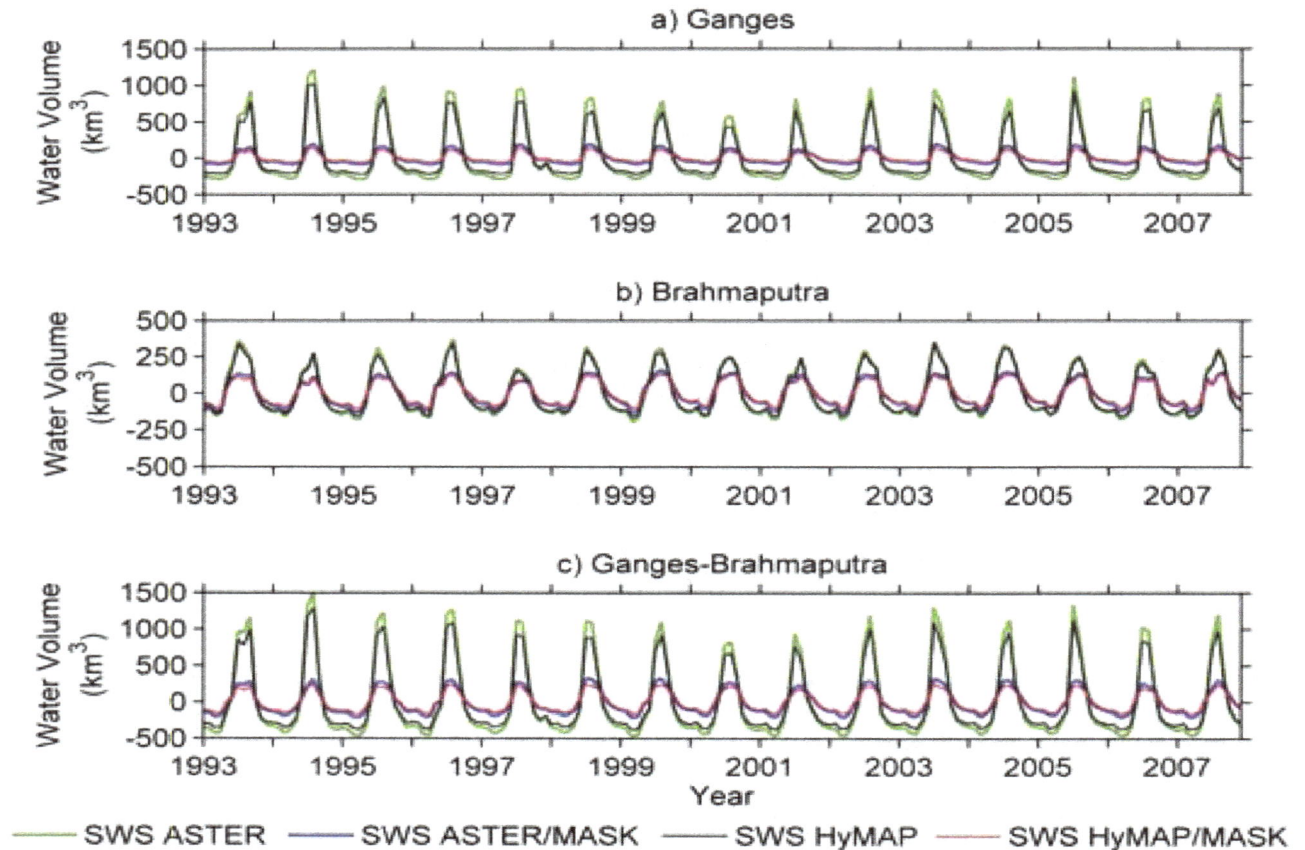

Figure 6. Monthly mean surface water storage variations for the period 1993–2007 estimated by the combination of GIEMS water extents (before and after the usage of a MODIS mask) and a GDEM (ASTER or HyMAP) over: (**a**) Ganges; (**b**) Brahmaputra, and (**c**) Ganges-Brahmaputra River basins.

The variations of SWS (Figure 6) after the use of the MODIS mask maintain a similar variation pattern (correlation coefficients higher than 0.94 over G, B and GB for ASTER and HyMAP) along with a significant decrease in amplitude. For the entire GB system, SWS ASTER mean amplitude decreases from ~1612 km^3 to ~496 km^3 and SWS HyMAP mean amplitude decreases from ~1339 km^3 to ~378 km^3. Mean annual amplitudes estimated using the two techniques (GIEMS-MODIS/GDEM and GIEMS-MODIS/Alt) over G, B and GB are given in Table 1. The magnitude of SWS corrections is higher in the Ganges than the Brahmaputra; this might be the result of relatively higher agricultural coverage in the Ganges basin. These results are consistent with the evaluation of SWS over the GB basin carried out by [13] as shown in Figure 7, giving a SWS mean amplitude of 410 km^3 and a SWS/TWS ratio ~0.45. The Pearson correlation coefficient calculated between SWS estimated by the two techniques is higher than 0.95 for the different couples of time series (SWS GIEMS/Alt–SWS GIEMS/ASTER and SWS GIEMS/Alt-SWS GIEMS/HyMAP) over G, B, and GB.

A strong seasonal cycle is observed with maximum surface water volume in August, one month before the TWS peak and a minimum volume in March–April one month before the TWS minima (Figure 7). The maximum lagged correlation coefficient between TWS and the SWS time series (ASTER and HyMAP) is always higher than 0.9, with a lag time of one month. The delayed seasonal phases of TWS relative to SWS can be explained by the lower flow velocities of water in the soil and in groundwater in comparison to the surface water movement, causing recharge and drainage to continue after the maximum and minimum of SWS. In agreement with the results from [13], these new datasets confirm that SWS contributes annually to ~50% of TWS variations for both river basins.

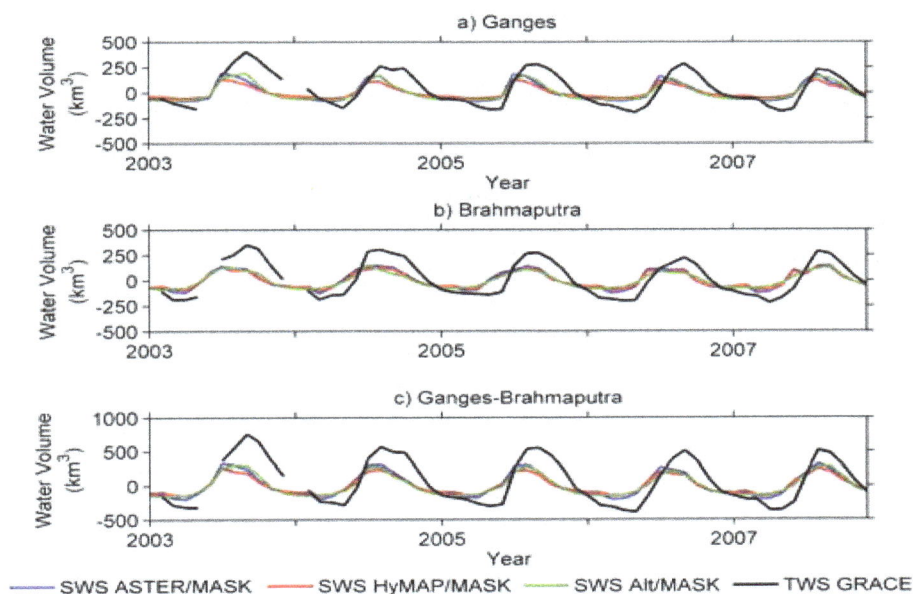

Figure 7. Monthly basin-scale SWS variations estimated using the GIEMS/GDEM approach (blue for ASTER and red for HyMAP), by the GIEMS/Altimetry technique (green) and total water storage (TWS) variations estimated using Gravity Recovery And Climate Experiment (GRACE) for the period 2003–2007 over: (**a**) Ganges; (**b**) Brahmaputra, and (**c**) Ganges-Brahmaputra River basins.

Table 1. Mean annual amplitudes of SWS variations estimated by GIEMS/GDEM and GIEMS/Altimetry techniques over the G, B, and GB system.

	Mean Annual Amplitude (km³)		
	Using GDEM		Using Altimetry
Basin	GIEMS/ASTER	GIEMS/HyMAP	GIEMS/Alt
Ganges	254	172	300
Brahmaputra	253	212	250
Ganges-Brahmaputra	496	378	410

In the following, SWS estimates are compared and evaluated with two other related hydrological variables: satellite altimeter-derived river discharge measured at Hardinge Bridge and Bahadurabad (see Figure 1 for locations) and basin-scale estimates of precipitation from GPCP. Figures 8 and 9 show the annual variations (a), the mean seasonal cycle (b) and the inter-annual variations (c) of SWS (blue), precipitation (gray), and discharge (red) over the Ganges (Figure 8) and the Brahmaputra (Figure 9) River basins. The normalized anomalies shown in Figures 8c and 9c are obtained by removing the mean seasonal cycle of 1993–2007 and normalizing by the corresponding standard deviation. Considering the similarity between SWS obtained by GIEMS/ASTER and GIEMS/HyMAP, only SWS GIEMS/ASTER time series are presented in the figures below.

SWS time series (GIEMS/ASTER and GIEMS/HyMAP) show high consistency (Figures 8a and 9a) with precipitation ($R > 0.87$) and discharge ($R > 0.90$) time series for the period 1993–2007. As shown in Figures 8 and 9, there is no delay between precipitation and SWS at basin scale as intense local rainfall during the annual monsoon results in fast soil saturation followed by quasi-instantaneous inundation of large extents and SWS variations. In contrast, SWS leads discharge by one month in Ganges and Brahmaputra: this time lag corresponds to the residence time of water in floodplains before flowing into the mainstream. The mean seasonal cycle (Figures 8b and 9b) in the GB watershed shows well the increase in the SWS as a consequence to the wet south-west monsoonal high rainfall rate between June and September. Over the period 2003–2007, there is also a close correspondance between SWS and TWS for both river basins ($R > 0.91$). Maximum cross-correlation coefficients (R_{max}) calculated for annual and inter-annual variations are given in Table 2 along with the corresponding time lags. Note that correlation coefficients between inter-annual time series of TWS and SWS were not computed due to the short period of observation.

   Inter-annual variations shown in Figures 8c and 9c, highlight the years when wetter and dryer events take place. For the Ganges River basin, extreme negative anomalies associated to SWS occur in 1993, 1994, 2001 and 2006; as for the Brahmaputra basin, the years 1994, 1997, 2006 and 2007 show significant negative anomalies. High positive anomalies lasting several consecutive months are especially observed in 1998 and 2004 over the Ganges, with many other years over the record showing smaller positive anomalies. The years 1998 and 2004 show strong positive anomalies for the Brahmaputra basin.

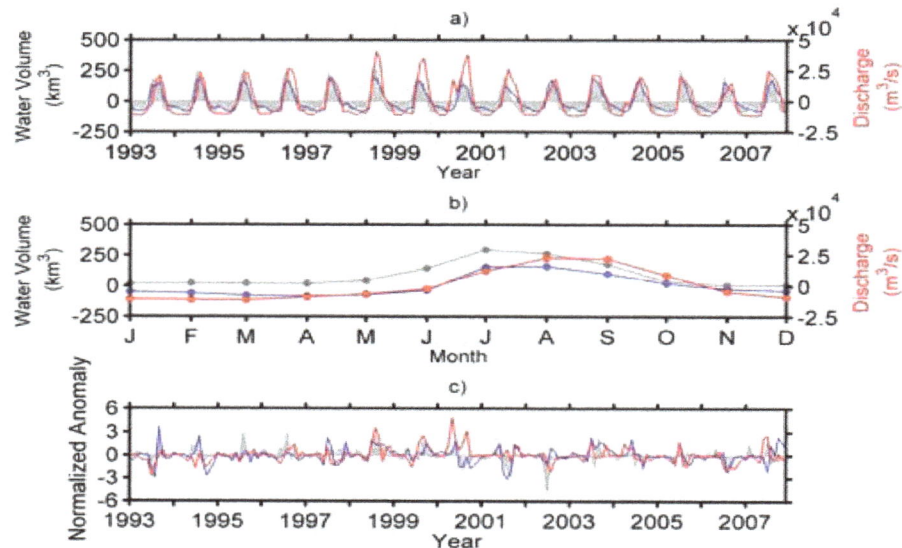

Figure 8. Surface water storage (blue) time series comparison with precipitation (gray) and discharges (red) over the Ganges catchment: (**a**) Annual variations; (**b**) Mean seasonal cycle; (**c**) Inter-annual variations.

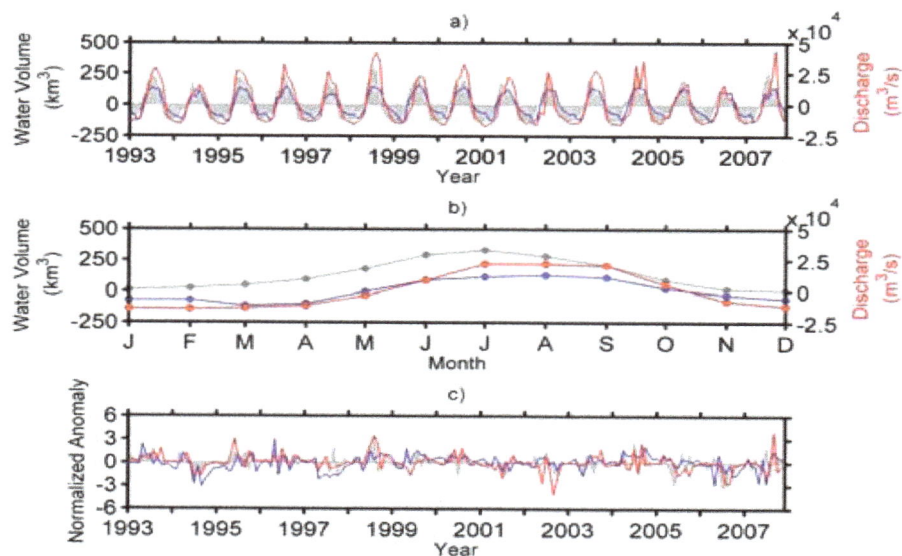

Figure 9. Surface water storage (blue) time series comparison with precipitation (gray) and discharges (red) over the Brahmaputra catchment: (**a**) Annual variations; (**b**) Mean seasonal cycle; (**c**) Inter-annual variations.

Table 2. Maximum cross correlation (R_{max}) coefficients and time lags calculated between SWS time series (GIEMS/ASTER and GIEMS/HyMAP) and the three related hydrological parameters: TWS, discharge and precipitation.

| | | | R_{max} (Time Lag in Month) | |
Basin	Technique	Parameter	Annual Time Series	Inter-Annual Time Series
Ganges	GIEMS/ASTER	TWS	0.91 (−1)	/
		Discharge	0.87 (0)	0.3 (0)
		Precipitation	0.87 (0)	0.51 (0)
	GIEMS/HyMAP	TWS	0.91 (−1)	/
		Discharge	0.86 (0)	0.26 (0)
		Precipitation	0.88 (0)	0.50 (0)
Brahmaputra	GIEMS/ASTER	TWS	0.94 (−1)	/
		Discharge	0.90 (0)	0.34 (0)
		Precipitation	0.89 (1)	0.38 (0)
	GIEMS/HyMAP	TWS	0.94 (−1)	/
		Discharge	0.91 (0)	0.38 (0)
		Precipitation	0.90 (1)	0.33 (0)

In order to further illustrate such drought/flood events in the Ganges-Brahmaputra River system, Figure 10 shows spatio-temporal patterns of SWS for the two contrasted years of 1994 and 1998. Figure 10a–c show the temporal variability of SWS (over the Ganges, the Brahmaputra, and the entire Ganges-Brahmaputra respectively, estimated by GIEMS/ASTER technique) for 1994 and 1998 as compared to their mean seasonal cycle (estimated over the period 1993–2007). For both the Ganges and Brahmaputra basins, SWS estimates in 1998 are larger than the mean seasonal cycle values, especially during the monsoon season. This is in agreement with several past studies that characterized the 1998 monsoon season as extremely "flooded", with for instance [15,63] reporting that during the summer of 1998 over 60% of Bangladesh was inundated for nearly three months. Figure 10e, showing the spatial distributions of SWS anomalies for July 1998, illustrates well the patterns of these major flood events with large positive anomalies over the entire GB system. On the other hand, it is interesting to note that SWS estimates in 1994, which is characterized as a drought year, are below the mean seasonal cycle only for the Brahmaputra basin as reported in [64]. Nevertheless, Figure 10d shows that major drought patterns are well observed over the Brahmaputra and the main channel of the Ganges. The drought observed over the main Ganges channels is compensated by positive anomalies distributed over the rest of the basin, which diminish the signature of the drought over the entire basin.

This case scenario highlights the importance of the new SWS dataset that helps spatialize large-scale drought/flood patterns. Nevertheless, it should be also noted that when estimating SWS during severe droughts (which involved the low end of the hypsographic curves), the proposed method assumes that we cannot have access to water storage below the minimum values that GIEMS/ASTER and GIEMS/HyMAP can provide. This can be a potential source of uncertainties when estimating the extreme low storage values of exceptional drought years. Indeed, in order to capture correctly the extreme low storage values during droughts, ASTER and SRTM-GDEM should have produced credible elevation data for those periods at the low end of the histograms. Unfortunately, it is not possible at this stage to verify such information.

Investigating the large-scale climate causes of these anomalous drought/flooding events in the Ganges-Brahmaputra is far beyond the scope of this paper, but the new availability of these long-term continuous estimates of SWS will help such future studies. For instance, these new observations are in accordance with the results of [65], which investigate how the occurrence (or co-occurrence) of different climate modes (El-Niño, La Niña and Indian Ocean Dipole (IOD) events) affects the variability of precipitation in the GB basin, as well as the occurrences of major flood and drought events: for instance, major droughts, such as the one observed in 1994, are linked to a positive Indian Ocean Dipole (pIOD) mode, whereas major floods, such as the one in 1998 for the entire GB system, might be linked to a negative Indian Ocean Dipole (nIOD) mode.

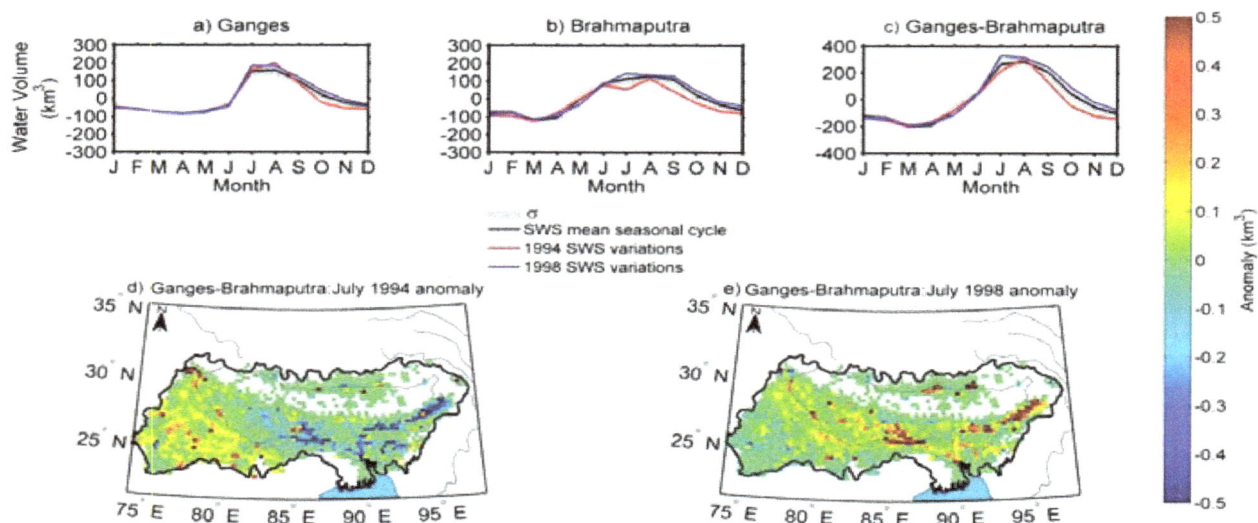

Figure 10. Spatio-temporal variations of SWS estimated by the GIEMS/ASTER technique for the two years of 1994 and 1998: (**a–c**) show the SWS mean seasonal cycle (black) and its standard deviation (shaded gray) for the 1993–2007 period along with SWS variations for 1994 (red) and 1998 (blue) over the Ganges, the Brahmaputra, and the Ganges-Brahmaputra respectively; (**d,e**) show the spatial distribution of SWS anomalies over the GB system for July 1994 and July 1998 respectively.

6. Conclusions

This study presents an estimation and evaluation of SWS variations over the Ganges-Brahmaputra system between 1993 and 2007. The technique used to determine the water volume variations consists in combining water extents from GIEMS with topographic information extracted from GDEMs (ASTER and HyMAP). It follows the method developed by [14] and previously applied in the Amazon basin. Due to the presence of extensive saturated soils (resulting from intense irrigation practices) in the GB basin, the inundated extent detected using GIEMS was overestimated. A MODIS-based static mask was applied to discriminate between flooded and water saturated soil, providing realistic surface water extent. Our results show realistic spatial distribution of surface water reservoirs over the GB compared with previous estimates based on GIEMS and altimetry-based water levels. Basin integrated time series of SWS (G, B and GB) exhibit strong annual and inter-annual variations. For the entire basin, a mean amplitude of ~496 km^3 of SWS is estimated by GIEMS/ASTER while GIEMS/HyMAP gives a mean amplitude of ~378 km^3, accounting for 41% and 51% of the seasonal amplitude of TWS respectively.

The monthly SWS estimates are evaluated against monthly SWS time series estimated by another technique (GIEMS/Alt) and other related hydrological variables such as satellite altimetry-derived river discharge, precipitation and GRACE-derived TWS. Correlations higher than 0.86 were observed among all variables. Lower correlations are calculated when subtracting the seasonal cycle (between 0.23 and 0.51).

In this study, we use ASTER and SRTM-GDEM, which as global satellite-derived DEMs show a series of characteristics, artifacts and anomalies that can cause significant problems or errors when used for hydrological applications [47,48]. It includes the influence of vegetation cover, man-made constructions and even errors due to cloud cover, such as very low but dense boundary layer clouds in tropical regions that are difficult to correct. These effects may introduce inaccurate elevation in the DEM with consequences on the hypsographic curve technique that we developed. These issues should be investigated in future studies and the future release of the DEM from TerraSAR-X for some key regions, such as the Ganges-Brahmaputra delta, might help to solve some issues.

This new data set provides valuable information on the hydrology of the Indian Sub-Continent. It can be used for a better understanding of the complex relationship between the water cycle, climate variability and human activities, for estimating the sub-surface water storage and discharge to the ocean and their impact on key parameters for oceanography of the Bay of Bengal such as salinity and temperature [65–67] and for the validation of regional/global hydrological models.

This methodology had already been tested in the Amazon basin. The present study shows its validity in a very different environment. As GIEMS and the DEMs are available globally, this study is also a first step towards the development of such a database at the global scale. There is also ongoing work to extend the GIEMS time series from 2007 to present. A consistent global SWS dataset from 1993 to present will play a key role in the definition and development of the future hydrology-oriented satellite missions such as the NASA-CNES SWOT (Surface Water and Ocean Topography) dedicated to surface hydrology [68,69].

Acknowledgments: This work was supported by the Centre National d'Etudes Spatiales (CNES) TOSCA (Terre Solide, Océan, Surfaces Continentales et Atmosphère) and OSTST (Ocean Surface Topography Science Team) grants "Variability of terrestrial freshwater storage in the Tropics from multi-satellite observations" managed by Selma Cherchali and by the Belmont Forum project BAND-AID (ANR-13-JCLI-0002, http://Belmont-BanDAiD.org or http://Belmont-SeaLevel.org) with PI C.K. Shum from Ohio State University. We thank Raffael Maurer for his involvement in processing ASTER-GDEM data during his work at NASA-GISS.

Author Contributions: Fabrice Papa and Frédéric Frappart conceived and designed the experiments; Fabrice Papa, Frédéric Frappart and Catherine Prigent performed the experiments; Fabrice Papa, Frédéric Frappart and Edward Salameh analyzed the data; Catherine Prigent, Filipe Aires, Andreas Güntner, Vuruputur Venugopal, David Labat and Benoit Laignel were involved in the analysis of the results; Edward Salameh, Fabrice Papa and Frédéric Frappart wrote the paper. All authors contributed to the discussion of results and the preparation of the manuscript.

References

1. Chahine, M.T. The hydrological cycle and its influence on climate. *Nature* **1992**, *359*, 373–380. [CrossRef]
2. Papa, F.; Prigent, C.; Aires, F.; Jimenez, C.; Rossow, W.B.; Matthews, E. Interannual variability of surface water extent at the global scale, 1993–2004. *J. Geophys. Res.* **2010**, *115*, 1–17. [CrossRef]
3. Alsdorf, D.E.; Rodriguez, E.; Lettenmaier, D.P. Measuring surface water from space. *Rev. Geophys.* **2007**, *45*, 1–24. [CrossRef]
4. Frappart, F.; Papa, F.; Santos da Silva, J.; Ramillien, G.; Prigent, C.; Seyler, F.; Calmant, S. Surface freshwater storage and dynamics in the Amazon basin during the 2005 exceptional drought. *Environ. Res. Lett.* **2012**, *7*, 044010. [CrossRef]
5. Prigent, C.; Papa, F.; Aires, F.; Jimenez, C.; Rossow, W.B.; Matthews, E. Changes in land surface water dynamics since the 1990s and relation to population pressure. *Geophys. Res. Lett.* **2012**, *39*, 2–6. [CrossRef]
6. Prigent, C.; Papa, F.; Aires, F.; Rossow, W.B.; Matthews, E. Global inundation dynamics inferred from multiple satellite observations, 1993–2000. *J. Geophys. Res.* **2007**, *112*. [CrossRef]
7. Frappart, F.; Do Minh, K.; L'Hermitte, J.; Cazenave, A.; Ramillien, G.; Le Toan, T.; Mognard-Campbell, N. Water volume change in the lower Mekong from satellite altimetry and imagery data. *Geophys. J. Int.* **2006**, *167*, 570–584. [CrossRef]
8. Birkett, C.M. The contribution of TOPEX/POSEIDON to the global monitoring of climatically sensitive lakes. *J. Geophys. Res.* **1995**, *100*, 25179–25204. [CrossRef]
9. Frappart, F.; Papa, F.; Famiglietti, J.S.; Prigent, C.; Rossow, W.B.; Seyler, F. Interannual variations of river water storage from a multiple satellite approach: A case study for the Rio Negro River basin. *J. Geophys. Res. Atmos.* **2008**, *113*, 1–12. [CrossRef]
10. Frappart, F.; Calmant, S.; Cauhopé, M.; Seyler, F.; Cazenave, A. Preliminary results of ENVISAT RA-2-derived water levels validation over the Amazon basin. *Remote Sens. Environ.* **2006**, *100*, 252–264. [CrossRef]

11. Frappart, F.; Papa, F.; Guntner, A.; Werth, S.; Ramillien, G.; Prigent, C.; Rossow, W.B.; Bonnet, M.P. Interannual variations of the terrestrial water storage in the lower Ob' basin from a multisatellite approach. *Hydrol. Earth Syst. Sci.* **2010**, *14*, 2443–2453. [CrossRef]

12. Frappart, F.; Papa, F.; Malbéteau, Y.; Leon, J.G.; Ramillien, G.; Prigent, C.; Seoane, L.; Seyler, F.; Calmant, S. Surface freshwater storage variations in the Orinoco floodplains using multi-satellite observations. *Remote Sens.* **2015**, *7*, 89–110. [CrossRef]

13. Papa, F.; Frappart, F.; Malbéteau, Y.; Shamsudduha, M.; Vuruputur, V.; Sekhar, M.; Ramillien, G.; Prigent, C.; Aires, F.; Pandey, R.K.; et al. Satellite-derived surface and sub-surface water storage in the Ganges-Brahmaputra River Basin. *J. Hydrol. Reg. Stud.* **2015**, *4*, 15–35. [CrossRef]

14. Papa, F.; Frappart, F.; Guntner, A.; Prigent, C.; Aires, F.; Getirana, A.C.V.; Maurer, R. Surface freshwater storage and variability in the Amazon basin from multi-satellite observations, 1993–2007. *J. Geophys. Res. Atmos.* **2013**, *118*, 11951–11965. [CrossRef]

15. Chowdhury, R.; Ward, N. Hydro-meteorological variability in the greater Ganges-Brahmaputra-Meghna basins. *Int. J. Climatol.* **2004**, *24*, 1495–1508. [CrossRef]

16. Gain, A.K.; Wada, Y. Assessment of future water scarcity at different spatial and temporal scales of the Brahmaputra River Basin. *Water Resour. Manag.* **2014**, *28*, 999–1012. [CrossRef]

17. Winkel, L.; Berg, M.; Amini, M.; Hug, S.J.; Johnson, C.A. Predicting groundwater arsenic contamination in Southeast Asia from surface parameters. *Nat. Geosci.* **2008**, *1*, 536–542. [CrossRef]

18. Singh, I.B. The Ganga River. In *Large Rivers Geomorphology and Management*; Gupta, A., Ed.; John Wiley & Sons Ltd: Chichester, UK, 2007.

19. Singh, S.K. Erosion and Weathering in the Brahmaputra River System. In *Large Rivers Geomorphology and Management*; Gupta, A., Ed.; John Wiley & Sons Ltd: Chichester, UK, 2007.

20. Prigent, C.; Matthews, E.; Aires, F.; Rossow, W.B. Remote sensing of global wetland dynamics with multiple satellite data sets. *Geophys. Res. Lett.* **2001**, *28*, 4631–4634. [CrossRef]

21. Papa, F.; Prigent, C.; Durand, F.; Rossow, W.B. Wetland dynamics using a suite of satellite observations: A case study of application and evaluation for the Indian Subcontinent. *Geophys. Res. Lett.* **2006**, *33*, 5–8.

22. Prigent, C.; Rossow, W.B.; Matthews, E. Microwave land surface emissivities estimated from SSM/I observations. *J. Geophys. Res.* **1997**, *102*, 21867–21890. [CrossRef]

23. Prigent, C.; Aires, F.; Rossow, W.B. Land Surface Microwave Emissivities over the Globe for a Decade. *Bull. Am. Meteorol. Soc.* **2006**, *87*, 1573–1584. [CrossRef]

24. Rossow, W.B.; Schiffer, R.A. Advances in Understanding Clouds from ISCCP. *Bull. Am. Meteorol. Soc.* **1999**, *80*, 2261–2287. [CrossRef]

25. Kalnay, E.; Kanamitsu, M.; Kistler, R.; Collins, W.; Deaven, D.; Gandin, L. The NCEP/NCAR 40-Year Reanalysis Project. *Bull. Am. Meteorol. Soc.* **1996**, *77*, 437–470. [CrossRef]

26. Papa, F.; Guntner, A.; Frappart, F.; Prigent, C.; Rossow, W.B. Variations of surface water extent and water storage in large river basins: A comparison of different global data sources. *Geophys. Res. Lett.* **2008**, *35*, L11401. [CrossRef]

27. Papa, F.; Prigent, C.; Rossow, W.B. Ob' River flood inundations from satellite observations: A relationship with winter snow parameters and river runoff. *J. Geophys. Res. Atmos.* **2007**, *112*, 1–11. [CrossRef]

28. Papa, F.; Prigent, C.; Rossow, W.B. Monitoring flood and discharge variations in the large siberian rivers from a multi-satellite technique. *Surv. Geophys.* **2008**, *29*, 297–317. [CrossRef]

29. Ringeval, B.; De Noblet-Ducoudré, N.; Ciais, P.; Bousquet, P.; Prigent, C.; Papa, F.; Rossow, W.B. An attempt to quantify the impact of changes in wetland extent on methane emissions on the seasonal and interannual time scales. *Glob. Biogeochem. Cycles* **2010**, *24*, 1–12. [CrossRef]

30. Bousquet, P.; Ciais, P.; Miller, J.B.; Dlugokencky, E.J.; Hauglustaine, D.A.; Prigent, C.; Van der Werf, G.R.; Peylin, P.; Brunke, E.; Carouge, C.; et al. Contribution of anthropogenic and natural sources to atmospheric methane variability. *Nature* **2006**, *443*, 439–443. [CrossRef] [PubMed]

31. Ringeval, B.; Decharme, B.; Piao, S.L.; Ciais, P.; Papa, F.; De Noblet-Ducoudré, N.; Prigent, C.; Friedlingstein, P.; Guttevin, I.; Koven, C.; et al. Modelling sub-grid wetland in the ORCHIDEE global land surface model: Evaluation against river discharges and remotely sensed data. *Geosci. Model Dev.* **2012**, *5*, 941–962. [CrossRef]

32. Pedinotti, V.; Boone, A.; Decharme, B.; Crétaux, J.F.; Mognard, N.; Panthou, G.; Papa, F.; Tanimoun, B.A. Evaluation of the ISBA-TRIP continental hydrologic system over the Niger basin using in situ and satellite derived datasets. *Hydrol. Earth Syst. Sci.* **2012**, *16*, 1745–1773. [CrossRef]

33. Getirana, A.; Boone, A.; Yamazaki, D.; Decharme, B.; Papa, F.; Mognard, N. The Hydrological Modeling and Analysis Platform (HyMAP): Evaluation in the Amazon basin. *J. Hydrometeorol.* **2012**, *13*, 1641–1665. [CrossRef]

34. Decharme, B.; Alkama, R.; Papa, F.; Faroux, S.; Douville, H.; Prigent, C. Global off-line evaluation of the ISBA-TRIP flood model. *Clim. Dyn.* **2012**, *38*, 1389–1412. [CrossRef]

35. Decharme, B.; Douville, H.; Prigent, C.; Papa, F.; Aires, F. A new river flooding scheme for global climate applications: Off-line evaluation over South America. *J. Geophys. Res. Atmos.* **2008**, *113*, 1–11. [CrossRef]

36. Aires, F.; Papa, F.; Prigent, C. A Long-Term, High-Resolution Wetland Dataset over the Amazon Basin, Downscaled from a Multiwavelength Retrieval Using SAR Data. *J. Hydrometeorol.* **2013**, *14*, 594–607. [CrossRef]

37. Lehner, B.; Doell, P. Development and validation of a global database of lakes, reservoirs and wetlands. *J. Hydrol.* **2004**, *296*, 1–22. [CrossRef]

38. Adam, L.; Döll, P.; Prigent, C.; Papa, F. Global-scale analysis of satellite-derived time series of naturally inundated areas as a basis for floodplain modeling. *Adv. Geosci.* **2010**, *27*, 45–50. [CrossRef]

39. Toutin, T. ASTER DEMs for geomatic and geoscientific applications: A review. *Int. J. Remote Sens.* **2008**, *29*, 1855–1875. [CrossRef]

40. Abrams, M.; Bailey, B.; Tsu, H.; Hato, M. The ASTER Global DEM. *Photogramm. Eng. Remote Sens.* **2010**, *76*, 344–348.

41. Li, P.; Shi, C.; Li, Z.; Muller, J.-P.; Drummond, J.; Li, X.; Li, T.; Li, Y.; Liu, J. Evaluation of ASTER GDEM using GPS benchmarks and SRTM in China. *Int. J. Remote Sens.* **2013**, *34*, 1744–1771. [CrossRef]

42. Hirano, A.; Welch, R.; Lang, H. Mapping from ASTER stereo image data: DEM validation and accuracy assessment. *ISPRS J. Photogramm. Remote. Sens.* **2003**, *57*, 356–370. [CrossRef]

43. Hayakawa, Y.S.; Oguchi, T.; Lin, Z. Comparison of new and existing global digital elevation models: ASTER G-DEM and SRTM-3. *Geophys. Res. Lett.* **2008**, *35*, 1–5. [CrossRef]

44. Fujisada, H.; Bailey, G.B.; Kelly, G.G.; Hara, S.; Abrams, M.J. ASTER DEM performance. *IEEE Trans. Geosci. Remote Sens.* **2005**, *43*, 2707–2713. [CrossRef]

45. Tachikawa, T.; Hato, M.; Kaku, M.; Iwasaki, A. The characteristics of ASTER GDEM version 2. In Proceedings of the 2011 IEEE International Geoscience and Remote Sensing Symposium (IGARSS), Vancouver, BC, Canada, 24–29 July 2011; pp. 3657–3660.

46. Farr, T.; Kobrick, M. The shuttle radar topography mission. *Eos Trans. AGU* **2007**. [CrossRef]

47. Yamazaki, D.; Kanae, S.; Kim, H.; Oki, T. A physically based description of floodplain inundation dynamics in a global river routing model. *Water Resour. Res.* **2011**, *47*, 1–21. [CrossRef]

48. Yamazaki, D.; Baugh, C.; Bates, P.D.; Kanae, S.; Alsdorf, D.E.; Oki, T. Adjustment of a spaceborne DEM for use in floodplain hydrodynamic modeling. *J. Hydrol.* **2012**, *436*, 81–91. [CrossRef]

49. Pavlis, N.K.; Holmes, S.A.; Kenyon, S.C.; Factor, J.K. Erratum: Correction to the development and evaluation of the earth gravitational model 2008 (EGM2008). *J. Geophys. Res. Solid Earth* **2012**, *118*, 2633. [CrossRef]

50. Center for Topographic Studies of the Ocean and Hydrosphere. Available online: http://ctoh.legos.obs-mip. fr (accessed on 28 March 2017).

51. Ramillien, G.; Frappart, F.; Cazenave, A.; Güntner, A. Time variations of land water storage from an inversion of 2 years of GRACE geoids. *Earth Planet. Sci. Lett.* **2005**, *235*, 283–301. [CrossRef]

52. Landerer, F.W.; Swenson, S.C. Accuracy of scaled GRACE terrestrial water storage estimates. *Water Resour. Res.* **2012**. [CrossRef]

53. Rodell, M.; Famiglietti, J.S. Detectability of variations in continental water storage from satellite observations of the time dependent gravity field. *Water Resour. Res.* **1999**. [CrossRef]

54. Frappart, F.; Ramillien, G.; Maisongrande, P.; Bonnet, M.-P. Denoising satellite gravity signals by Independent Component Analysis. *IEEE Geosci. Remote Sens. Lett.* **2010**, *7*, 421–425. [CrossRef]

55. Frappart, F.; Ramillien, G.; Leblanc, M.; Tweed, S.; Bonnet, M.; Maisongrande, P. An independent component analysis filtering approach for estimating continental hydrology in the GRACE gravity data. *Remote Sens. Environ.* **2011**, *115*, 187–204. [CrossRef]

56.   Adler, R.F.; Huffman, G.J.; Chang, A.; Ferraro, R.; Xie, P.; Janowiak, J.; Rudolf, B.; Schneider, U.; Curtis, S.; Bolvin, D.; et al. The Version-2 Global Precipitation Climatology Project (GPCP) Monthly Precipitation Analysis (1979–Present). *J. Hydrometeorol.* **2003**, *4*, 1147–1167. [CrossRef]

57.   Bangladesh Water Development Board. Available online: http://www.bwdb.gov.bd/ (accessed on 28 March 2017).

58.   Papa, F.; Biancamaria, S.; Lion, C.; Rossow, W.B. Uncertainties in mean river discharge estimates associated with satellite altimeter temporal sampling intervals: A case study for the annual peak flow in the context of the future SWOT hydrology mission. *IEEE Geosci. Remote Sens. Lett.* **2012**, *9*, 569–573. [CrossRef]

59.   Dartmouth Flood Observatory. Available online: http://floodobservatory.colorado.edu (accessed on 28 March 2017).

60.   Brakenridge, G.R.; Anderson, E. MODIS-based flood detection, mapping, and measurement: The potential for operational hydrological applications. In *Transboundary Floods: Reducing the Risks through Flood Management*; Marsalek, J., Stancalie, G., Balint, G., Eds.; Springer: Dordrecht, The Netherlands, 2006.

61.   Gianinetto, M.; Villa, P.; Lechi, G. Postflood damage evaluation using Landsat TM and ETM+ data integrated with DEM. *IEEE Trans. Geosci. Remote Sens.* **2006**, *44*, 236–243. [CrossRef]

62.   Frappart, F.; Bourrel, L.; Brodu, N.; Riofrío Salazar, X.; Baup, F.; Darrozes, J.; Pombosa, R. Monitoring of the Spatio-Temporal Dynamics of the Floods in the Guayas Watershed (Ecuadorian Pacific Coast) Using Global Monitoring ENVISAT ASAR Images and Rainfall Data. *Water* **2017**, *9*, 12. [CrossRef]

63.   Mirza, M.M.Q.; Warrick, R.A.; Ericksen, N.J. The implications of climate change on floods of the Ganges, Brahmaputra and Meghna rivers in Bangladesh. *Clim. Chang.* **2003**, *57*, 287–318. [CrossRef]

64.   Pervez, M.S.; Henebry, G.M. Spatial and seasonal responses of precipitation in the Ganges and Brahmaputra river basins to ENSO and Indian Ocean dipole modes: Implications for flooding and drought. *Nat. Hazards Earth Syst. Sci.* **2015**, *15*, 147–162. [CrossRef]

65.   Akhil, V.P.; Durand, F.; Lengaigne, M.; Vialard, J.; Keerthi, M.G.; Gopalakrishna, V.V.; Deltel, C.; Papa, F.; de Boyer Montégut, C. A modeling study of the processes of surface salinity seasonal cycle in the Bay of Bengal. *J. Geophys. Res. Ocean.* **2014**, *119*, 3926–3947. [CrossRef]

66.   Sengupta, D.; Bharath Raj, G.N.; Ravichandran, M.; Sree Lekha, J.; Papa, F. Near-surface salinity and stratification in the north Bay of Bengal from moored observations. *Geophys. Res. Lett.* **2016**, *43*, 4448–4456. [CrossRef]

67.   Pant, V.; Girishkumar, M.S.; Udaya Bhaskar, T.V.S.; Ravichandran, M.; Papa, F.; Thangaprakash, V.P. Observed interannual variability of near-surface salinity in the Bay of Bengal. *J. Geophys. Res. C Ocean.* **2015**, *120*, 3315–3329. [CrossRef]

68.   Prigent, C.; Lettenmaier, D.P.; Aires, F.; Papa, F. Toward a High-Resolution Monitoring of Continental Surface Water Extent and Dynamics, at Global Scale: From GIEMS (Global Inundation Extent from Multi-Satellites) to SWOT (Surface Water Ocean Topography). *Surv. Geophys.* **2015**, *37*, 339–355. [CrossRef]

69.   Biancamaria, S.; Lettenmaier, D.P.; Pavelsky, T.M. The SWOT Mission and Its Capabilities for Land Hydrology. *Surv. Geophys.* **2015**, *37*, 307–337. [CrossRef]

4

Characterization of Terrestrial Discharges into Coastal Waters with Thermal Imagery from a Hierarchical Monitoring Program

Claudia Ferrara [1,2], Massimiliano Lega [3] (iD), Giannetta Fusco [1,2] (iD), Paul Bishop [4] and Theodore Endreny [5,*] (iD)

[1] Department of Science and Technologies, University of Naples Parthenope, Centro Direzionale di Napoli, Isola C4, 80143 Napoli, Italy; claudia.ferrara@uniparthenope.it (C.F.); giannetta.fusco@uniparthenope.it (G.F.)
[2] CINFAI, (Consorzio Interuniversitario Nazionale per la Fisica delle Atmosfere e delle Idrosfere), 62029 Tolentino (MC), Italy
[3] Department of Engineering, University of Naples Parthenope, Centro Direzionale di Napoli, Isola C4, 80143 Napoli, Italy; lega@uniparthenope.it
[4] College of Engineering, University of Rhode Island, 102 Bliss Hall 1 Lippitt Rd, Kingston, RI 02881, USA; bishop@mail.uri.edu
[5] Department of Environmental Resources Engineering, College of Environmental Science and Forestry, SUNY, 402 Baker Labs, 1 Forestry Drive, Syracuse, NY 13244, USA
* Correspondence: te@esf.edu

Abstract: Background: The hierarchical use of remotely-sensed imagery from satellites, and then proximally-sensed imagery from helicopter sand drones, can provide a range of spatial and temporal coverage that supports water quality monitoring of complex pollution scenarios. Methods: The study used hierarchical satellite-, helicopter-, and drone-acquired thermal imagery of coastal plumes ranging from 3 to 300 m, near Naples, Italy, and captured temporally- and spatially-overlapping in situ samples to correlate thermal and water quality parameters in each plume and the seawater. Results: In situ sampling determined that between-plume salinity varied by 37%, chlorophyll-a varied by 356%, dissolved oxygen varied by 81%, and turbidity varied by 232%. The radiometric temperature, T_{rad}, for the plume area of interest had a correlation of 0.81 with salinity, 0.74 with chlorophyll-a, 0.98 with dissolved oxygen, and -0.61 with turbidity. Conclusion: This study established hierarchical use of remote and proximal thermal imagery can provide monitoring of complex coastal areas.

Keywords: remote sensing; hydrology; drones; environmental forensics

1. Introduction

Rapid water quality monitoring of receiving waters is important for the protection and preservation of water and related terrestrial resources. However, in natural systems water quality pollution phenomena can occur across a range of spatial scales, and involve a variety of chemical pollutants, making them difficult and costly to rapidly monitor with in situ retrieval of samples and with a fixed spatial or temporal scale remote sensing approach. Ideally, water quality monitoring of pollutant target areas is spatially and temporally flexible to facilitate the environmental forensics process of characterizing the path of the pollutant between the source and target. To monitor pollution phenomena with extensive spatial or temporal scales, remotely-sensed imagery provides distinct benefits not easily achieved by in situ techniques [1,2] particularly when the pollution interface is affected by dispersion generated by terrestrial inflows to coastal zones [3]. A hierarchical monitoring program is proposed in this manuscript to extend the benefits of water quality monitoring to sites where there are spatial, temporal, financial, and radiometric constraints prohibiting the use of more traditional monitoring with in situ sampling, in situ sensor networks, and remote sensing. A hierarchical monitoring program can use a combination of satellite, helicopter, and drone imagery, as well as in situ sampling (Figure 1), to cover the spatial and temporal scales, and radiometric needs, of the pollution phenomena.

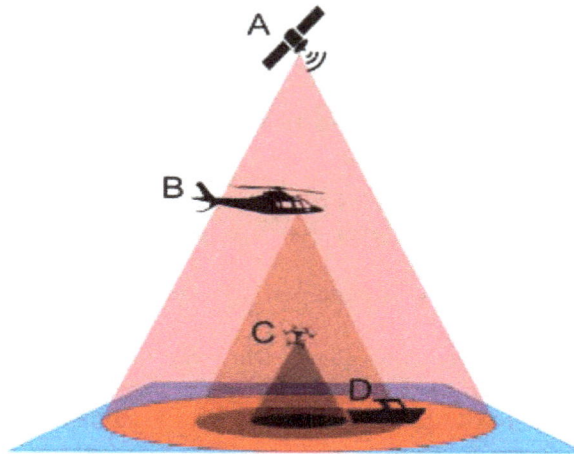

Figure 1. Illustration of the hierarchical monitoring program using (**A**) satellite, (**B**) helicopter, (**C**) drone and (**D**) field boat to study the mixing of channel plumes in the coastal zone.

Prior to the emergence of remote sensing hydrology and the evolution of distributed sensor networks, water quality monitoring was traditionally performed using an in situ sample of the water column. In situ samples are either analyzed in real-time by the sampling instrument, such as placing the dissolved oxygen meter into the target water body, or preserved for subsequent analysis in a laboratory. The temporal, spatial, and financial limitations of this approach have encouraged development of remote sensing and in situ sensor-based monitoring. Remote sensing has used airborne and spaceborne hyperspectral and thermal imagery to characterize water quality parameters, establishing relationships with the spectral or temperature signal and the water quality parameter [4]. Hyperspectral sensors detect tens to hundreds of narrow spectral bands throughout the visible, near-infrared, and mid-infrared portions of the electromagnetic spectrum in order to better discriminate between different targets and, as such, can contain the temperature signal [5]. However, hyperspectral sensors generate large volumes of data that create challenges for data storage, manipulation, and water quality analysis.

Nearly all airborne sensors are considered high spatial resolution, ranging between 25 and 0.5 m, and include the Airborne Visible Infrared Imaging Spectrometer manufactured by NASA that provides 17 m resolution, 224 bands, across a 12 km swath width [6]. Spaceborne sensors that are high-resolution (e.g., 20 to 0.5 m) are typically limited to eight bands, and are often operated by commercial firms providing contracted monitoring services [6]. Moderate-resolution sensors include the government-operated Landsat-8 (30 m resolution, 10 bands, 16 day revisit interval) and the Hyperspectral Imager for the Coastal Ocean (100 m resolution, 128 bands, 10 day revisit interval) [6–8]. Satellite microwave radiometers used for sea surface temperature water quality studies include the Advanced Microwave Scanning Radiometer-2, with resolution ranging from 5×3 km to 62×35 km [6].

To summarize the constraints a water quality monitoring program may face with these sensors, at high spatial resolution the monitoring is typically limited to pre-arranged, fee-based campaigns, and, if publicly-available moderate-resolution sensor data are sufficient for the site, the 10 to 16 day revisit interval may become a constraint. As an alternative to radiometric monitoring, finer spatial-and temporal-scale monitoring is available via in situ wireless sensor networks, which relay auto-sampled water quality data to a base station [9]. For some sites, the constraint of these systems include installation, operation, and maintenance costs, their inability to detect pollution extents due to spatial gaps in coverage, and potential interference of the in situ sensor with other activities in those coastal waters.

Thermal infrared imagery, or thermographic data, compared with hyperspectral imagery, are relatively inexpensive and, as such, can allow more resources for increasing spatial and temporal resolution. To capture the peak spectral emission of objects on Earth's surface, and thereby obtain a suitable signal to noise ratio, the thermographic image should be captured between 9 and 11 μm, which is near the middle of the long-wave infrared electromagnetic spectrum. The thermographic data offer a limited radiometric signal, but contain valuable information on surface properties affecting the energy flux characteristics and dynamics. The radiant energy detected by thermal sensors is a composite of energy emitted by the investigated surface that is transmitted through the atmosphere and energy that is emitted by the atmosphere. While the water-atmosphere coupling complicates interpretation of thermographic data, with proper signal processing researchers can estimate a number of environmental variables important to Earth system science modelling [10].

The thermographic image measures the radiometric temperature, T_{rad} (K), which is related to the thermodynamic, or kinetic, temperature, T_{kin} (K), is typically measured with a thermometer. The principle of thermographic data analysis is based on the physical phenomenon that all objects at a temperature >0 K emit thermal radiation as a function of the body's T_{kin}, and emissivity, ε, which typically ranges $0 \geq \varepsilon \leq 1$ depending on properties of the material.

As described by the Stefan-Boltzman law [5], the spectral radiant flux of a black body object, M_b, (W/m^2) is:

$$M_b = \sigma T_{kin}^4 \tag{1}$$

where σ is the Stefan-Boltzman constant, 5.6697×10^{-8} W m^{-2} K^{-4}. As explained by Kirchoff's Law [5], Equation (1) presumes the black body is a perfect absorber and emitter, with an $\varepsilon = 1$, but otherwise the spectral radiant flux of a real-world object, M_r, is:

$$M_r = \varepsilon \sigma T_{kin}^4 \tag{2}$$

where $\varepsilon < 1$, which is the case for most substances [11]. In our work we use the Stefan-Boltzman law to relate the apparent radiant temperature, T_{rad} (K) of the real world object to M_r as:

$$M_r = \sigma T_{rad}^4 \tag{3}$$

where T_{rad} is measured by thermal remote sensing. By combining Equations (2) and (3), we can then obtain:

$$\varepsilon = T_{rad}^4 / T_{kin}^4 \tag{4}$$

and establish emissivity as the property relating the remotely sensed T_{rad} and the in situ measured T_{kin}. This same relationship was established by Equation 1.7 in Kuenzer and Dech [12].

In standard processing of multi-pixel infrared thermographic image, it is possible to set only one emissivity value for the whole image. If the observed surface is heterogeneous material (e.g., spatially-varying water chemistry, soil moisture, lithology, or vegetation), the homogenous emissivity will generate erroneous kinetic temperatures for some pixels. However, these discrepancies are diminished when emissivity values approach 1, which is the case for some natural surfaces, such as water. The detection of possible water quality anomalies in coastal waters is made possible by the processing of the multi-pixel thermographic image using a homogeneous emissivity value; for seawater an emissivity value of 0.986 is recommended [13,14]. In monitoring of seawater, when a thermographic pixel captures the radiometric temperature of non-seawater material with the same kinetic temperature but different emissivity than the seawater, the thermographic sensor estimates an erroneous kinetic temperature. In situ measurement of the kinetic temperature in parallel with remote sensing of the radiometric temperature would allow for the derivation of the emissivity. With advances in technologies for in situ and remote sensing data capture, monitoring campaigns that use both techniques for data capture are increasingly effective in spatially extending inferences about water quality [15,16]. Spatial variation of emissivity across a thermographic image can generate noticeable spatial variation in kinetic temperature [17,18]. In order to utilize anomalies in the thermographic image, it is important to obtain images with an appropriate point of view, spatial resolution, and sensor accuracy [19].

This goal of this research is to determine if coastal water quality monitoring across an area with multiple channel inflows can be achieved using infrared thermographic imagery collected from a hierarchical monitoring program. The research question is whether T_{rad} from remote or proximal imagery has a strong correlation (>60) with water quality parameters within channel plumes entering a coastal area. There are a range of spatial scales in ecosystem management, and hierarchical monitoring is developed to cross that range using a variety of thermal monitoring platforms. Applied ecosystem science requires monitoring across variable spatial, temporal, and organizational scales [20], and scientific knowledge guiding data management and fusion across time and space [21]. For the study of a coastal area receiving inputs from rivers of varying spatial extent, temporal flows, and water quality characteristics, a hierarchical monitoring program will utilize various tools. Typically, environmental coastal monitoring actions have been provided by an in situ aquatic vessel that collect a low spatial and temporal density of sample data, are time consuming, and are accomplished only with significant advance planning. The hierarchical monitoring program with T_{rad} might be used to efficaciously deploy monitoring with more sophisticated and precise analytical tools, such as in situ sampling or remote and proximal sensing with hyperspectral instruments.

2. Materials and Methods

The study area was the coastal zone northwest of Naples, Italy where four channels deliver terrestrial discharges that can jeopardize coastal water quality. The four channels discharging to this section of coast are the Volturno with a ~300 m wide channel at the outlet, Regi Lagni with a ~100 m wide channel, Agnena with two ~15 m wide channels bifurcating around a seawall at the outlet, and Cuma with a ~3 m wide channel at the outlet. This coastal zone is a critical area that requires monitoring due its important ecological value and the risk of pollution from discharge draining the adjacent terrestrial area. The principal water quality concerns are discharges from wastewater treatment plants and discharges from factory agricultural activities. The coastal bathymetry, warm season intensification of currents, and diurnal reversal in winds act together to create a very complex surface dynamic, resulting in different flushing mechanisms and exchange patterns between the coastal zone with the river plumes discharging to the coast and the outer Tyrrhenian waters [22]. Vertical mixing of the water is less pronounced in the warm season due to a stable thermocline [22]. This thermocline is disrupted by the cooler weather and increase in precipitation during the winter months, but in the warm season stratification of the water column allows for formation of a surface mixed layer 30 to 40 m thick [23]. Water quality is particularly important during the warm season when there is more recreational contact with the water due to sport and tourist activities, and for this reason the study was conducted in July during peak use of the coastal area.

The hierarchical monitoring used temporally and spatially overlapping satellite, helicopter, and drone platforms (Figure 1) in order to simultaneously obtain thermal images of the targets (Figure 1). The selection of an imaging platform for water quality monitoring is based on the suitability of platform characteristics (Table 1) to detect, recognize, and identify criteria for characterizing the thermal anomalies of the plume (Figure 2). These criteria are specific to each sensor and target, and may involve: ratios of image spatial resolution or image swath size to the target size; matching of image spectral resolution to target thermal properties; and time to achieve target image relative to target time constraints. In cases where a thermal imaging platform is suitable, targeted, overlapping in situ monitoring may be deployed; otherwise un-targeted, wide in situ sampling might be used.

The first step of the hierarchical monitoring was coordinating with flight dates for Landsat-8. After each flight, the Landsat imagery was retrieved and reviewed for evidence of discharging plumes, noted by different reflectance than the seawater, from the channels into the coastal zone. Evidence of discharging plumes was identified in the Landsat-8 images obtained on 18 June 2013, both in the natural look and TIRS band 10 thermal imagery. Given plume detection with Landsat-8, the second step in hierarchical monitoring was deployment of the helicopter, drone, and field boat, either immediately, or to coincide with the next Landsat-8 flight. In this project all monitoring was coordinated to temporally overlap with the Landsat-8 flight date of 27 June 2013. The helicopter was a rotorcraft AW139 (Augusta Westland, Rome, Italy), with a pilot and photographer, and acquired imagery at a flight acquisition of 300 m, with Star SAFIRE QWIP (FLIR, USA) + FLIR T620 camera (FLIR, USA), providing images with ~10 cm resolution. The unmanned drone was a StillFly 6–R Natural Drone (San Diego, CA, USA), and acquired imagery at a flight acquisition of 50 m with a FLIR T620, providing resolution with ~2 cm resolution [24,25]. The thermal cameras used by the helicopter and drone missions used a constant emissivity value of 0.986, and were set for ambient parameters in the thermal sensor control software, including atmospheric temperature, relative humidity, and distance, which was used by the T620 camera to compensate for the atmospheric interference and provide an accuracy of 2 mK in the measurements. Images were rejected if they had poor view angles (e.g., too oblique) and if they had extreme solar reflection off the water surface.

Table 1. Hierarchical monitoring platforms of satellite, helicopter, and drone, with the associated thermal image sensor, target distance, swath size, spatial resolution, and spectral resolution.

Hierarchical Monitoring Platform	Thermal Image Sensor	Target Distance (km)	Image Swath Size (km)	Image Spatial Resolution (m)	Image Spectral Resolution (μm)
Satellite	Landsat 8 TIRS	705	190	100	Band 10: 10.60–11.19
Helicopter	FLIR T620	0.3	~4.7	0.12	7–14
Drone	FLIR T620	0.05	~0.13	0.02	7–14

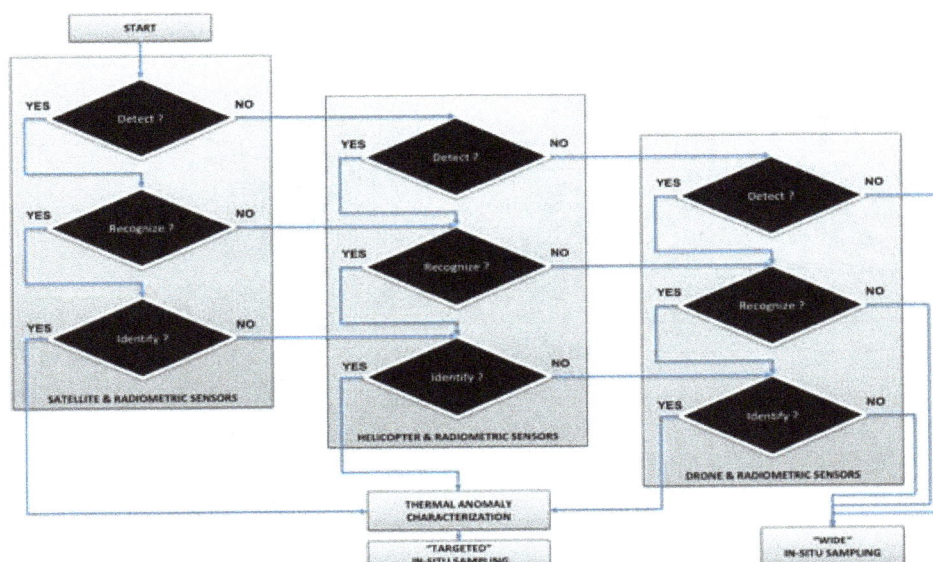

Figure 2. Flowchart illustrating the choice of platform in hierarchical monitoring, related to target detect, recognize, and identify criteria. All three platforms may be needed for a set of distinct targets.

The in situ samples were coordinated to spatially and temporally overlap with the thermal imagery about the discharge plumes (Figure 3). A field boat was provided by the Regional Agency for the Environmental Protection and it carried an Ocean Seven 320 Plus multi-parameter Conductivity Temperature Depth (CTD) probe. The boat was used for in situ sampling to test the correlation with T_{rad}, however, once this was performed subsequent monitoring would not require the costly concurrent collection of in situ samples. To coordinate the spatial congruence of the in situ sampling and proximal sensing, the operator of the field boat navigated to the coordinates of the areas of interest where thermal anomalies in the water were detected, with coordinates sent by radio between the helicopter and boat crews. The CTD in situ samples provided measurement of water kinetic temperature, T_{kin}, salinity, dissolved oxygen, chlorophyll-a, and turbidity. The CTD records were acquired at a frequency of 24 Hz, the highest allowed by the probes, taking the surficial measurements at 50 cm from the surface. During the CTD acquisitions the wind speeds measured 6 m/s at the weather station close to the sampling area, and this condition suggests the surface layer of the water was completely mixed with a homogeneous T_{kin} sampled at the 50 cm depth [26].

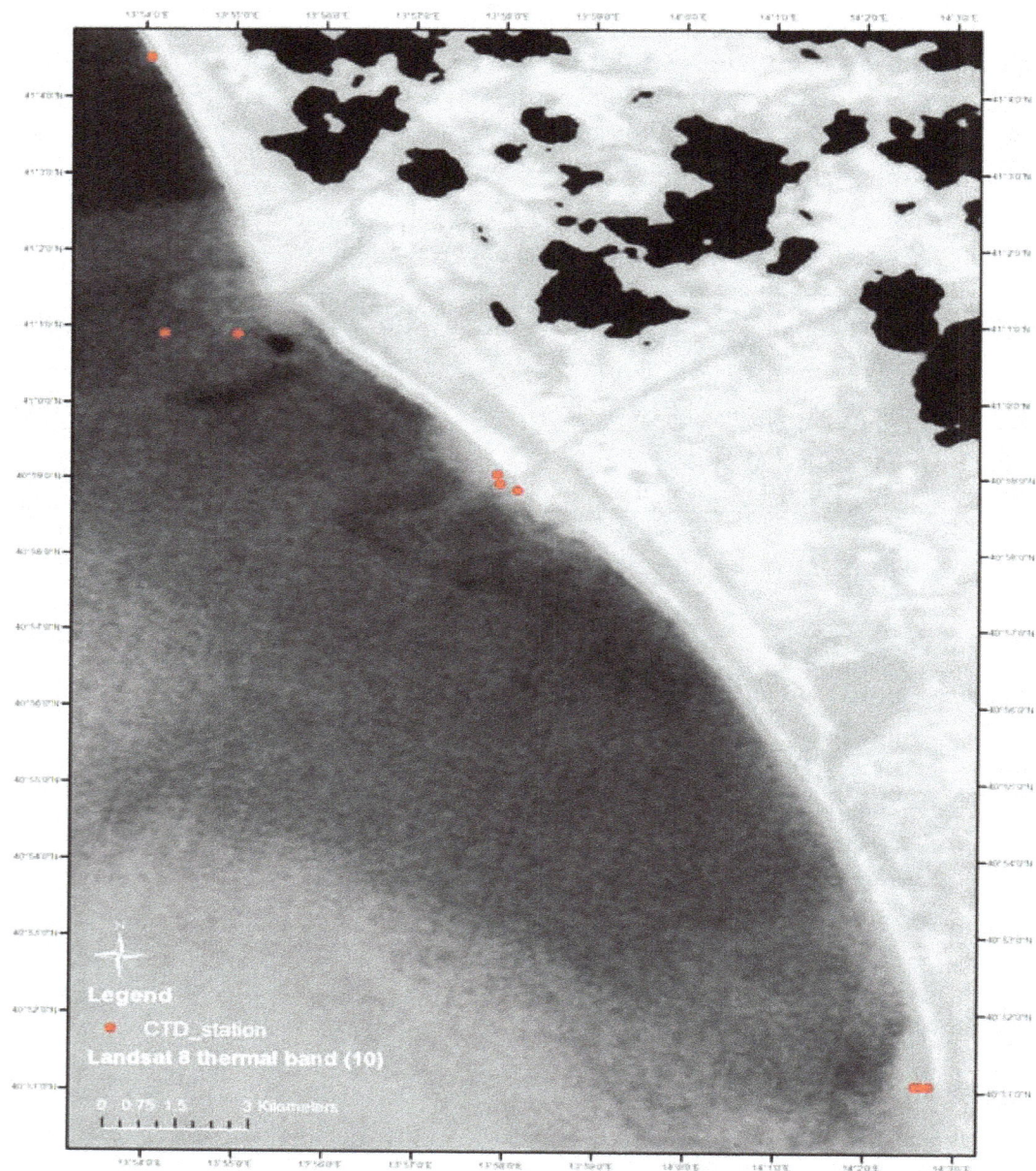

Figure 3. Landsat-8 image of the study area with dots indicating the location of the field boat sampling sites, within the channel discharge areas of interest.

The third step of the monitoring program was to post-process the data and obtain the dataset of water quality and temperature. The thermographic imagery was post-processed to delineate the area of interest and obtain the average T_{rad} value from that area, as well as other statistical data on the distribution of the T_{rad} values. The in situ sampling provided one temperature parameter, T_{kin}, and the proximal sensing provided one temperature parameter, T_{rad}, the combination of the two parameters allowed for a third parameter equal to the difference $T_{kin} - T_{rad}$, and a fourth parameter of emissivity, ε, derived using the relationship between T_{kin} and T_{rad}. An optional fourth step of the monitoring program was to post-process the imagery to define the edges of the areas of interest, where the plumes met with the coastal water. This involves extracting statistical information about the IR temperature spectra, defining standard thermal patterns related to the phenomenology of water pollution [27]. In this study, the consistent temperature difference along the perimeter suggested the edge of the channel plumes, which was confirmed with using Canny edge detection [28] (see image Figure 4 for the Volturno channel).

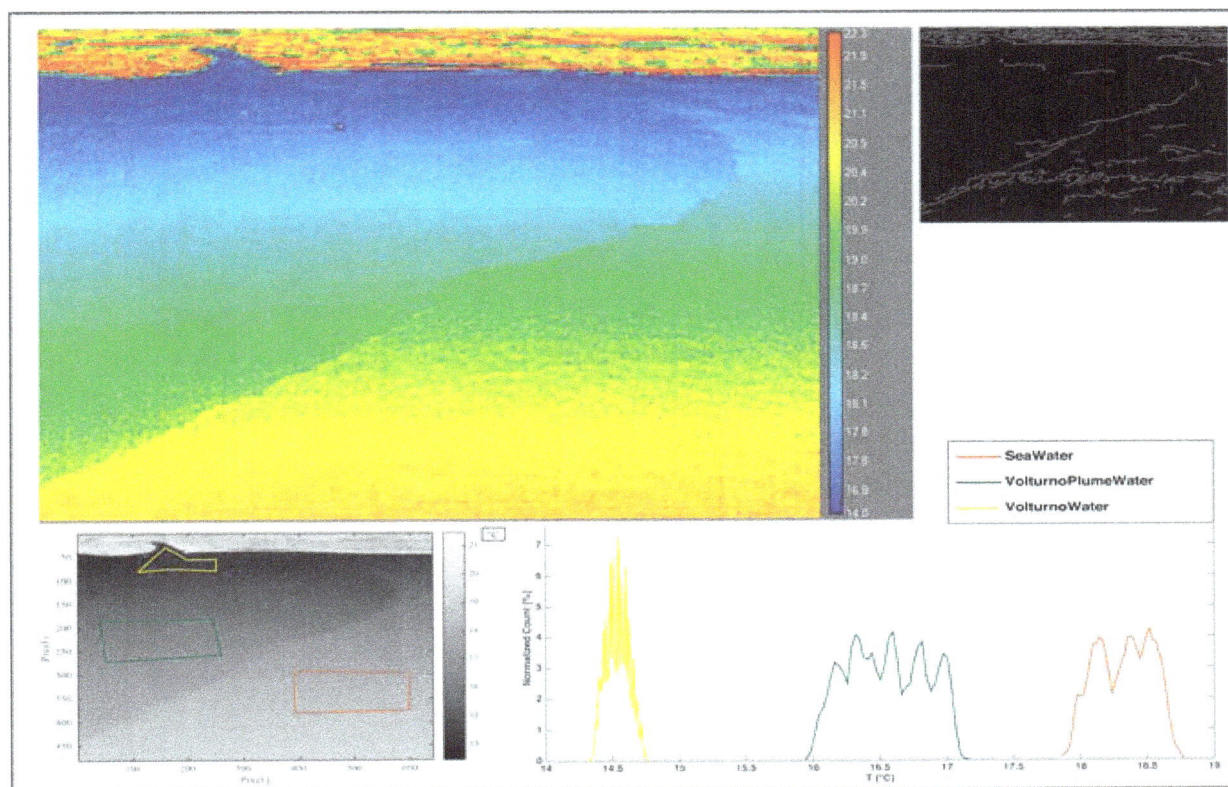

Figure 4. Thermal image of the Volturno plume with polygons over the channel, seawater, and plume areas of interest, which correspond to histograms of radiometric temperature; and (upper right side) edges of the Volturno channel plume as white lines to the right of the image.

3. Results and Discussion

Plumes discharging from the four channels into the coastal waters were observed and captured with the helicopter thermographic camera. To capture the largest plume from the Volturno channel, with a channel width of ~300 m, the helicopter was used to take an oblique image (Figure 4), while less oblique images were used for the Agnena plume (Figure 5), Regi-Lagni plume (Figure 6), and the Cuma plume (Figure 7). The Cuma channel width was ~3 m, sufficient to allow for an area of interest to be delineated in the channel using the helicopter-acquired ~10 cm thermographic image. For more detailed analysis of the mixing zone between the Cuma channel discharge and coastal water, the drone-acquired ~2 cm thermographic image provides excellent detail (Figure 7). The Landsat-8 imagery performed its function as a first step of the hierarchical monitoring, and was sufficient to confirm that spectral differences existed along the coastal zone, and initiated the subsequent monitoring step 2, proximal image collection, and step 3, image post-processing, in the hierarchical program.

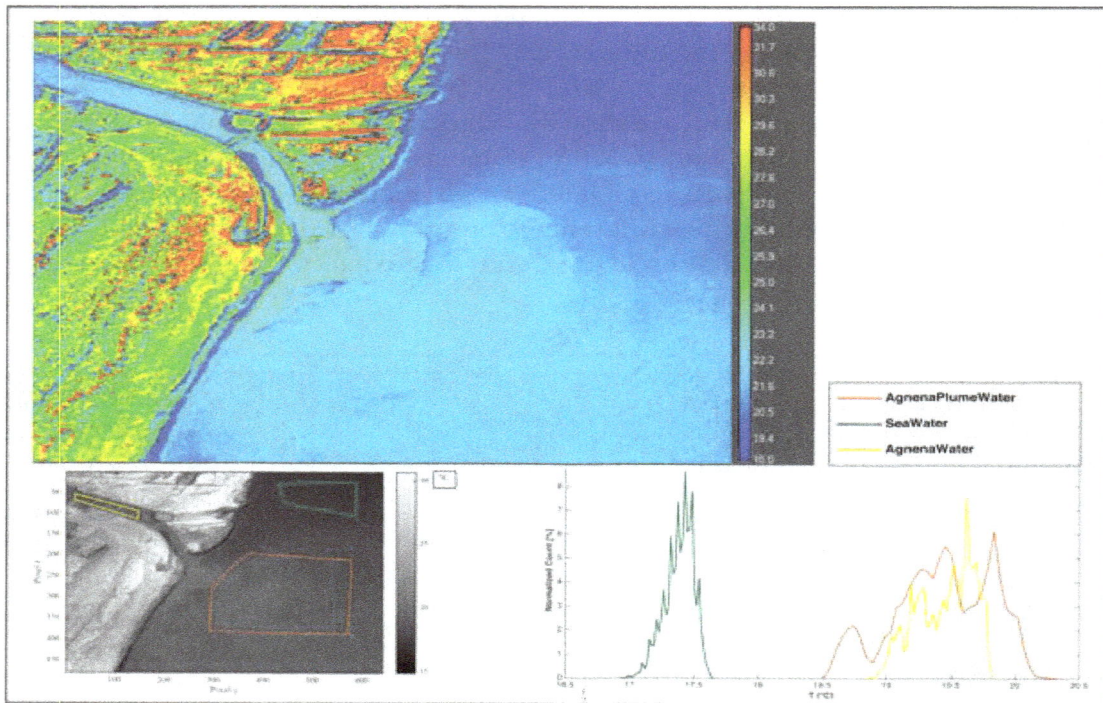

Figure 5. Thermal image of Agnena plume with histogram of radiometric temperature for areas of the channel, seawater and area of the plume.

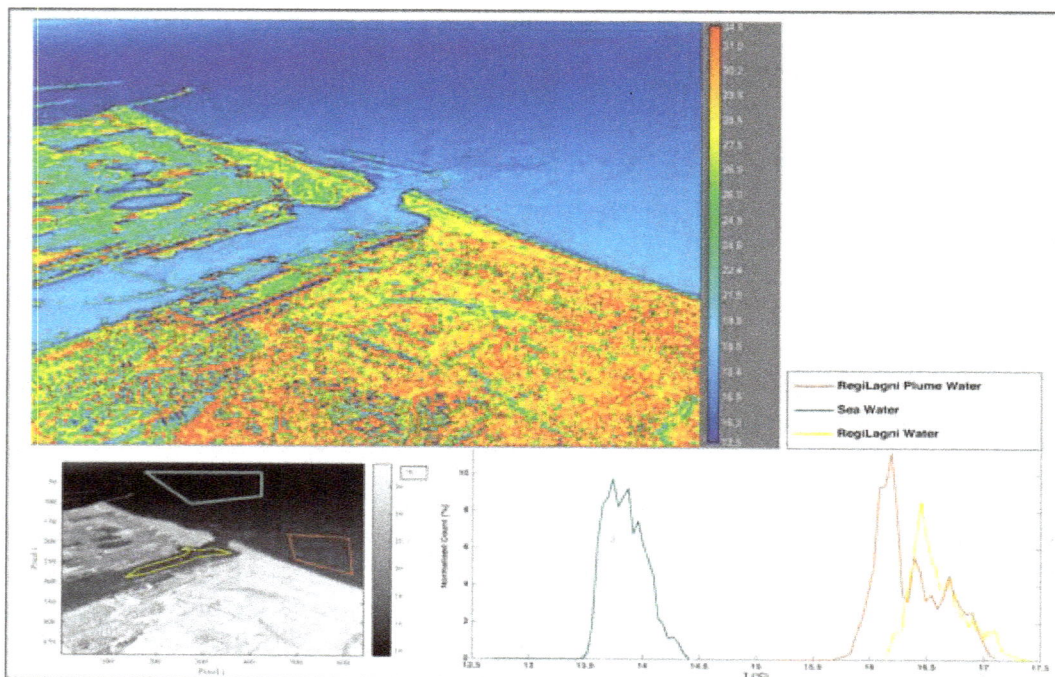

Figure 6. Thermal image of Regi Lagni plume with histogram of radiometric temperature for areas of the channel, seawater, and area of the plume.

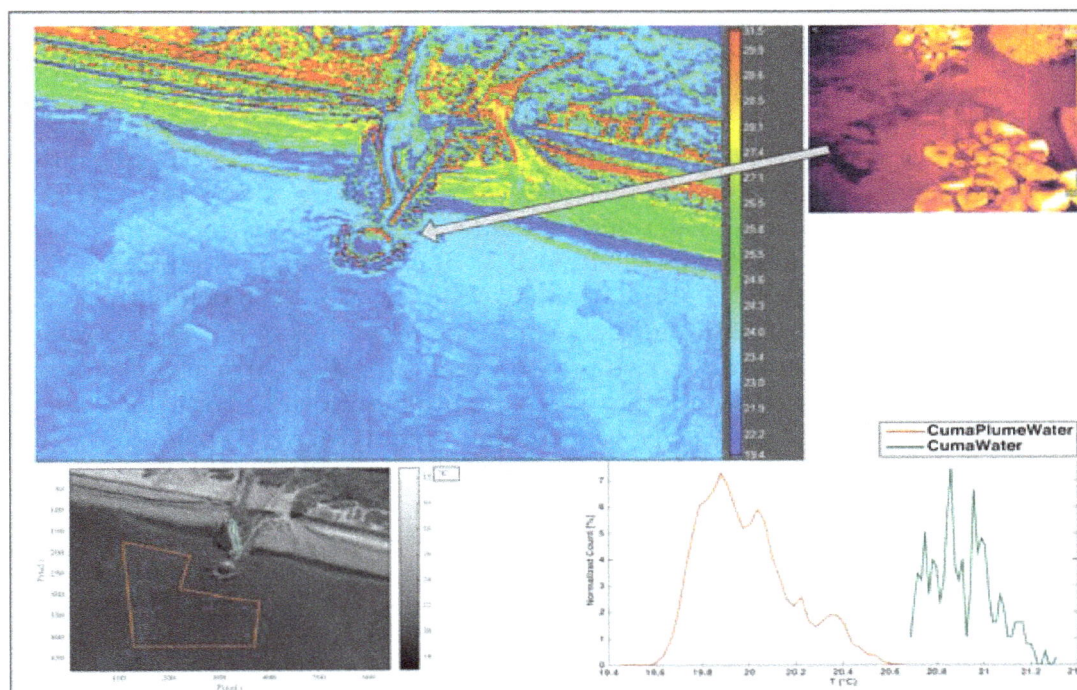

Figure 7. Thermal image of Cuma plume with histogram of radiometric temperature for areas of the channel, seawater, and area of the plume; and (in upper right side) drone-acquired thermal image of the mixing zone in the Cuma channel outlet.

The in situ water quality measurements were obtained for one sample of coastal seawater beyond the plumes, one sample in the Volterno plume, one sample in the Agnena plume, two samples in the Regi Lagni plume, and two samples in the Cuma plume (Figure 3, Table 2). The salinity of the coastal seawater was 3.4% (i.e., 34 parts per thousand), while the salinity of plumes had a maximum of 3.7% and a minimum of 2.7%. The chlorophyll-a of the coastal seawater was 0.93 µg/L, while the chlorophyll-a of plumes had a maximum of 2.5 µg/L and a minimum of 0.97 µg/L. The dissolved oxygen of the coastal seawater was 87% of saturation, while the dissolved oxygen of plumes had a maximum of 107% and a minimum of 59.3%. The detection of dissolved oxygen above 100% saturation is relatively common in coastal sites due to the production of pure oxygen by photosynthetically-active organisms, as well as a momentary lack of equilibrium of dissolved oxygen between the water column and air column. The turbidity of the coastal seawater was 0.93 FTU (Formazin turbidity unit), while the turbidity of plumes had a maximum of 5.6 FTU and a minimum of 0.9 FTU. The kinetic temperature of the coastal seawater was 22.1 °C, while the kinetic temperature of plumes had a maximum of 23.7 °C and a minimum of 21.9 °C.

Table 2. Water quality parameters measured during the in situ field campaign, with salinity in %, chlorophyll-a in µg/L, dissolved oxygen (DO) in % saturation, turbidity in Formazin turbidity units (FTU), and kinetic temperature in °C.

Locations	Salinity %	Chl-a µg/L	DO% Sat	Turbidity FTU	T_{kin} °C
Volturno Plume	3.4	0.97	107	1.70	21.90
Agnena Plume	3.7	0.55	87	2.05	22.10
Regi Lagni Plume 1	3.3	2.20	65	2.29	22.48
Regi Lagni Plume 2	2.7	1.66	59	5.65	22.90
Cuma Plume 1	3.7	2.25	100	1.94	23.06
Cuma Plume 2	3.7	2.51	102	1.94	23.65
Coastal Seawater	3.4	0.93	87	0.93	22.10

The kinetic temperature, T_{kin}, measured by the CTD in situ, and the radiometric temperature, T_{rad}, measured from IR imagery, were used with Equation (4) to derive the emissivity, ε, for the areas of interest (Table 3). The derived ε values ranged from 0.93 to 0.97, with seawater at 0.97; this seawater ε value deviates from the standard of 0.98 and illustrates that ε can vary around a common value due to the variation in environmental and viewing conditions [29,30]. The T_{kin}, T_{rad}, the difference $T_{kin} - T_{rad}$, and the ε for each area of interest was correlated with the water quality parameters (Table 4). The correlations with in situ water quality data were consistently the best for the parameters of T_{kin}, T_{rad} and ε, which were identical, or within 0.01 of each other, and worst for the parameter of T_{kin}. The absolute values of the T_{rad} correlations were just 0.02 to 0.06 below the absolute values of the ε correlations, and 0.21 to 0.36 higher than the absolute values of the T_{kin} correlations. Given that parameterizing ε and T_{kin} requires in situ measurement, the T_{rad} correlations were of particular interest because T_{rad} can be collected remotely and expedite monitoring. For in situ salinity T_{rad} had a correlation of 0.81, for chlorophyll-a T_{rad} had a correlation of 0.74, for dissolved oxygen T_{rad} had a correlation of 0.98, and for turbidity T_{rad} had a correlation of -0.61.

Table 3. Water in situ kinetic temperature, T_{kin} ($^\circ$C), IR thermographic radiometric temperature, T_{rad} ($^\circ$C), the difference of T_{kin} and T_{rad}, and the derived emissivity, ε, for the areas of interest.

Locations	T_{kin} $^\circ$C	T_{rad} $^\circ$C	$T_{kin} - T_{rad}$ $^\circ$C	ε
Volturno Plume	21.90	17.50	4.40	0.942
Agnena Plume	22.10	19.90	2.20	0.971
Regi Lagni Plume 1	22.48	17.10	5.38	0.929
Regi Lagni Plume 2	22.90	17.30	5.60	0.926
Cuma Plume 1	23.06	20.08	2.98	0.960
Cuma Plume 2	23.65	20.90	2.75	0.963
Coastal Seawater	22.10	19.90	2.20	0.971

Table 4. Correlation coefficients between water quality parameters salinity (%), chlorophyll-a (μg/L), dissolved oxygen (DO, % saturation), turbidity (Formazin turbidity units, FTU), and in situ kinetic temperature, T_{kin} ($^\circ$C), IR thermographic radiometric temperature, T_{rad} ($^\circ$C), the difference of T_{kin} and T_{rad}, and the derived emissivity, ε, for the areas of interest.

Variables	Salinity %	Chl-a μg/L	DO %Sat	Turbidity FTU
T_{kin}	0.48	0.51	0.77	-0.25
T_{rad}	0.81	0.74	0.98	-0.61
$T_{kin} - T_{rad}$	-0.87	-0.78	-1.00	0.69
ε	0.86	0.77	1.00	-0.67

In the Regi Lagni channel there was no detected difference between the channel continental water and the seawater mixing zone water, attributed to a more complete mixing along this section of the coast. All channels had a relatively low discharge during the warm season, with water temperatures warmer than the seawater for all channels, but the larger Volturno, which drains a larger river basin with high-altitude tributaries that may feed cooler water. The smallest channel, the ~3 m wide Cuma, had the highest levels of chlorophyll-a, while the medium sized channel, the ~100 m wide Regi Lagni, had the next highest levels of chlorophyll-a, suggesting they may be the most polluted channels. The reported values of oxygen under saturation suggest the presence of a high level of photosynthesis, and the Cuma channel had the highest chlorophyll-a and second highest dissolved oxygen values. The high turbidity values from all channels, above the 0.93 FTU of the seawater, suggest the channels are carrying a high concentration of particulate matter, which may carry additional contamination.

The two temperature variables that correlated best with water quality parameters, the difference T_{kin} and T_{rad} and ε, which is from a ratio of T_{kin} and T_{rad}, are based on using both kinetic and radiometric temperatures. However, in typical applications of this monitoring program the T_{kin} and ε will not be available. Only with in situ measurements will T_{kin} be available, and only with T_{kin} can ε be derived. As intended in this study, T_{rad} is the most available temperature parameter from this monitoring program, and in this application it had a much stronger correlation with the water quality parameters than the T_{kin}. The T_{rad} correlations were, on average, 71% better than the T_{kin} correlations, and ranged from 144% better for turbidity to 27% better for oxygen. The T_{rad} correlations were all above 0.6, and three were above 0.74, while three of the T_{kin} correlations were below 0.51.

The study demonstrates how thermographic data can support a coastal monitoring program that targets the water quality impact of channel plumes discharge along the coast. This study addressed the research question by establishing a strong correlation between T_{rad} and four common water quality parameters. In a subsequent application of this hierarchical monitoring program, the second step would not require a concurrent field boat with in situ sampling and, instead, would just collect T_{rad} for areas of interest within the thermographic imagery acquired by the helicopter and/or drone. If the plumes and seawater areas had different values of T_{rad} for any one channel, then in situ samples could subsequently be obtained to characterize the water quality. The limitations of the monitoring program include the inability to estimate the water quality parameter concentration using the T_{rad}, and predictive equations are difficult to establish given the sensitivity of T_{rad} to variation in ambient air temperature, water temperature, and emissivity of the pollutant. At sites where thermographic data alone are insufficient for the monitoring program, other approaches have been used. The fusion of optical data and synthetic aperture radar have been used for feature-based detection of environmental hazards [31,32], and ratios of multi-spectral bands have been used to detect surface contamination of soil and water [33,34]. Monitoring programs can also use remote sensing-based detection cyanobacteria together with knowledge of flow paths to make inferences of the impact and source of water pollution [35].

Without the hierarchical monitoring via helicopter, prior coastal monitoring for this region limited its focus to pollution from the largest channels, including the Volturno, and another large channel further north, called the Garigliano [36–39]. Due to the larger discharge plumes from these rivers, they have been considered to be the principal cause of water quality impairment. This research revealed that the smallest channel, the Cuma, and the medium-sized channel, the Regi Lagni, had the highest concentrations of chlorophyll-a, which can lead to significant local degradation of coastal water quality. Indeed, the contaminant concentrations from the small channels originate with wastewater treatment and agricultural runoff, and can be higher than those of the larger rivers. The lower flow rates from these small channels do not generate significant plume dispersion, and constrain the dilution and degradation of the pollutant during transport. During the warm season the coastal currents are mostly onshore, due to local land-sea breezes, and the river discharge and pollutants are retained in the littoral area.

In summary, the hierarchical sampling protocol might search for thermal anomalies first using satellite data, if the channel width is large enough, and then proceed to helicopter or drone data depending on channel size and distance between channels. Collaboration between teams with environmental expertise and teams with access to helicopters and drones may be critical to combine resources and complete the monitoring program. The collaboration with governmental authorities for access to helicopters can satisfy their needs for pollution monitoring, and ideally fit within their operational requirements in terms of both flight regulations and the mission goals, while satisfying the scientific aims and requirements to analyze the data. The application of this methodology produces multi-resolution data that can be processed to highlight thermal anomalies, and the inferences with respect to water quality are enhanced using local knowledge of pollutant sources. Indeed, with a proper knowledge of the environmental dynamics, such as the interaction of the channels and coastal currents, this application can link thermal anomalies and environmental criticalities.

4. Conclusions

This research demonstrated that a hierarchical use of remotely-sensed imagery from satellites, then helicopter, and then proximal sensed imagery from drones, provides a range of spatial and temporal coverage to support water quality monitoring of complex pollution scenarios. The research established that the thermal infrared cameras can be used in the monitoring of water quality anomalies, with the radiometric temperature, T_{rad}, strongly correlating with water quality parameters of salinity, chlorophyll-a, dissolved oxygen, and turbidity. The Landsat-8 remotely-sensed imagery was used as a first step to identify that plumes were discharging into the coastal water. The helicopter was used as a second step to obtain proximal imagery with a spatial resolution of ~10 cm, able to sample the plumes discharging from ~300 to ~3 m channels. The area of interest in the proximal thermal imagery captured T_{rad} values that had a correlation of 0.81 with salinity, of 0.74 with chlorophyll-a, 0.98 with dissolved oxygen, and −0.61 with turbidity. This study demonstrates the utility of using thermal imagery in cases where more advanced monitoring is unable due to spatial, temporal, and financial constraints.

Acknowledgments: We acknowledge the Italian Governmental Authorities, and the ARPA Campania and the director of the Marine Protection and Oceanography division, Lucio De Maio, for providing their instrumentation for in situ measurements. We recognize the support of CINFAI and RITMARE Project, funded by the Italian Ministry of University and Research. The last two authors are grateful to the US and Italian Fulbright Commission and Parthenope University of Naples for sponsoring their collaboration with the Italian research partners. Massimiliano Lega and Giannetta Fusco acknowledge the financial support received by Università degli Studi di Napoli 'Parthenope' under "Bando di sostegno alla ricerca individuale per il triennio 2015–2017".

Author Contributions: M.L. and P.B. conceived and designed the experiments; G.F. coordinated the in situ sampling; M.L. coordinated the proximal sensing; C.F. and T.E. post-processed and analyzed the data; and C.F., M.L., and T.E. wrote the paper.

References

1.   Ahn, Y.H.; Shanmugam, P.; Lee, J.H.; Kang, Y.Q. Application of satellite infrared data for mapping of thermal plume contamination in coastal ecosystem of Korea. *Mar. Environ. Res.* **2006**, *61*, 186–201. [CrossRef] [PubMed]

2.   Hook, S.J.; Chander, G.; Barsi, J.A.; Alley, R.E.; Abtahi, A.; Palluconi, F.D.; Markham, B.L.; Richards, R.C.; Schladow, S.G.; Helder, D.L. In-flight validation and recovery of water surface temperature with Landsat-5 thermal infrared data using an automated high-altitude lake validation site at Lake Tahoe. *IEEE Trans. Geosci. Remote Sens.* **2004**, *42*, 2767–2776. [CrossRef]

3.   Hedger, R.D.; Malthus, T.J.; Folkard, A.M. Estimation of velocity fields at the estuary-coastal interface through statistical analysis of successive airborne remotely sensed images. *Int. J. Remote Sens.* **2001**, *22*, 3901–3906. [CrossRef]

4.   El Din, E.S.; Zhang, Y.; Suliman, A. Mapping concentrations of surface water quality parameters using a novel remote sensing and artificial intelligence framework. *Int. J. Remote Sens.* **2017**, *38*, 1023–1042. [CrossRef]

5.   Jansen, J.R. *Remote Sensing of the Environment: An Earth Resource Perspective*, 2nd ed.; Pearson: New York, NY, USA, 2006; p. 608.

6.   Gholizadeh, M.H.; Melesse, A.M.; Reddi, L. Spaceborne and airborne sensors in water quality assessment. *Int. J. Remote Sens.* **2016**, *37*, 3143–3180. [CrossRef]

7.   Brando, V.E.; Dekker, A.G. Satellite hyperspectral remote sensing for estimating estuarine and coastal water quality. *IEEE Trans. Geosci. Remote Sens.* **2003**, *41*, 1378–1387. [CrossRef]

8.   Keith, D.J.; Schaeffer, B.A.; Lunetta, R.S.; Rocha, K.; Cobb, D.J.; Gould, R.W., Jr. Remote sensing of selected water-quality indicators with the hyperspectral imager for the coastal ocean (HICO) sensor. *Int. J. Remote Sens.* **2014**, *35*, 2927–2962. [CrossRef]

9.   Chung, W.Y.; Yoo, J.H. Remote water quality monitoring in wide area. *Sens. Actuators B Chem.* **2015**, *217*, 51–57. [CrossRef]

10. Czajkowski, K.P.; Goward, S.N.; Stadler, S.J.; Walz, A. Thermal remote sensing of near surface environmental variables: Application over the Oklahoma Mesonet. *Prof. Geogr.* **2000**, *52*, 345–357. [CrossRef]

11. Wooster, M.J.; Roberts, G.; Smith, A.; Johnston, J.; Freeborn, P.; Amici, S.; Hudak, A. Thermal remote sensing of active vegetation fires and biomass burning events. In *Thermal Infrared Remote Sensing: Sensors, Methods, Applications*; Kuenzer, C., Dech, S., Eds.; Springer: Dordrecht, The Netherlands; Heidelberg, Germany; New York, NY, USA, 2013; pp. 347–390.

12. Kuenzer, C.; Dech, S. Theoretical background of thermal infrared remote sensing. In *Thermal Infrared Remote Sensing: Sensors, Methods, Applications*; Kuenzer, C., Dech, S., Eds.; Springer: Dordrecht, The Netherlands; Heidelberg, Germany; New York, NY, USA, 2013; pp. 1–26.

13. Branch, R.; Chickadel, C.C.; Jessup, A.T. Infrared emissivity of seawater and foam at large incidence angles in the 3–14 mu m wavelength range. *Remote Sens. Environ.* **2016**, *184*, 15–24. [CrossRef]

14. Niclos, R.; Caselles, V.; Coll, C.; Valor, E.; Rubio, E. Autonomous measurements of sea surface temperature using in situ thermal infrared data. *J. Atmos. Ocean. Technol.* **2004**, *21*, 683–692. [CrossRef]

15. Cotroneo, Y.; Aulicino, G.; Ruiz, S.; Pascual, A.; Budillon, G.; Fusco, G.; Tintoré, J. Glider and satellite high resolution monitoring of a mesoscale eddy in the algerian basin: Effects on the mixed layer depth and biochemistry. *J. Mar. Syst.* **2016**, *162*, 73–88. [CrossRef]

16. Aulicino, G.; Cotroneo, Y.; Lacava, T.; Sileo, G.; Fusco, G.; Carlon, R.; Budillon, G. Results of the first Wave Glider experiment in the southern Tyrrhenian Sea. *Adv. Oceanogr. Limnol.* **2016**, *7*. [CrossRef]

17. Norman, J.M.; Becker, F. Terminology in Thermal Infrared Remote-sensing of Natural Surfaces. *Agric. For. Meteorol.* **1995**, *77*, 153–166. [CrossRef]

18. Li, Z.L.; Wu, H.; Wang, N.; Qiu, S.; Sobrino, J.A.; Wan, Z.; Tang, B.H.; Yan, G. Land surface emissivity retrieval from satellite data. *Int. J. Remote Sens.* **2013**, *34*, 3084–3127. [CrossRef]

19. Lega, M.; Kosmatka, J.; Ferrara, C.; Russo, F.; Napoli, R.M.A.; Persechino, G. Using Advanced Aerial Platforms and Infrared Thermography to Track Environmental Contamination. *Environ. Forensics* **2012**, *13*, 332–338. [CrossRef]

20. Lindenmayer, D.B.; Likens, G.E. The science and application of ecological monitoring. *Biol. Conserv.* **2010**, *143*, 1317–1328. [CrossRef]

21. Wikle, C.K. Hierarchical models in environmental science. *Int. Stat. Rev.* **2003**, *71*, 181–199. [CrossRef]

22. De Maio, A.; Moretti, M.; Sansone, E.; Spezie, G.; Vultaggio, M. Outline of Marine Currents in the Bay of Naples and Some Considerations on Pollutant Transport. *Nuovo Cimento Della Societa Ital. Di Fisica C-Geophys. Space Phys.* **1985**, *8*, 955–969. [CrossRef]

23. Carrada, G.C.; Hopkins, T.S.; Bonaduce, G.; Ianora, A.; Marino, D.; Modigh, M.; Ribera, D.A.M.; Scotto, C.B. Variability in the Hydrographic and Biological Features of the Gulf of Naples. *Mar. Ecol.* **1980**, *1*, 105–120. [CrossRef]

24. Lega, M.; Napoli, R.M.A. Aerial infrared thermography in the surface waters contamination monitoring. *Desalination Water Treat.* **2010**, *23*, 141–151. [CrossRef]

25. Lega, M.; Ferrara, C.; Persechino, G.; Bishop, P. Remote sensing in environmental police investigations: Aerial platforms and an innovative application of thermography to detect several illegal activities. *Environ. Monit. Assess.* **2014**, *186*, 8291–8301. [CrossRef] [PubMed]

26. Brainerd, K.E.; Gregg, M.C. Surface Mixed and Mixing Layer Depths. *Deep-Sea Res. Part I-Oceanogr. Res. Pap.* **1995**, *42*, 1521–1543. [CrossRef]

27. Lega, M.; Persechino, G. GIS and Infrared Aerial View: Advanced Tools for the Early Detection of Environmental Violations. *WIT Trans. Ecol. Environ.* **2014**, *180*, 225–235.

28. Canny, J. A Computational Approach to Edge-detection. *IEEE Trans. Pattern Anal. Mach. Intell.* **1986**, *8*, 679–698. [CrossRef] [PubMed]

29. Sidran, M. Broadband reflectance and emissivity of specular and rough water surfaces. *Appl. Opt.* **1981**, *20*, 3176–3183. [CrossRef] [PubMed]

30. Smith, W.L.; Knuteson, R.O.; Revercomb, H.E.; Feltz, W.; Nalli, N.R.; Howell, H.B.; Menzel, W.P.; Brown, O.; Brown, J.; Minnett, P.; et al. Observations of the Infrared Radiative Properties of the Ocean—Implications for the Measurement of Sea Surface Temperature via Satellite Remote Sensing. *Bull. Am. Meteorol. Soc.* **1996**, *77*, 41–51. [CrossRef]

31. Errico, A.; Angelino, C.V.; Cicala, L.; Podobinski, D.P.; Persechino, G.; Ferrara, C.; Lega, M.; Vallario, A.; Parente, C.; Masi, G.; et al. SAR/multispectral image fusion for the detection of environmental hazards with a GIS. In Proceedings of the SPIE Remote Sensing, Amsterdam, The Netherlands, 23–25 September 2014; International Society for Optics and Photonics: Bellingham, WA, USA; p. 924503. [CrossRef]

32. Errico, A.; Angelino, C.V.; Cicala, L.; Persechino, G.; Ferrara, C.; Lega, M.; Vallario, A.; Parente, C.; Masi, G.; Gaetano, R.; et al. Detection of environmental hazards through the feature-based fusion of optical and SAR data: A case study in southern Italy. *Int. J. Remote Sens.* **2015**, *36*, 3345–3367. [CrossRef]

33. Lega, M.; D'antonio, L.; Napoli, R.M. Cultural Heritage and Waste Heritage: Advanced techniques to preserve cultural heritage, exploring just in time the ruins produced by disasters and natural calamities. In *Waste Management and the Environment V*; Popov, V., Itoh, H., Mander, U., Brebbia, C.A., Eds.; WIT Press: Southampton, UK; Billerica, MA, USA, 2010; pp. 123–134.

34. Lega, M.; Endreny, T. Quantifying the environmental impact of pollutant plumes from coastal rivers with remote sensing and river basin modelling. *Int. J. Sustain. Dev. Plan.* **2016**, *11*, 651–662. [CrossRef]

35. Teta, R.; Romano, V.; Della Sala, G.; Picchio, S.; De Sterlich, C.; Mangoni, A.; Di Tullio, G.; Costantino, V.; Lega, M. Cyanobacteria as indicators of water quality in Campania coasts, Italy: A monitoring strategy combining remote/proximal sensing and in situ data. *Environ. Res. Lett.* **2017**, *12*, 024001. [CrossRef]

36. Isidori, M.; Lavorgna, M.; Nardelli, A.; Parrella, A. Integrated environmental assessment of Volturno River in South Italy. *Sci. Total Environ.* **2004**, *327*, 123–134. [CrossRef] [PubMed]

37. De Pippo, T.; Donadio, C.; Pennetta, M.; Petrosino, C.; Terlizzi, F.; Valente, A. Coastal hazard assessment and mapping in Northern Campania. *Italy Geomorphol* **2008**, *97*, 451–466. [CrossRef]

38. Infascelli, R.; Pelorosso, R.; Boccia, L. Spatial assessment of animal manure spreading and groundwater nitrate pollution. *Geospat. Health* **2009**, *4*, 27–38. [CrossRef] [PubMed]

39. Amodiococchieri, R.; Arnese, A. Organochlorine Pesticide-residues in Fish from Southern Italian Rivers. *Bull. Environ. Contam. Toxicol.* **1988**, *40*, 233–239. [CrossRef]

5

Mapping Palaeohydrography in Deserts: Contribution from Space-Borne Imaging Radar

Philippe Paillou

Laboratoire d'Astrophysique de Bordeaux, Université de Bordeaux, UMR 5804-CNRS, 33600 Pessac, France;
philippe.paillou@u-bordeaux.fr

Academic Editor: Frédéric Frappart

Abstract: Space-borne Synthetic Aperture Radar (SAR) has the capability to image subsurface features down to several meters in arid regions. A first demonstration of this capability was performed in the Egyptian desert during the early eighties, thanks to the first Shuttle Imaging Radar mission. Global coverage provided by recent SARs, such as the Japanese ALOS/PALSAR sensor, allowed the mapping of vast ancient hydrographic systems in Northern Africa. We present a summary of palaeohydrography results obtained using PALSAR data over large deserts such as the Sahara and the Gobi. An ancient river system was discovered in eastern Lybia, connecting in the past the Kufrah oasis to the Mediterranean Sea, and the terminal part of the Tamanrasett river was mapped in western Mauritania, ending with a large submarine canyon. In southern Mongolia, PALSAR images combined with topography analysis allowed the mapping of the ancient Ulaan Nuur lake. We finally show the potentials of future low frequency SAR sensors by comparing L-band (1.25 GHz) and P-band (435 MHz) airborne SAR acquisitions over a desert site in southern Tunisia.

Keywords: SAR; radar; deserts; palaeohydrography; Sahara; Gobi

1. Introduction

Space-borne Synthetic Aperture Radar (SAR) allows the mapping of continental surfaces at centimetre-scale wavelengths. It is an active remote sensing technique, producing high resolution images sensitive to surface topography and roughness, and to soil water content [1]. In very dry soils, SAR is able to probe the subsurface down to several meters: it was shown that L-band (1.25 GHz) radar is able to penetrate meters of low electrical loss material such as sand [2,3]. Thanks to the first Shuttle Imaging Radar SIR-A mission, McCauley et al. [4] demonstrated radar subsurface imaging capabilities for a site located in the Selima Sand Sheet, in southern Egypt. Radar images revealed buried and previously unknown palaeodrainage channels (see Figure 1), which afterwards were confirmed by field studies [5,6]. Later in 1995, SIR-C radar was used to map the subsurface basement structures that control the Nile's course in northeastern Sudan [7]. More recent studies have shown that combining Shuttle Radar Topography Mission (SRTM) data [8] with SAR images better reveals subsurface features that still present a topographic signature. New palaeodrainage flow directions have thus been mapped in eastern Sahara [9], allowing better definition of drainage lines leading to oases and valleys, as well as a better mapping of the past aquifers [10,11].

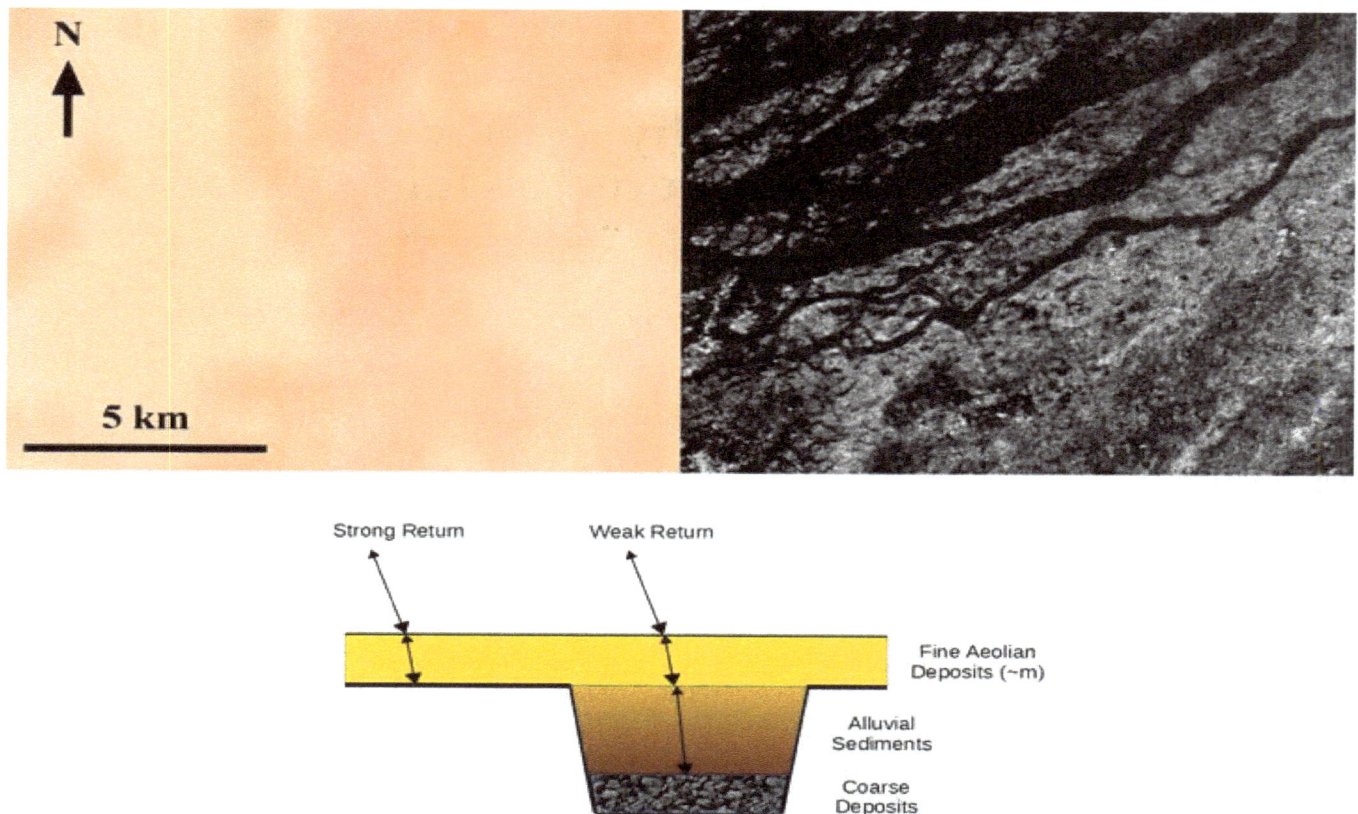

Figure 1. (**Top left**) SPOT (Satellite Pour l'Observation de la Terre) image of part of the Selima Sand Sheet located in southern Egypt (22.55° N–29.28° E). (**Top right**) Radar image from Japanese polarimetric L-band synthetic aperture radar (PALSAR) of the same area, revealing numerous palaeochannels hidden by a meter-thick layer of aeolian sand. (**Bottom**) Sketch of radar wave interaction with surface sand, subsurface bedrock and palaeochannel. The radar wave is absorbed by the alluvial sediments, leading to a weak return, while it is backscattered by the rough bedrock under the thin sand cover, leading to a stronger return.

While the geographical coverage of the Shuttle Imaging Radar missions was limited, a more complete L-band radar coverage of the eastern Sahara was acquired by the JERS-1 satellite of the Japanese space agency (JAXA). It was used to produce the first regional-scale radar mosaic covering Egypt, northern Sudan, eastern Libya, and northern Chad. This data set helped discover numerous unknown crater structures in eastern Sahara [12]. Later in 2006, JAXA successfully launched the Advanced Land Observing Satellite (ALOS), carrying a full polarimetric L-band SAR, named PALSAR, which offered higher resolution imagery and a much improved signal to noise ratio as compared to JERS-1 [13]. Full coverage of the Sahara and Arabia was acquired during June and July 2007, delivering more than 400 PALSAR strips at a resolution of 50 m (see Figure 2). A fully automated data processing chain allowed to produce geocoded 1° × 1° SAR scenes that can be superposed to the corresponding 1° × 1° SRTM squares, covering latitudes between 17° N and 37° N and longitudes between 17° W and 60° E. The whole dataset is managed with the help of a web map server, allowing the import and display of PALSAR data using Google Earth. It is freely accessible through a dedicated web site [14]. This PALSAR dataset constitutes a unique tool for the scientific community to study the palaeo-environment and palaeoclimate of North Africa and Arabia. It also helps in the building of more complete geological maps and in support of future water prospecting in arid and semi-arid regions [15]. Recently, the PALSAR dataset was extended to the Gobi Desert in Central Asia, covering latitudes between 34° N and 52° N and longitudes between 73° E and 120° E.

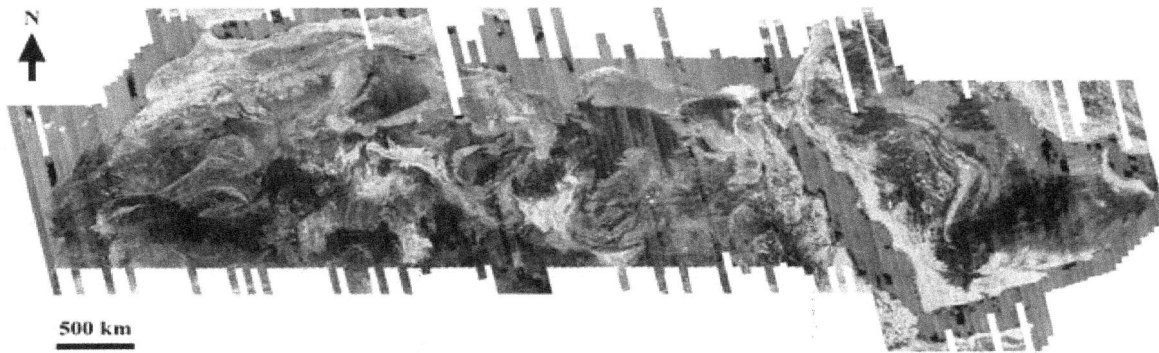

Figure 2. PALSAR strips acquired by the Japanese space agency (JAXA) over the Sahara and Arabia during the summer of 2017.

2. Case Studies Using PALSAR L-Band Sensor

2.1. The Kufrah Palaeoriver in Eastern Sahara

Many of the major drainage basins in North Africa were influenced by the Messinian salinity crisis in the late Miocene, when desiccation of the Mediterranean Sea promoted deep landscape incision. In central Sahara, extensive drainage systems originating in the Tibesti mountains were flowing northward to the Mediterranean Sea and southward to the Chad Basin. While this region is now hyperarid, remains of past river systems have been detected using remote sensing imagery, leading some authors to propose palaeodrainage pathways between south Libya and the Mediterranean Sea [9,16].

For the first time, PALSAR L-band images allowed an accurate mapping of a continuous 900 km-long palaeodrainage system, named the Kufrah River (see Figure 3). Its headwaters are mainly in southern Libya with observed tributaries arising in three main areas: El Fayoud and El Akdamin hamadas in northeastern Tibesti (Wadi Al Kufrah), northern Uweinat close to the Sudanese border (Uweinat tributary), and the western Gilf Kebir and Abu Ras plateaux on the Egyptian border [17]. The end of the Kufrah River disperses as a network of small shallow channels across the surface of the broad Sarir Dalmah alluvial fan, that covers more than 15,000 km^2, possibly constituting an inland delta. It is not possible to follow the river course to the north because the large and thick sand dunes of the Calanscio Sand Sea preclude radar mapping of the subsurface. However, about 300 km away to the northwest and emerging from beneath the Calanscio Sand Sea, lies the major, 2 to 4 km-wide, alluvium-filled Wadi Sahabi palaeochannel that incised more than 300 m into bedrock. Analysis of SRTM topography combined with PALSAR scenes allowed the mapping of several additional palaeochannels located west of the Kufrah River, each of which is likely to have formed a tributary that supplied water and sediment to the main palaeodrainage system. SRTM topography also revealed local depressions which allow to connect the western palaeochannels and the terminal alluvial fan of the Kufrah River to the Wadi Sahabi palaeochannel, through a 400 km-long palaeocorridor [18].

The Kufrah River is then a major palaeodrainage system in eastern Sahara, which at its maximum extent would have drained an area of more than 400,000 km^2 between the Tibesti, Al Haruj, and Gilf Kebir massifs and connected to the Mediterranean Sea in the Sirt Basin through the Wadi Sahabi palaeochannel, possibly discharging a comparable amount of water as does the present-day Nile. Despite the fact we have no direct indication about the age of initiation and history of the Kufrah River, it is very likely to be have been active during in recent (Holocene) times as proposed by Pachur and Altmann [19]: even though L-band radar does not allow to see deeper than a couple of meters, the palaeochannels are clearly visible in PALSAR images, suggesting that they are only at shallow depths. Earlier (Pleistocene) phases of activity are also likely: Osborne et al. [20] proposes a "humid corridor" that was connecting the Kufrah Basin to the Mediterranean coast 120,000 years ago. The Kufrah River system is then clearly a major palaeohydrological feature to take into account when studying the past environments and climates of northern Africa, from the middle Miocene to the Holocene. It also represents a likely corridor for fauna and human dispersal in the eastern Sahara, and thus indicates locations where further palae-ontological, palaeo-anthropological, and archaeological field exploration should be conducted.

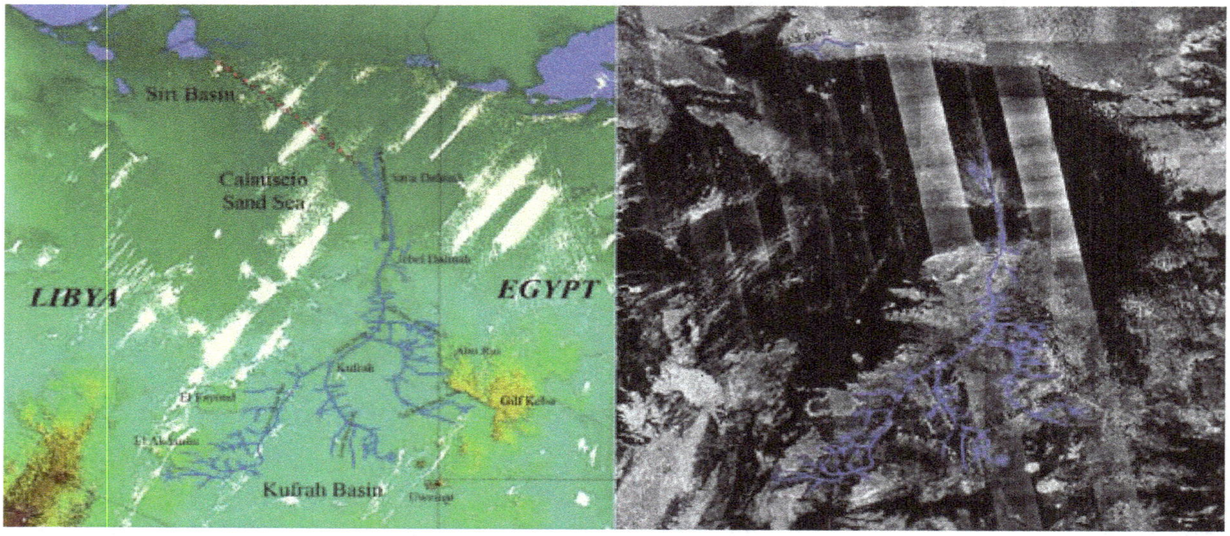

Figure 3. The Kufrah palaeoriver system (in blue) on top of Shuttle Radar Topography Mission (SRTM) topography (**left**) and on top of the PALSAR mosaic (**right**). The red dash line indicates a possible corridor connecting the terminal fan of the Kufrah River to the Mediterranean coast, through the Wadi Sahabi palaeochannel.

2.2. The Tamanrasett Palaeoriver in Western Sahara

The Green Sahara Periods (GSPs) are due to astronomically related changes and are the consequence of the transformation of the hydrological cycle over North Africa, with the intensification of the African summer monsoon. Changes in the position of this rain belt led to development of important fluvial networks over the Sahara, which resulted in an increase of freshwater delivery to surrounding oceans. Marine sediment records from the Mediterranean and Atlantic margins have provided consistent evidence of monsoon variability in northern Africa since the middle Pleistocene. The most recent GSP, which spans from 12,000 to 5000 years BP, is commonly referred to as the early Holocene African Humid Period (AHP). It is well recorded in marine sedimentary archives from the Gulf of Guinea to the northeastern Tropical Atlantic Ocean, the Mediterranean margin, and eastern Africa.

Off the western African margin, fluvial signals have been identified in deep-sea sediments dated from the early Holocene [21]. Recently, a large 400 km-long submarine channel system, the Cap Timiris Canyon, has been discovered on the western Sahara margin off Mauritania [22]. The Cap Timiris Canyon was very likely connected to a major river system in the past, and potential flow pathways simulated from present-day topography indicates the existence of a large river system in western Sahara, taking its sources in the Hoggar highlands and the southern Atlas Mountains. This so-called Tamanrasett River valley has been described as a possible vast and ancient hydrographic system [23]. Although a possible link between the Tamanrasett River and the Cap Timiris Canyon has already been suggested, no direct evidence of any fluvial activity and of a connection to the canyon has ever been found. PALSAR images of the Mauritanian coast provide geomorphological evidence for the existence of a palaeodrainage system located in the Arguin Bay, between Cap Blanc and Cap Timiris (see Figure 4). This newly identified palaeodrainage system is about 500 km-long and overlaps very well with the coastal section of the course of the Tamanrasett River inferred from topography. The reconstruction of the complete system was not possible using PALSAR images, due to the presence of thick sand dunes. However, the branch of the palaeodrainage network identified using radar represents a fifth of the total length of the Tamanrasett palaeoriver. The palaeochannels detected in radar images are also perfectly aligned with palaeovalleys identified in the Arguin Basin, as well as with the proximal tributaries of the submarine Cap Timiris Canyon system [24].

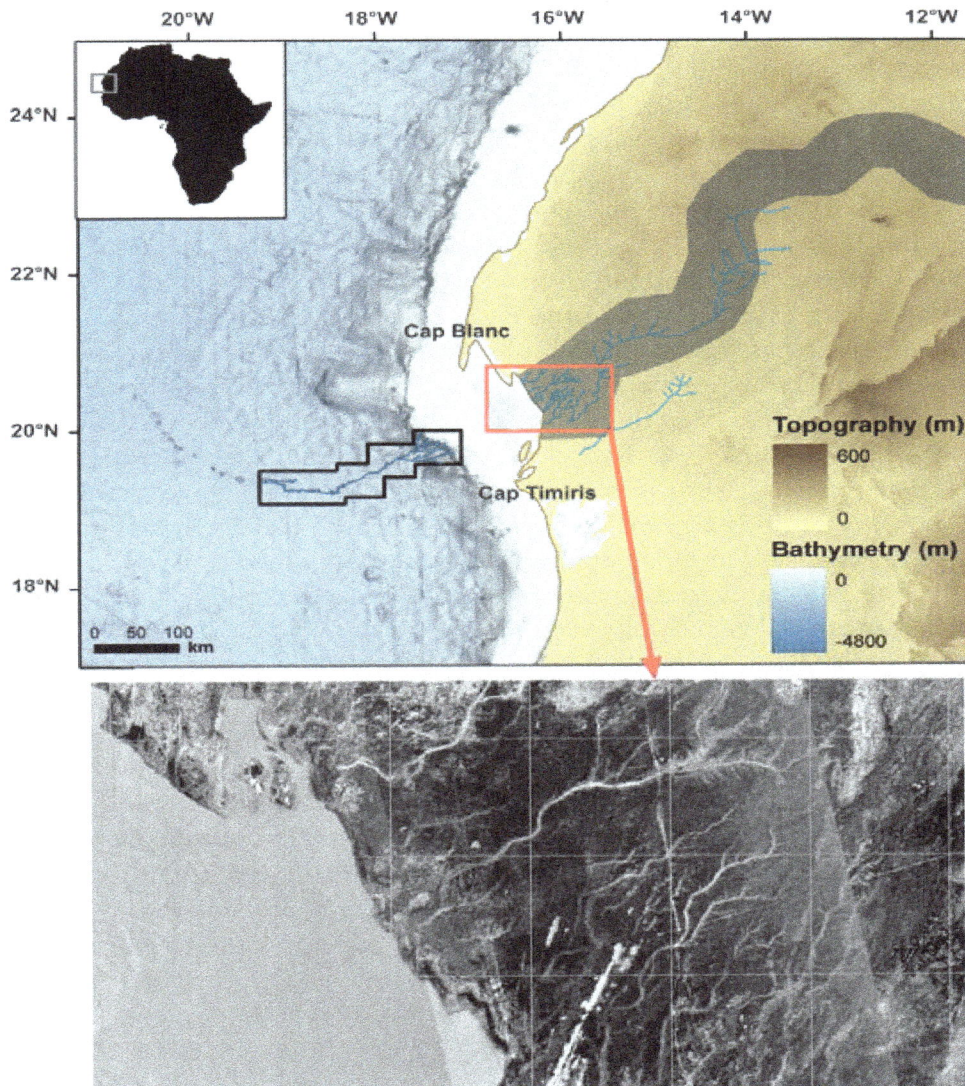

Figure 4. (**Top**) Topographic map of Mauritanian coast, showing locations of the Timiris Canyon (black square), palaeochannels detected in PALSAR data (blue lines) and Tamanrasett River valley (dark brown area), after [24]. (**Bottom**) PALSAR mosaic showing the discovered palaeodrainage system. "Radar rivers" appear here as bright features, due to the accumulation of coarse gravels in the terminal part of the channels, producing a higher radar return.

Thanks to orbital space-borne imaging radar, it was possible to establish the continuity of a past giant drainage system in western Africa, from the continent (Tamanrasett River) to the shelf (Arguin basin), and then to the bottom ocean (Cap Timiris Canyon). Overall, the identification of this palaeodrainage system in the Arguin Bay area is very coherent with the hydrological landscape of the western Sahara, especially during the most recent GSPs. The evidence of a fluvial activity on the Mauritanian coast during the recent past provides a missing link between the development of lakes over Algeria and Mauritania and fluvial evidences in Algeria. The presence of a vast fluvial system in the past in the place of present-day major dust sources also provides a new light on the interplay between aeolian and fluvial supplies in the making of the terrigenous signal off western Africa. This finding also provides valuable constraints for numerical simulations of Saharan climate throughout the late Quaternary.

2.3. The Ulaan Nuur Palaeolake in Central Asia

Arid Inner Asia has experienced dramatic climate fluctuations over the last millennia [25] and remains highly sensitive to ongoing climatic change. As a consequence, lakes in the Gobi Desert underwent major level changes during the Late Quaternary, with high stages being associated with wetter climates. The early Holocene in Mongolia is characterized by increasing temperature and humidity, followed by a humid early-mid Holocene stage, when lakes were at high volume. Enhanced aridity occurred during the mid-Holocene, but the beginning and end of the dry interval differs from location to location. In the late Holocene the humidity increased due to decreased evaporation when temperatures dropped in Mongolia. Due to its location, Mongolia is influenced by both the North Atlantic Oscillation and the East Asian Monsoon, associated with El Nino effect, making the region an important source for establishing Holocene climatic signals [26].

Remains of past wetter climates can be found in today's topography in the form of palaeochannels and palaeoshorelines, in particular in the desert of southern Mongolia [27,28]. We studied the Ulaan Nuur depression (44.53° N–103.73° E), located in a large and flat expanse in the Omnogov province, which was in the recent past the southern terminus for the Ongi River. We conducted the analysis of present-day topography provided by SRTM data, coupled with PALSAR imaging. PALSAR allowed the mapping of palaeochannels and palaeofans of ancient rivers feeding the Ulaan Nuur depression, while SRTM topography was used to simulate various palaeolake levels and map the location of possible palaeoshorelines [29]. We actually identified ten potential palaeochannels, strongly indicative of watercourses feeding Ulaan Nuur in the past. Two main streams are represented by the Ongi River (Figure 5A) and a southeastern channel (Figure 5C) that correspond to strong incision in the bedrock. They appear as bright linear structures in PALSAR images, because coarse alluvial gravel filling the channel bed increase the surface roughness and volume scattering effects, leading to a strong radar return. Secondary tributes also appear to have fed the Ulaan Nuur depression from the northwest, but with a somewhat less important water flow, creating numerous alluvial fans (Figure 5B). Palaeoshoreline morphologies can be clearly observed in present day topography: the Ulaan Nuur depression is bordered by wave-cut terraces in the north (Figure 5D) and in the south (Figure 5E,F).

The observed terraces actually form sequences that can be accounted for by varying lake levels. Figure 5 shows three main lake levels that we reconstructed by artificially flooding the present-day topography, choosing water levels in order to match the observed palaeoshorelines and channel fans. The bigger outer lake covers an area of 19,500 km^2 and is limited in the south by a sharp west-east trending palaeoshoreline. It also corresponds in the north to the limit where the Ongi River flows out of higher relief, and shows a bed transition from narrow to wide, indicating that the flow competence has declined markedly downstream. A medium-size lake, corresponding to a surface of about 6900 km^2, can then be defined and limited by northwestern and southeastern alluvial fans and shorter palaeoshorelines. The intermediate-size lake is likely to have been fed by low competence flows from the northwest and from the southeast, which left numerous, and wide alluvial fan remnants. It is coherent with a decrease in water level, since the Ongi River also becomes less competent between the large and medium lake limits, as shown by a wider and shallower bed morphology. Finally, a small-size lake, probably the most recent one, occupied an area close to 1700 km^2 and was limited in the north by a plateau and by being the termination of all mapped palaeochannels, including the present day Ongi River. This is likely to have been the final stage of the Ulaan Nuur, by the end of the progressive drying Holocene phase, before the present-day dry and desiccated depression.

Considering the surface area covered by the bigger extent of Ulaan Nuur lake, and taking the present day topography as lake floor, we estimated that an amount of more than 3000 km^3 of fresh water was present in the past, filling the Ulaan Nuur depression. Considering the Holocene palaeoclimatology of southern Mongolia, this lake is likely to have last several thousands of years, leading to significant amount of water infiltrating in the shallow subsurface. This suggests that shallow water resources may be located at multiple sites in the Ulaan Nuur depression: this is a suitable resource for small-scale community, accessible using inexpensive shallow drilling techniques.

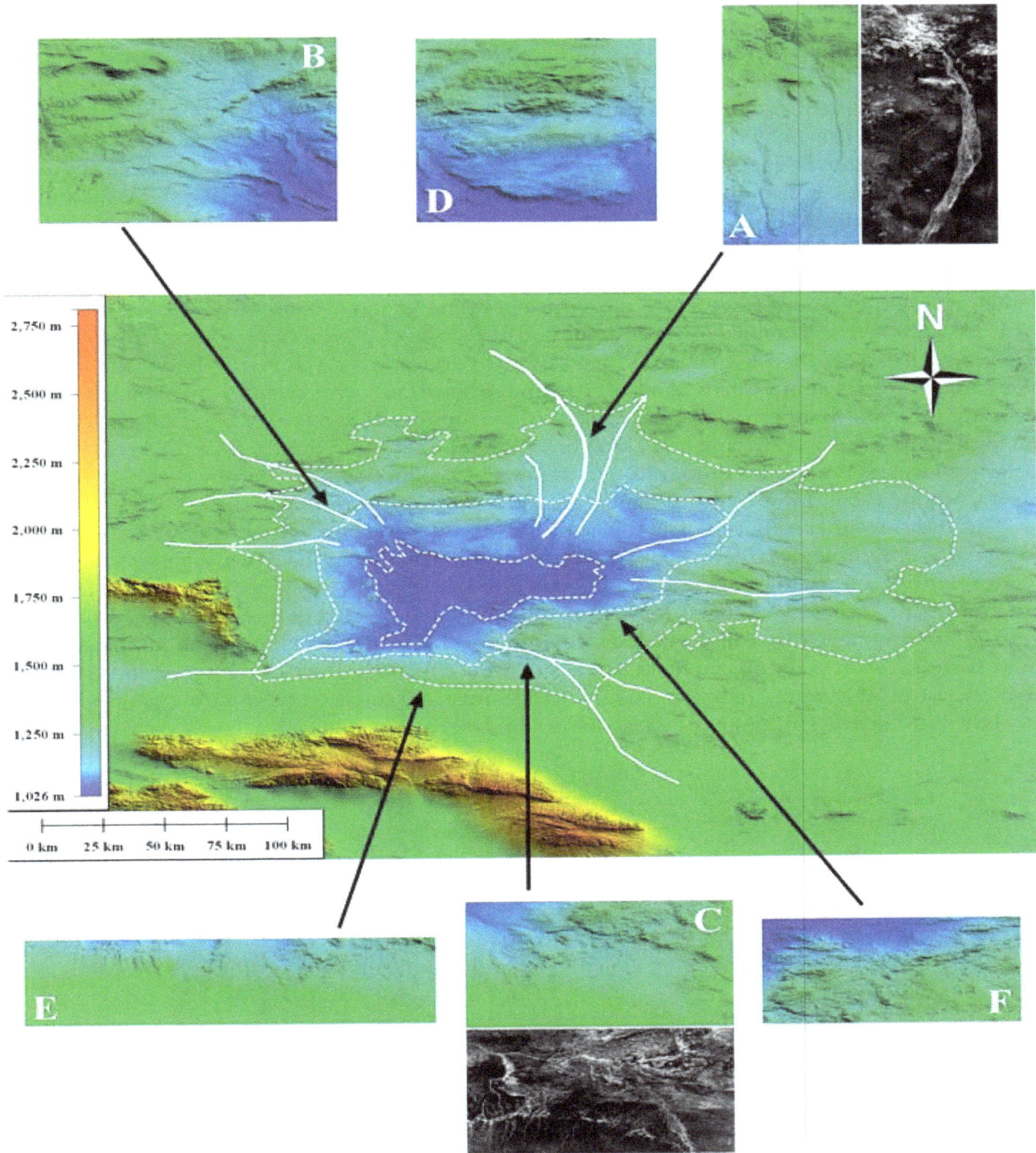

Figure 5. Ulaan Nuur depression on top of SRTM topography, showing the three past lake levels (dotted white lines) and the ten palaeochannels (solid white lines). Ulaan Nuur's main streams are the Ongi River (**A**) and a southeastern channel (**C**). Terminal fans of secondary tributaries (**B**) and palaeoshoreline morphologies (**D–F**) remain visible in the present day topography.

3. The Future of Space-Borne Imaging Radar: Low Frequency Sensors

Previous cases have shown the benefit of using L-band space-borne imaging radar for mapping the shallow subsurface in arid environments: even a shallow investigation depth of a couple of meters is enough to obtain significant and new information about palaeohydrography. An easy way to probe the subsurface deeper is to go for longer radar wavelengths: while a L-band (1.25 GHz) radar can penetrate 1–2 m of dry sand, a P-band system (435 MHz) should be able to probe the subsurface down to more than 5 m [30]. In June 2010, we conducted an airborne P-band SAR campaign over a desert site in southern Tunisia, using the SETHI system developed by ONERA [31]. This is the first time a low frequency P-band radar was flown over the Sahara. We acquired several radar scenes over the Ksar Ghilane oasis (32.98° N–9.63° E), an arid area at the limit between past alluvial plains and present day sand dunes. Figure 6 shows the comparison between a L-band radar scene acquired by the ALOS-2 Japanese sensor and a P-band radar scene acquired by the SETHI system: P-band radar better reveals the subsurface features under the superficial sand layer because of its higher penetration depth. A lower frequency radar is also less sensitive to the covering sand surface, leading to a lower contribution of the superficial layer. Using a two-layers scattering model for the surface and subsurface geometry shown in Figure 1, we could reproduce both the L- and P-band measured scattering levels, which are actually comparable. At L-band, the subsurface layer produces a backscattering component about 30 times lower than the one produced by the surface layer, while at P-band, the subsurface layer contribution is about thirty times higher than the surface layer component. The lower surface scattering term at P-band, due to a smoother surface roughness at a longer wavelength, is balanced by a higher subsurface scattering term, due to a higher penetration depth. As a final result, the total scattering level at P-band is comparable to the one at L-band, as observed by ALOS-2 and SETHI sensors, but the P-band return is dominated by the subsurface layer [32].

This indicates that a space-borne P-band SAR should be able to very efficiently map subsurface geological and hydrological features in arid areas. In 2021, the European Space Agency will launch its seventh Earth Explorer mission, named BIOMASS [33]. It is being designed to provide, for the first time from space, P-band SAR measurements to determine the amount of biomass and carbon stored in forests by combining a low radar frequency, polarimetric, interferometric, and tomographic techniques. It will also, as a secondary objective, map the subsurface geology in large desert regions such as the Sahara, Central Asia, and Australia. The BIOMASS mission will then offer a unique opportunity to reveal the hidden and still unknown past history of deserts.

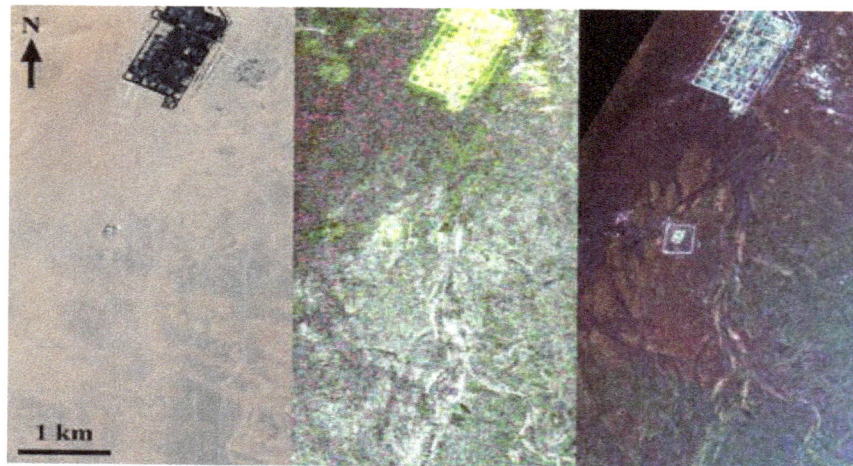

Figure 6. (**Left**) SPOT image of the Ksar Ghilane oasis region in southern Tunisia, palaeochannels are hidden by aeolian sand deposits. (**Middle**) ALOS-2 L-band radar image, showing some subsurface features still blurred by the radar return of the superficial sand layer. (**Right**) SETHI P-band radar image, revealing subsurface hydrological features in a very efficient way.

Acknowledgments: The author would like to thank the JAXA for providing PALSAR data in the framework of the Kyoto & Carbon Protocol project, and the ONERA for providing SETHI data in the framework of the TUNISAR project. This work was financially supported by the Centre National d'Etudes Spatiales (CNES) and by the European Space Agency (ESA).

References

1. Lee, J.-S.; Pottier, E. *Polarimetric Radar Imaging: From Basics to Applications*; CRC Press: Boca Raton, FL, USA, 2009; p. 398.

2. Elachi, C.; Roth, L.E.; Schaber, G.G. Spaceborne radar subsurface imaging in hyperarid regions. *IEEE Trans. Geosci. Remote Sens.* **1984**, GE-22, 383–388. [CrossRef]

3. Farr, T.G.; Elachi, C.; Hartl, P.; Chowdhury, K. Microwave penetration and attenuation in desert soil: A field experiment with the Shuttle Imaging Radar. *IEEE Trans. Geosci. Remote Sens.* **1986**, GE-24, 590–594. [CrossRef]

4. McCauley, J.F.; Schaber, G.G.; Breed, C.S.; Grolier, M.J.; Haynes, C.V.; Issawi, B.; Elachi, C.; Blom, R. Subsurface valleys and geoarchaeology of the eastern Sahara revealed by Shuttle Radar. *Science* **1982**, *218*, 1004–1020. [CrossRef] [PubMed]

5. Paillou, P.; Grandjean, G.; Baghdadi, N.; Heggy, E.; August-Bernex, T.; Achache, J. Sub-surface imaging in central-southern Egypt using low frequency radar: Bir Safsaf revisited. *IEEE Trans. Geosci. Remote Sens.* **2003**, *41*, 1672–1684. [CrossRef]

6. Schaber, G.G.; McCauley, J.F.; Breed, C.S.; Olhoeft, G.R. Shuttle Imaging Radar: Physical controls on signal penetration and subsurface scattering in the Eastern Sahara. *IEEE Trans. Geosci. Remote Sens.* **1986**, GE-24, 603–623. [CrossRef]

7. Abdelsalam, M.G.; Stern, R.J. Mapping precambrian structures in the Sahara Desert with SIR-C/X-SAR radar: The neoproterozoic Keraf suture, NE Sudan. *J. Geophys. Res.* **1996**, *101*, 23063–23076. [CrossRef]

8. Farr, T.G.; Rosen, P.A.; Caro, E.; Crippen, R.; Duren, R.; Hensley, S.; Kobrick, M.; Paller, M.; Rodriguez, E.; Roth, L.; et al. The Shuttle Radar Topography Mission. *Rev. Geophys.* **2007**, *45*. [CrossRef]

9. Drake, N.A.; El-Hawat, A.S.; Turner, P.; Armitage, S.J.; Salem, M.J.; White, K.H.; McLaren, S. Palaeohydrology of the Fazzan Basin and surrounding regions: The last 7 million years. *Palaeogeogr. Palaeoclimatol. Palaeoecol.* **2008**, *263*, 131–145. [CrossRef]

10. Ghoneim, E.; El-Baz, F. The application of radar topographic data to mapping of a mega-paleodrainage in the eastern Sahara. *J. Arid Environ.* **2007**, *69*, 658–675. [CrossRef]

11. Robinson, C.A.; Werwer, A.; El-Baz, F.; El-Shazly, M.; Fritch, T.; Kusky, T. The Nubian aquifer in Southwest Egypt. *Hydrogeol. J.* **2007**, *15*, 33–45. [CrossRef]

12. Paillou, P.; Reynard, B.; Malézieux, J.-M.; Dejax, J.; Heggy, E.; Rochette, P.; Reimold, W.U.; Michel, P.; Baratoux, D.; Razin, P.; et al. An extended field of crater-shaped structures in the Gilf Kebir region—Egypt: Observations and hypotheses about their origin. *J. Afr. Earth Sci.* **2006**, *46*, 281–299. [CrossRef]

13. Rosenqvist, A.; Shimada, M.; Ito, N.; Watanabe, M. ALOS PALSAR: A pathfinder mission for global-scale monitoring of the environment. *IEEE Trans. Geosci. Remote Sens.* **2007**, *45*, 3307–3316. [CrossRef]

14. SAHARASAR. Available online: http://saharasar.obs.u-bordeaux1.fr/ (accessed on 8 March 2017).

15. Paillou, P.; Lopez, S.; Farr, T.; Rosenqvist, A. Mapping Subsurface Geology in Sahara using L-band SAR: First Results from the ALOS/PALSAR Imaging Radar. *IEEE J. Select. Top. Earth Obs. Remote Sens.* **2010**, *3*, 632–636. [CrossRef]

16. Pachur, H.J.; Hoelzmann, P. Late Quaternary palaeoecology and palaeoclimates of the eastern Sahara. *J. Afr. Earth Sci.* **2000**, *30*, 929–939. [CrossRef]

17. Paillou, P.; Schuster, M.; Tooth, S.; Farr, T.; Rosenqvist, A.; Lopez, S.; Malézieux, J.-M. Mapping of a major paleodrainage system in Eastern Libya using orbital imaging Radar: The Kufrah River. *Earth Planet. Sci. Lett.* **2009**, *277*, 327–333. [CrossRef]

18. Paillou, P.; Tooth, S.; Lopez, S. The Kufrah Paleodrainage System in Libya: A Past Connection to the Mediterranean Sea? *C.R. Geosci.* **2012**, *344*, 406–414. [CrossRef]

19. Pachur, H.J.; Altmann, N. *Die Ostsahara im Spätquartär*; Springer: Berlin/Heidelberg, Germany; New York, NY, USA, 2006; p. 662.

20. Osborne, H.A.; Vance, D.; Rohling, E.J.; Barton, N.; Rogerson, M.; Fello, N. A humid corridor across the Sahara for the migration of early modern humans out of Africa 120,000 years ago. *Proc. Natl. Acad. Sci. USA* **2008**, *105*, 16444–16447. [CrossRef] [PubMed]

21. McGee, D.; Demenocal, P.B.; Winckler, G.; Stuut, W.; Stuur, J.B.; Bradtmiller, L.I. The magnitude, timing and abruptness of changes in North African dust deposition over the last 20,000 year. *Earth Planet. Sci. Lett.* **2013**, *371–372*, 163–176. [CrossRef]

22. Krastel, S.; Hanebuth, T.J.H.; Antobreh, A.A.; Henrich, R.; Holz, C.; Kolling, M.; Schulz, H.D.; Wien, K.; Wynn, R.B. Cap Timiris canyon: A newly discovered channel system offshore of Mauritania. *EOS Trans. Am. Geophys.* **2004**, *85*, 414–423.

23. Vörösmarty, C.J.; Fekete, B.M.; Meybeck, M.; Lammers, R.B. Global system of rivers. Its role in organizing continental land mass and defining land-to-ocean linkage. *Glob. Biogeochem. Cycles* **2000**, *14*, 599–621. [CrossRef]

24. Skonieczny, C.; Paillou, P.; Bory, A.; Bayon, G.; Biscara, L.; Crosta, X.; Eynaud, F.; Malaizé, B.; Revel, M.; Aleman, N.; et al. African Humid periods triggered the reactivation of a large river system in Western Sahara. *Nat. Commun.* **2015**, *6*. [CrossRef] [PubMed]

25. Yang, X.; Rost, K.T.; Lehmkuhl, F.; Zhenda, Z.; Dodson, J. The evolution of dry lands in northern China and in the Republic of Mongolia since the Last Glacial Maximum. *Quat. Int.* **2004**, *118*, 69–85. [CrossRef]

26. Lee, M.; Lee, K.; Lim, S.; Lee, J.; Yoon, H. Late Pleistocene–Holocene records from Lake Ulaan, southern Mongolia: Implications foreast Asian palaeomonsoonal climate changes. *J. Quat. Sci.* **2013**, *28*, 370–378. [CrossRef]

27. Komatsu, G.; Brantingham, P.; Olsen, J.; Baker, V. Paleoshoreline geomorphology of Boon Tsagaan Nuur, Tsagaan Nuur and Orog Nuur: The Valley of Lakes, Mongolia. *Geomorphology* **2011**, *39*, 83–98. [CrossRef]

28. Lehmkuhl, F.; Lang, A. Geomorphological investigations and luminescence dating in the southern part of the Khangay and the Valley of the Gobi Lakes (Central Mongolia). *J. Quat. Sci.* **2001**, *16*, 69–87. [CrossRef]

29. Sternberg, T.; Paillou, P. Mapping potential shallow groundwater in the Gobi using remote sensing: Lake Ulaan Nuur. *J. Arid Environ.* **2015**, *118*, 21–27. [CrossRef]

30. Paillou, P.; Dreuillet, P. The PYLA'01 experiment: Flying the new RAMSES P-band facility. In *AIRSAR Earth Applications Workshop*; JPL Publication: Pasadena, CA, USA, 2002.

31. Paillou, P.; Ruault du Plessis, O.; Coulombeix, C.; Dubois-Fernandez, P.; Bacha, S.; Sayah, N.; Ezzine, A. The TUNISAR experiment: Flying an airborne P-Band SAR over southern Tunisia to map subsurface geology and soil salinity. In Proceedings of the Progress Electromagnetics Research Symposium Abstracts, Marrakesh, Marocco, 20–23 March 2011.

32. Paillou, P.; Dubois-Fernandez, P.; Lopez, S.; Touzi, R. SAR polarimetric scattering processes over desert areas: Ksar Ghilane, Tunisia. In Proceedings of the POLINSAR Programme, Frascati, Italy, 23–28 January 2017.

33. Le Toan, T.; Quegan, S.; Davidson, M.; Baltzer, H.; Paillou, P.; Papathanassiou, K.; Plummer, S.; Rocca, F.; Saatchi, S.; Shugart, H.; et al. The BIOMASS mission: Mapping global forest biomass to better understand the terrestrial carbon cycle. *Remote Sens. Environ.* **2011**, *115*, 2850–2860. [CrossRef]

6

Evaluation of the Water Cycle in the European COSMO-REA6 Reanalysis using GRACE

Anne Springer [1,*], **Annette Eicker** [1,2], **Anika Bettge** [1], **Jürgen Kusche** [1] and **Andreas Hense** [3]

[1] Institute of Geodesy and Geoinformation, Bonn University, 53115 Bonn, Germany; annette.eicker@hcu-hamburg.de (A.E.); s7anbett@uni-bonn.de (A.B.); kusche@geod.uni-bonn.de (J.K.)

[2] Hafen-City University, 20457 Hamburg, Germany

[3] Meteorological Institute, Bonn Univesrity, 52121 Bonn, Germany; ahense@uni-bonn.de

* Correspondence: springer@geod.uni-bonn.de

Academic Editor: Frédéric Frappart

Abstract: Precipitation and evapotranspiration, and in particular the precipitation minus evapotranspiration deficit ($P - E$), are climate variables that may be better represented in reanalyses based on numerical weather prediction (NWP) models than in other datasets. $P - E$ provides essential information on the interaction of the atmosphere with the land surface, which is of fundamental importance for understanding climate change in response to anthropogenic impacts. However, the skill of models in closing the atmospheric-terrestrial water budget is limited. Here, total water storage estimates from the Gravity Recovery and Climate Experiment (GRACE) mission are used in combination with discharge data for assessing the closure of the water budget in the recent high-resolution Consortium for Small-Scale Modelling 6-km Reanalysis (COSMO-REA6) while comparing to global reanalyses (Interim ECMWF Reanalysis (ERA-Interim), Modern-Era Retrospective Analysis for Research and Applications, Version 2 (MERRA-2)) and observation-based datasets (Global Precipitation Climatology Centre (GPCC), Global Land Evaporation Amsterdam Model (GLEAM)). All 26 major European river basins are included in this study and aggregated to 17 catchments. Discharge data are obtained from the Global Runoff Data Centre (GRDC), and insufficiently long time series are extended by calibrating the monthly Génie Rural rainfall-runoff model (GR2M) against the existing discharge observations, subsequently generating consistent model discharge time series for the GRACE period. We find that for most catchments, COSMO-REA6 closes the water budget within the error estimates. In contrast, the global reanalyses underestimate $P - E$ with up to 20 mm/month. For all models and catchments, short-term (below the seasonal timescale) variability of atmospheric terrestrial flux agrees well with GRACE and discharge data with correlations of about 0.6. Our large study area allows identifying regional patterns like negative trends of $P - E$ in eastern Europe and positive trends in northwestern Europe.

Keywords: water cycle; GRACE; water budget equation; atmospheric reanalyses; numerical weather prediction models; precipitation minus evapotranspiration; COSMO-REA6

1. Introduction

Precipitation (P) minus evapotranspiration (E) represents the net flux of water between the atmosphere and the Earth's surface. Globally, $P - E$ is close to zero since land and ocean areas balance each other nearly, but not perfectly, as can be deduced from the present gain of water stored on continents [1,2]. Over land areas, in the temporal mean, $P - E$ is slightly positive since average evapotranspiration should not be greater than precipitation. Here, precipitation minus evapotranspiration links the terrestrial water budget to atmospheric moisture transports and, through latent heat flux, to the Earth's surface energy budget. The precipitation versus evapotranspiration deficit thus represents an important boundary condition for climate modeling and hydrological studies. Its temporal evolution can be traced to changes in climate forcing (temperature, precipitation, wind, CO_2 levels, etc.), the direct and indirect impacts of anthropogenic activities such as groundwater abstraction and land use change and the hydrological response of the system. Simulations have shown that while $P - E$ over oceans follows thermodynamical scaling with global warming, the "wet-get-wetter, dry-get-drier" scaling does not seem to apply over land [3]. Thus, investigating $P - E$ is very important to develop more elaborate scaling laws.

Various measurement techniques and observation datasets for precipitation and evapotranspiration exist [4,5]. In reanalyses, precipitation and evapotranspiration are generated based on numerical weather prediction (NWP) modeling and the assimilation of many datasets, with the result that $P - E$ is usually better simulated compared to P and E individually. It is known that P, E and $P - E$ have biases in reanalyses and do not close budgets [6–8]. Regional reanalyses seek to improve over global reanalyses through improved process representation and high-resolution modeling, but it is difficult to quantify improvements with independent datasets.

Reanalyses usually do not close the water budgets within error bars due to assimilation increments and since data errors and model inconsistencies necessarily propagate into the estimates. Assessing the closure of water and energy cycles enables diagnosing the error level and also understanding how well numerical weather prediction modeling can derive unobserved fields as residuals. The objective of this study is to validate the net flux of water between the atmosphere and the land surface in global and regional reanalyses including the regional European Consortium for Small-Scale Modelling 6-km Reanalysis (COSMO-REA6, [9]), differentiating per river basin and utilizing total water storage (TWS) measurements obtained from the Gravity Recovery and Climate Experiment (GRACE) satellites and discharge observations. To this end, we equate areal averages of $P - E$ for 26 river catchments grouped into 17 target areas through the terrestrial water budget equation to the observations of TWS change ΔS and discharge R.

GRACE data, commonly expressed as monthly gridded fields of total vertically-integrated water storage change, provide an opportunity to close the terrestrial water balance and thus replace the assumption that storage does not change over long time intervals. The GRACE mission consists of two satellites in tandem formation, chasing each other at about a 450-km altitude and equipped with a highly precise K-band ranging system. Variations of the inter-satellite distance can then be converted to gravitational potential change and further to mass change that is typically expressed as total water storage change per area. Several studies have combined GRACE-derived river basin averaged water storage change with discharge measurements [10,11] for either constraining $P - E$ [12] or combining with observed precipitation data for area-wide estimates of evapotranspiration [13–15]. However, challenges related to the use of GRACE data are the need to average over large regions, signal attenuation related to the peculiar filtering required to smooth out spatially-correlated data noise and signal leakage from neighboring regions.

Our study follows essentially the approach outlined in [8]; yet the availability of 20 years of COSMO-REA6 reanalyses enables us to assess trends over the 11-year period (2003–2013) common with the GRACE mission data. Moreover, the present study now covers all major European river basins with 26 rivers aggregated to 17 catchment areas. For this, we had to reconstruct discharge data missing in the Global Runoff Data Centre (GRDC) database using a simple model-based approach. This allows evaluating, for the first time, regional patterns of consistency over the whole of Europe for $P - E$ datasets from the regional reanalysis COSMO-REA6, the global Interim ECMWF Reanalysis (ERA-Interim [16]) and the global Modern-Era Retrospective Analysis for Research and Applications, Version 2 (MERRA-2, [17]). For comparison, observation-based precipitation from the Global Precipitation Climatology Centre (GPCC) Version 7 dataset [18] and evapotranspiration from the Global Land Evaporation Amsterdam Model (GLEAM [19,20]) are evaluated. We anticipate that this study will be useful for (1) COSMO modelers, to diagnose model deficiencies and eventually bring forecast and reanalysis closer to observations and (2) scientists that study the evolution of the water cycle over Europe, using reanalysis data including (but not limited) to COSMO-REA6.

We found that for all but one of the investigated catchments, discharge time series could not be retrieved complete from the Global Runoff Data Centre (GRDC). While several methods exist to fill or continue discharge time series based on satellite altimetry and rating curves [21,22], upon remotely-sensed water surface or inundation and water area–discharge rating curves [23], and upon hydraulic equations and using remote sensing measurement of other hydraulic variables such as river width or flow velocities (e.g., [24]), here, we resort to simulating discharge with the monthly Génie Rural rainfall-runoff model (GR2M, [25]) while calibrating it against discharge observations from GRDC using the most recently available data. We test this approach, and all results are accompanied by a thorough error assessment.

Our results suggest that the high-resolution COSMO-REA6 reanalysis indeed improves the closure of the water budget compared to the global reanalyses. Due to the large study area, regional patterns were visible, e.g., positive trends of $P - E$ over western and central Europe and negative trends of $P - E$ over eastern Europe.

The paper is organized as follows. Section 2 describes the datasets used in this study and introduces the study area. Section 3 outlines our method for generating consistent time series of P, E, R and ΔS. In Section 4, first the individual components of the water budget equation are evaluated for selected river basins. Then, the closure of the water budget equation is assessed, followed by more detailed analyses of the two sides of the water budget equation, by contrasting mean values, amplitudes and trends. Additional discussions follow in Section 5.

2. Data and Models

2.1. Study Area

Twenty-six major European river basins were considered in this study (Figure 1). The border of each basin is defined by the catchment associated with the most downstream gauging station. Small neighboring catchments are merged together for water budget analyses taking into account the main European watersheds (marked by colors in Figure 1). This is necessary because of the limited spatial resolution of GRACE-derived TWS. The 17 combined catchments are listed in Table 1 together with their size ranging from ~70,000 km^2 for the Po to ~800,000 km^2 for the Danube.

Table 1. The target catchments and their size.

Catchment	Size (km^2)	Catchment	Size (km^2)
Danube	807,000	Meuse-Rhine	180,601
Daugava-Narva	120,500	Neman	81,200
Dnepr	463,000	Neva	281,000
Don	378,000	Oder	109,729
Douro-Tagus	158,981	Po	70,091
Ebro	84,230	Rhone	95,590
Elbe-Ems-Weser	178,039	Southern Bug-Dniester	112,300
Garonne-Loire-Seine	227,000	Vistula	194,376
Guadalquivir-Guadiana	107,878		

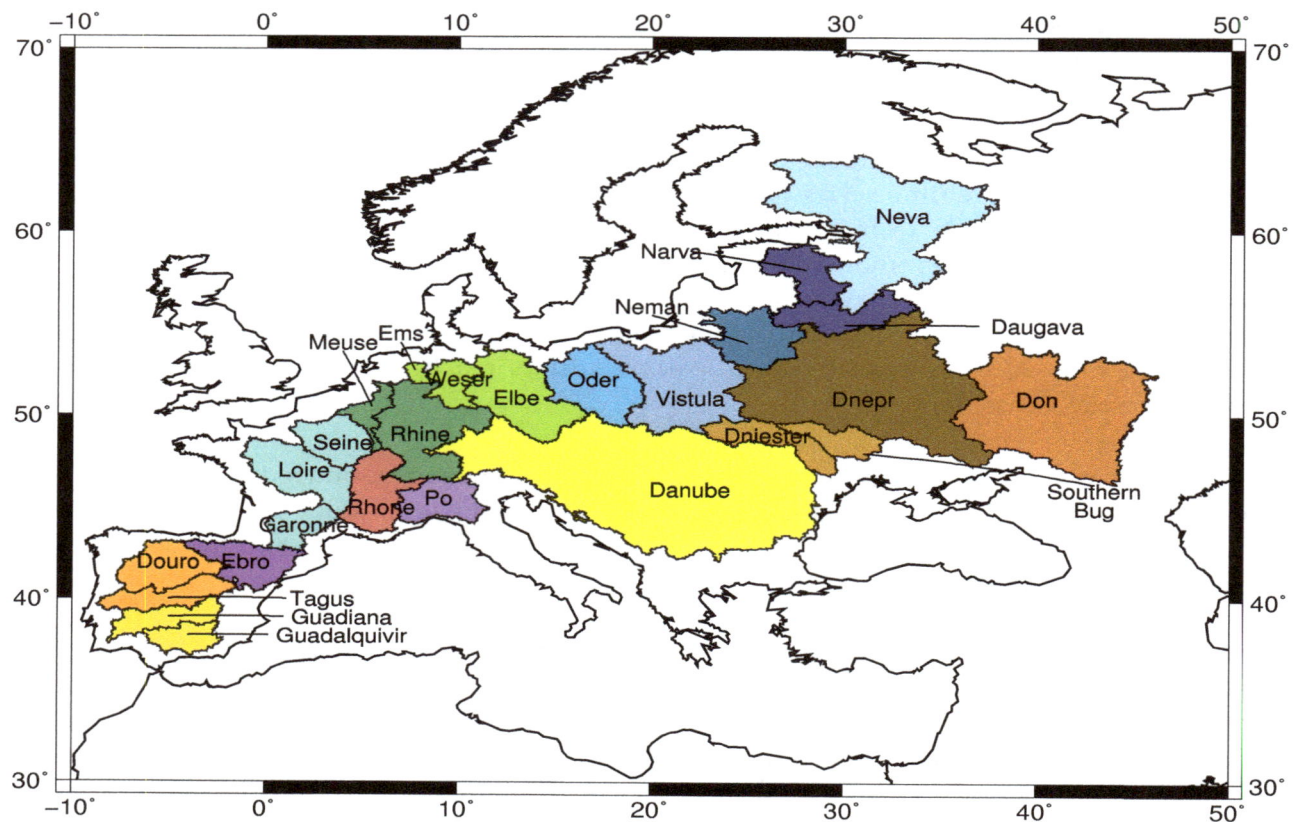

Figure 1. Study area: 26 European river basins, which are aggregated to 17 catchments as indicated by the different colors.

2.2. GRACE

The twin satellite mission GRACE (Gravity Field and Climate Experiment [26]) has been measuring spatial and temporal variations of the Earth's gravity field since 2002. GRACE consists of two satellites following each other on the same orbit. An inter-satellite K-band microwave link observes orbit variations caused by an inhomogeneous mass distribution on Earth. Temporal variations in gravity can then be converted to mass changes in terms of equivalent water height according to [27] taking into account the elastic loading effect.

In this study, we use GRACE release 05 (RL05) time series [28] provided by the German Research Center of Geosciences (GeoForschungsZentrum, GFZ). The monthly GRACE solutions were smoothed using a DDK4 filter [29] to account for the typical anisotropic error striping patterns. To consider the effect of seasonal and secular geo-center variations, we added the degree-one spherical harmonic coefficients provided by [30,31] to the GRACE solutions. The C20 coefficient, which cannot be well determined by GRACE, was replaced by a result obtained from satellite laser ranging [32]. Already during the GRACE processing chain, temporal gravity field variations caused by tides (ocean, Earth and pole tides), as well as by atmospheric and non-tidal ocean mass variations were subtracted from the observations prior to the gravity field estimation step. Additionally, the mass trend caused by glacial isostatic adjustment (GIA) [33] was removed in post-processing. Therefore, the resulting mass variations primarily reflect hydrological water storage changes. For a realistic estimation of the GRACE measurement accuracy, we used the calibrated errors provided by GFZ, as well as the standard deviations provided together with the degree-one and of the C20 coefficients in our variance propagation procedure (see Section 3.4).

2.3 Numerical Weather Prediction Models

2.3.1. COSMO-REA6

COSMO-REA6 is among the first reanalysis results covering only a part of the globe, but at a much larger spatial and temporal resolution than all available global reanalyses (see the next subsections). Other regional reanalyses are the North American Regional Reanalysis (NARR) [34], the arctic system reanalysis [35], or the recent European reanalyses from the U.K. MetOffice [36], or the Swedish Meteorological and Hydrological Institute (SMHI) [37].

The general goal of these regional reanalyses is to provide a physically consistent and homogeneous climate dataset of the atmosphere and land surface that resolves atmospheric patterns on spatial scales from 500 km down to about 10 km (also known as atmospheric mesoscale). Important weather and climate phenomena are situated in the mesoscale range e.g., frontal passages, land sea circulations or deep convection events. Furthermore, the effects of orography and land sea distribution (especially with respect to precipitation) are expected to be much better represented than in models with grid sizes of 50–100 km, which are common in global models due to the computational burden. Global reanalyses still play an important role because they provide the atmospheric boundary conditions to the regional high-resolution models.

The COSMO-REA6 uses the regional assimilation and forecasting system of the German National Meteorological Service (Deutscher Wetterdienst, DWD) [38]. It combines the non-hydrostatic forecast model COSMO with the continuous data assimilation method nudging [39]. The region covered by COSMO-REA6 is identical to the European Coordinated Regional Climate Downscaling Experiment at 0.11° resolution (CORDEX-EU11) domain, but with a grid size of 0.055° (~6 km), on a rotated latitude-longitude grid with 40 vertical layers in the atmosphere, between the surface and about a 20-km height. Prognostic variables are the three-component wind velocity vector, temperature, pressure and various water substance concentrations to account for the generation of clouds and precipitation in liquid and solid phases. Observations come from different observing systems like radiosondes, air planes, vertically pointing wind profilers, surface stations, buoys and ships. Depending on the observing system and the observed variable, windows of influence are defined that spread the actual observations to the neighboring grid cells in space and time. Nudging means that each equation for the prognostic variables temperature, pressure, wind velocity and moisture concentration is supplemented by an additional forcing term. This forcing term drives the model solution at any time and at all grid points to the respectively spread observations using a prescribed time scale. Note that this procedure will provide consistent prognostic variables through their physical connection dictated by the equations, but it cannot provide consistent budgets of energy and water because the additional forcing terms always add energy, water and momentum to the system proportional to the difference between the simulated and the observed values. Fluxes like precipitation and evapotranspiration are derived internally from the necessary state variables, but are in general not assimilated. Therefore, an evaluation of precipitation, evapotranspiration and other energy or water fluxes with independent data as is presented here is an important contribution to the overall quality assessment of the COSMO-REA6 or any other regional reanalysis.

The actual setup of COSMO-REA6, including the necessary preparation of land surface variables, like soil moisture, and oceanic variables, like sea surface temperatures, can be found in [9]. Here also, a first quality assessment of COSMO-REA6 is presented. Precipitation as one of the most interesting climate variables is further evaluated in [40], where also a comparison over Germany of the global reanalysis ERA-Interim, the U.K. Met Office and SMHI European reanalyses and an even higher resolution COSMO reanalysis at a 2-km grid size is presented, showing among other features the importance of resolution in representing precipitation.

2.3.2. ERA-Interim

The global Interim ECMWF Reanalysis (ERA-Interim) provides gridded fields of atmospheric and land surface variables with approximately 79-km grid spacing considering 60 vertical levels. The model output covers the time span from 1979 onwards at three-hourly resolution. The main objectives of the ERA-Interim reanalysis are a realistic representation of the hydrological cycle and temporal consistency in order to estimate reliable trends [16]. The sequential data assimilation system is based on the 4D-Var method [41]. In a 12-hourly cycle, the current state of the atmosphere is estimated from a forecast model combined with an important number of observations originating in the majority from satellites.

In this study, synoptic monthly means of precipitation and latent heat flux are evaluated, which are computed by the model from temperature, humidity, wind speed and radiometer observations. Compared to [8], here, we use a finer grid of $0.75° \times 0.75°$.

2.3.3. MERRA-2

While in [8], we assessed the performance of the Modern-Era Retrospective analysis for Research and Applications (MERRA) model, here we moved on to its successor MERRA-2. For data assimilation, an incremental analysis update scheme is applied by the Goddard Earth Observing System Version 5 (GEOS-5). MERRA-2 assimilates new observation types such as hyperspectral radiance, microwave, GPS radio occultation, ozone and aerosol datasets [17]. One particular improvement of MERRA-2 is the correction of precipitation biases using observation-based precipitation from different sources [42,43]. However, in this study, we assess the uncorrected precipitation in order to compare with the other two reanalyses.

MERRA-2 is provided by the Goddard Earth Sciences Data and Information Service Center (DISC) from 1980 onwards with one-hourly temporal resolution on a $0.625° \times 0.5°$ grid. Here, monthly outputs of modeled precipitation and surface latent heat flux are assessed.

2.4. Observational Datasets

In this study, the ability of P and E from NWP models to close the water budget equation is compared to the performance observation-based datasets, i.e., the Global Precipitation Climatology Centre (GPCC) dataset for P and the Global Land Evaporation Amsterdam Model (GLEAM) for E.

2.4.1. GPCC

We used the recent version of the GPCC Full Data Reanalysis (V7) that covers the period 1901–2013 and is available at 0.5° resolution from the GPCC web site [18]. The precipitation dataset is based on 75,000 gauging stations world-wide, which are subject to strict quality control. The monthly product is optimized for water budget studies [44]. GPCC data were also used as a reference by [9] for assessing the quality of P generated by COSMO-REA6.

2.4.2. GLEAM

GLEAM estimates the individual components of evapotranspiration using satellite observations together with model data and a set of algorithms [19,20]. First, the Priestley and Taylor equation [45] is used to calculate potential evaporation from net radiation and air temperature observations. Then, actual evaporation is derived by applying a multiplicative stress factor based on microwave vegetation optical depth and soil moisture observations. Three datasets exist with different forcings and spatio-temporal coverages. Of these, two datasets of limited geographical extent are exclusively based on satellite data. Here, we use the only global dataset, GLEAM_v3.0a. For GLEAM_v3.0a, net radiation and air temperature are obtained from ERA-Interim, but all other required input datasets (precipitation, soil moisture, vegetation optical depth, snow water equivalent) are observation based. GLEAM_v3.0a has a spatial resolution of 0.25° and is a daily dataset, which spans the period 1980–2014.

In [8], we used a gridded E dataset based on the current global network of eddy covariance towers (FLUXNET) provided by the Max Planck Institute (MPI) Jena. The MPI dataset is available only until 2011 and was therefore replaced by GLEAM. However, the authors of [20] found good agreement of GLEAM and FLUXNET data.

2.5. Discharge

Monthly discharge data from the Global Runoff Data Centre (GRDC) are available for limited periods of time only (Figure 2). Merely 10 out of 26 stations used in this study provide data covering parts of the GRACE time span. Approaches for extending discharge data in time became ever more relevant during the last few years [46], e.g., satellite altimetry, runoff-precipitation ratio and runoff-storage relationships. Here, we created a consistent discharge time series by calibrating the lumped rainfall-runoff model GR2M [25]. GR2M (a) is a simple empirical monthly model, (b) achieves good results for river basins of different sizes and with different hydrological conditions [47] and (c) allows for easy inclusion of error estimation. Further model improvement was achieved by extending GR2M using a distributed Hydrologiska Byrans Vattenavdelning (HBV) type snow model.

For each of the 26 river basins, the GR2M-snow model was calibrated against GRDC data using the 10 most recent continuous years available (marked in light blue in Figure 2). Then, the model was run for the GRACE time span. In addition, error estimates were obtained by running GR2M-snow several times with disturbed forcings and disturbed parameters. Finally, discharge from GR2M-snow was merged for the 17 aggregated catchments including error propagation. The resulting discharge time series were used in the further course of this study. While time series of P, E and ΔS are averaged over each basin, discharge is available at the most downstream gauging station, which also defines the border of the catchment.

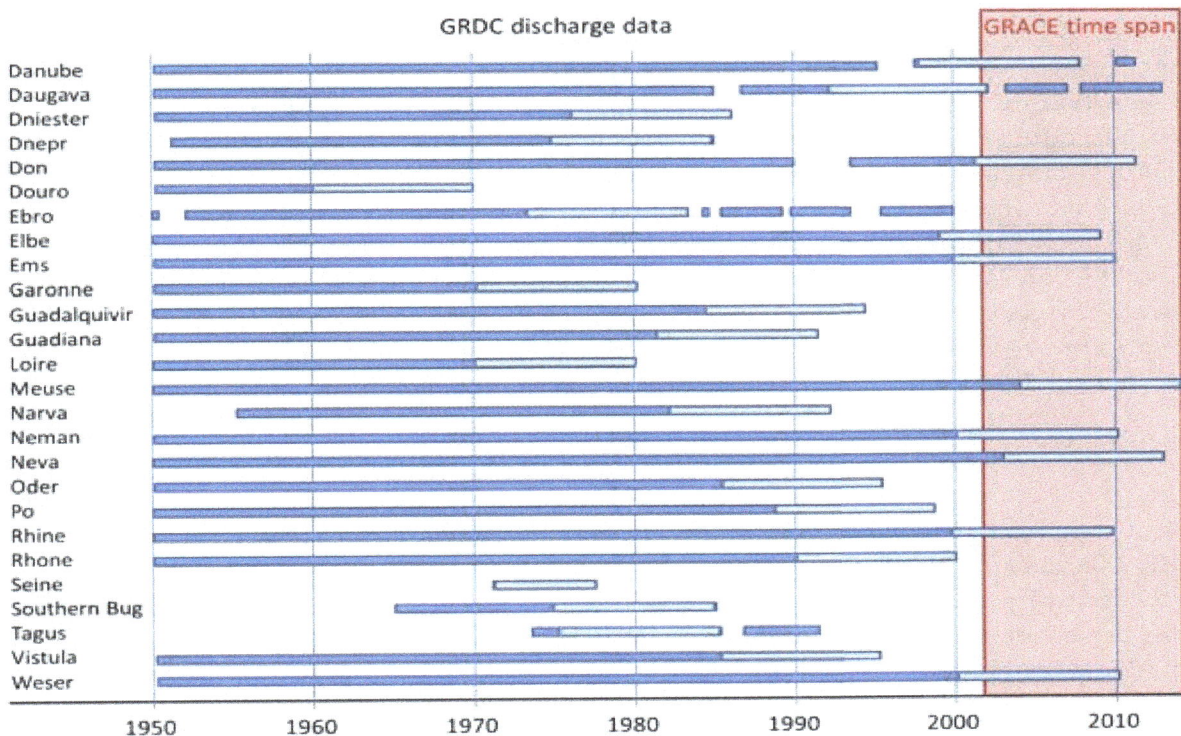

Figure 2. Discharge available from the Global Runoff Data Centre (GRDC) for the most downstream gauging stations of the rivers in our study region. The monthly Génie Rural rainfall-runoff model with snow extension (GR2M-snow) is calibrated against the 10 most recent continuous years of each basin (marked by light blue). In red, the Gravity Recovery and Climate Experiment (GRACE) time span is indicated.

3. Methodology

Water storage changes ΔS and discharge R are linked to net atmospheric-terrestrial flux $P - E$ via the water budget equation:

$$\Delta S + R = P - E. \tag{1}$$

Consistent time series of water flux and storage are required for assessing the closure of the water budget equation. In this section, we first describe the derivation of the time series of $P - E$ and ΔS. Afterwards, the set up of the rainfall-runoff model GR2M is presented aiming at the computation of discharge time series R covering the whole GRACE period. Finally, evaluation measures and the error assessment strategy are described.

3.1. Consistent Time Series of Atmospheric-Terrestrial Flux $P - E$ and Storage Change ΔS

First, GRACE-derived gridded total water storage (TWS) anomalies S were centered to the 11-year study period by removing the mean static field computed from 2003–2013. Next, monthly time series of P, E and S were computed by spatially averaging over all grid cells of the target area. Finally, storage changes ΔS were obtained for each month τ as central differences from the GRACE storage time series according to:

$$\Delta S(\tau) = \frac{1}{2}(S_{\tau+1} - S_{\tau-1}), \tag{2}$$

where $S_{\tau-1}$ is the previous month and $S_{\tau+1}$ the next month. Central differences in contrast to backward or forward difference operators avoid introducing a phase shift in the TWS change time series. Spatial averaging of TWS involves attenuation of the signal due to the spectral characteristics of GRACE data and further distortion due to the filtering procedure, known as the leakage effect [48]. Depending on the mass distribution, filter properties, and target area, mass is transported either into the basin (leakage in) or out of the basin (leakage-out). This effect becomes particularly large if the basin is very small or if the mass distribution outside and inside of the basin differs significantly [49]. Consequently, GRACE-derived time series of TWS change need to be rescaled. To this end, time-variable rescaling factors were derived for each target area separately. Assuming that the spatial distribution of $P - E$ approximately corresponds to the spatial distribution of ΔS, a number of different P and E datasets were used to simulate the effects from spectral resolution, filtering and leakage. A robust estimate for the multiplicative rescaling factor of each month was obtained from the median of the analyzed $P - E$ datasets. More detailed information on the derivation of consistent time series are provided by [8].

3.2. Using GR2M-Snow for Generating Modeled Discharge Time Series

The 2-parameter rainfall-runoff model GR2M requires as input monthly precipitation and potential evapotranspiration and includes two stores, a production store of variable size and a routing store with a fixed capacity of 60 mm. Monthly discharge is simulated by calibrating the capacity of the production store and an exchange coefficient, which accounts for the exchange of water with the outside of the basin. GR2M was extended by a distributed HBV-type snow model, which requires gridded temperature as input and adds three more calibration parameters (melting temperature, melting coefficient and temperature separating rain and snow). If the temperature is smaller than a defined value, snow is added to the corresponding grid cell, and if the temperature exceeds the melting temperature, the melting coefficient defines the amount of snow that is subtracted from the grid cell. Then, snow loss is accumulated over all grid cells and considered as additional precipitation. The precipitation and temperature datasets were obtained from the European daily high-resolution gridded dataset (E-OBS [50]) and potential evapotranspiration from the Climate Research Unit (CRU) data set TS3.22 [51]).

GR2M-snow was calibrated for each of the basins against observed discharge for the 10 most recent continuous years available from GRDC (Figure 2) minimizing the squared difference of observed and simulated discharge. As GRDC data are erroneous for Neman and Vistula watersheds, those basins were calibrated against data from E-RUN (observational gridded runoff estimates for Europe) [52]. Then, the model was run for the time span 1950–2013. As the validation period for each basin, the next 10 years available before the calibration period were chosen. In addition, error estimates were derived by generating an ensemble of simulated discharge by disturbing both the parameters and the input datasets. The input datasets were disturbed with noise following a Gaussian distribution, applying a standard deviation of 1 °C for temperature, 20% for precipitation and 5% for potential evapotranspiration. The variabilities of the five calibration parameters follow uniform distributions with appropriate uncertainty assumptions chosen for each parameter individually. For each river basin, 1000 model runs were performed, and from the ensemble of modeled discharge time series, the standard deviation was computed. The resulting error band fits well to the difference between GRDC and GR2M-snow (see Section 4.1). Furthermore, we were able to reproduce the results from [8] using discharge simulated by GR2M-snow within the error estimates. However, we are aware of limitations of this approach, e.g., when river dynamics changed over time due to river management and/or construction.

The skill of GR2M-snow in simulating discharge is evaluated by assessing different measures for the validation period, i.e., the mean bias and the root mean squared error (RMSE) between observed and simulated discharge, and Nash–Sutcliffe (NS) coefficients [53] for the time series with seasonal cycle and for de-trended and de-seasoned time series. NS coefficients are computed according to:

$$NS = 1 - \frac{\sum_\tau (\sqrt{R_{obs,\tau}} - \sqrt{R_{sim,\tau}})^2}{\sum_\tau (\sqrt{R_{obs,\tau}} - \sqrt{\overline{R_{obs}}})^2}, \tag{3}$$

where R_{obs} is observed discharge from GRDC, $\overline{R_{obs}}$ the temporal mean of observed discharge and R_{sim} simulated discharge from GR2M-snow. Here, NS coefficients are computed from the root of discharge in order to avoid too much weight on high discharge values. An NS coefficient of one means perfect agreement between observed and modeled discharge; values between zero and one imply that the model better simulates discharge than the mean of observed discharge. In the case of NS coefficients smaller than zero, the mean of observed discharge better represents actual discharge than the model.

3.3. Evaluation of the Water Budget Equation

The performance of the individual $P - E$ datasets is assessed with respect to $\Delta S + R$ using the following measures computed for the whole study period (2003–2013). A four-parameter model was fitted to the time series of basin averages for both sides of the water budget equation:

$$\begin{Bmatrix} P - E \\ \Delta S + R \end{Bmatrix} = a + b\tau + c\sin\left(\frac{2\pi}{T}\tau\right) + d\cos\left(\frac{2\pi}{T}\tau\right) + \epsilon, \tag{4}$$

where T is the annual period and ϵ the residuals. A difference between the mean a of the time series of $\Delta S + R$ and $P - E$ is defined as the bias of atmospheric-terrestrial flux. Parameter b corresponds to the trend, and amplitude and phase shift are derived from c and d. Additionally, correlations of the time series and of de-seasoned and de-trended time series $\hat{e}$ were investigated.

3.4. Error Assessment

For assessing the significance of our results, we performed a thorough error assessment for $\Delta S + R$. Calibrated errors given for monthly GRACE Stokes coefficients were propagated to the basin-averaged TWS changes ΔS while taking into account errors from degree-1 and -2 coefficients. However, the resulting error estimates might be too optimistic, as we neglect errors from the GIA model and from the rescaling procedure. The uncertainty of simulated discharge was obtained from ensemble runs with the GR2M-snow model by disturbing model parameters and input datasets. Then, errors were propagated further to $\Delta S + R$ for each of the aggregated catchments.

4. Results

Overall, the results presented below show the skill of the individual $P - E$ datasets in closing the water budget equation over Europe. First, the individual components of the water budget equation are analyzed. Then, both sides of the water budget equation are contrasted for selected target areas. Finally, biases, amplitudes and trends are illustrated and discussed for the whole study area.

4.1. Modeled Discharge from GR2M-Snow

Discharge simulated from GR2M-snow generally fits well to available discharge from GRDC within the error bounds. Figure 3 shows discharge for the Rhine basin from GRDC (blue) and GR2M-snow (red). Only a few peaks from GRDC exceed the ensemble-based standard deviation (grey).

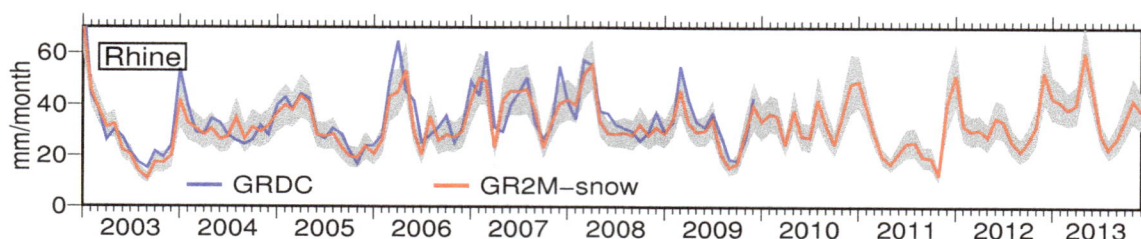

Figure 3. Discharge R for the Rhine from Global Runoff Data Centre (GRDC) (blue) and simulated from Génie Rural rainfall-runoff model (GR2M)-snow (red) including the standard deviation (grey).

The influence of discharge on the closure of the water budget equation depends not only on its magnitude, but also on the magnitude of TWS change. The magnitude of discharge and its standard deviation vary for the individual river basins with temporal means between 3 and 61 mm/month (Table 2). The largest values are found in central Europe where mean discharge rates are larger than 40 mm/month and corresponding standard deviations larger than 7 mm/month (Rhine, Rhone and Po). In contrast, many basins in eastern Europe, in France and on the Iberian peninsula have mean discharge rates of about 10–20 mm/month and standard deviations of only 2–5 mm/month. In general, the standard deviation derived from ensemble runs amounts to 20%–30% of simulated mean monthly discharge (Table 2). An exception is the Daugava, where the standard deviation is about 40% of simulated discharge. Root mean squared errors (RMSE) between observed and simulated discharge have in most basins approximately the size of the simulated standard deviations.

The bias of mean discharge for the validation period is smaller than 2 mm/month (Table 2) except for Po (4.0 mm/month) and Rhone (2.5 mm/month), and thus, we neglect its contribution to the closure of the water budget equation. Nash–Sutcliffe coefficients (NS) were computed for the validation phase and confirm that the calibration of GR2M-snow was successful for most basins. In western and central Europe, the NS coefficients reach values between 0.6 and 0.9. Only some basins in eastern Europe (Narva, Dnepr, Southern Bug and Don) have NS coefficients smaller than 0.3. Discharge from Don contains some extreme peaks during the validation time span that are not simulated by GR2M-snow, resulting in a negative NS coefficient. NS coefficients for de-seasoned and de-trended time series (computed according to Equation (4), extended by semiannual signals) are mostly between 0.2 and 0.6. Again, the largest values are obtained for western and central Europe with NS coefficients between 0.35 and 0.65. Negative NS coefficients of de-seasoned and de-trended time series are found for Danube, Don, Neva and Po indicating changes in the short-term behavior of discharge between the calibration and the validation time span. However, correlations are between 0.8 and 0.9 for the original time series and between 0.7 and 0.8 for de-seasoned and de-trended time series. Exceptions are the rivers Don and Neva, where correlations of de-seasoned and de-trended time series amount only to 0.3 and 0.4.

Table 2. Evaluation of simulated discharge from GR2M-snow for the validation period of each river basin: mean values, mean standard deviation (Std.), root mean squared errors (RMSE), bias of the mean and Nash–Sutcliffe (NS) coefficients for time series with seasonal cycle and de-seasoned and de-trended (des., det.) time series are computed as defined in Section 3.2. Due to missing observations, no validation could be performed for Seine and Tagus (* Neman and Vistula are calibrated and validated using E-RUN (observational gridded runoff estimates for Europe) because of erroneous GRDC data).

Catchment	Mean ($\frac{mm}{month}$)	Std. ($\frac{mm}{month}$)	RMSE ($\frac{mm}{month}$)	Bias ($\frac{mm}{month}$)	NS	NS des., det.
Danube	19.1	4.8	4.2	0.4	0.67	−0.34
Daugava	18.0	7.9	9.1	0.9	0.60	0.28
Dniester	13.8	4.0	5.9	−0.9	0.65	0.38
Dnepr	8.2	2.6	4.1	−0.5	0.42	0.22
Don	4.8	1.7	2.1	0.2	−0.12	−0.22
Douro	11.0	2.9	6.2	−1.0	0.61	0.45
Ebro	14.9	3.6	6.2	0.23	0.73	0.25
Elbe	11.9	3.1	4.2	−0.1	0.69	0.17
Ems	28.3	5.6	9.8	1.6	0.84	0.49
Garonne	33.7	6.7	11.2	2.5	0.81	0.38
Narva	16.2	4.1	4.0	1.2	0.62	0.42
Guadalquivir	3.4	1.2	4.6	−1.2	0.63	0.53
Guadiana	4.0	1.4	6.1	−1.8	0.66	0.65
Loire	23.8	5.2	7.4	1.8	0.84	0.35
Meuse	35.2	7.4	11.0	1.7	0.84	0.50
Neman *	14.6	4.7	5.8	−3.2	0.69	0.58
Neva	22.3	4.5	3.5	−0.5	0.66	−0.73
Oder	15.3	3.7	4.4	0.5	0.59	0.27
Po	61.4	16.4	20.4	4.0	0.51	−0.11
Rhine	38.2	7.3	7.6	1.1	0.77	0.36
Rhone	51.1	11.8	10.6	1.7	0.76	0.45
Southern Bug	5.7	2.1	4.5	−0.3	0.21	0.07
Vistula *	15.3	4.3	4.8	−1.0	0.54	0.29
Weser	22.5	5.1	7.3	0.1	0.83	0.51

4.2. Time Series of Fluxes and Storage Change

The noise of ΔS depends mainly on the size and the shape of the target areas [49]. For example, the Danube has a smooth time series with standard deviations of about 10 mm/month, while the time series of the Po is very noisy, and the standard deviations reach up to 30 mm/month, which is half of the annual amplitude (Figure 4). In fact, with a size of only 70,000 km^2, the Po basin is likely too small for obtaining reliable TWS information from GRACE.

In the following, time series of fluxes for five catchments from different regions in Europe and of different sizes are discussed in more detail: (a) Guadiana-Guadalquivir on the Iberian peninsula, which is a small region and the most southern target area with very dry climatic conditions, (b) Rhine-Meuse in central Europe with a medium size and large precipitation events in the Alpine region, (c) Neva, the most northern basin, (d) Dnepr, a catchment with particularly low precipitation rates, and (e) Danube, the largest river basin in Europe. Some, but not all findings from these regions can be transferred to basins with similar climatic conditions (see Section 4.3).

The magnitude of river discharge varies in the individual basins depending on the climatic conditions (Figure 5, black dashed line). In Guadiana-Guadalquivir and Dnepr, discharge is small with mean annual values of 7–9 mm/month and 0–14 mm/month. In contrast, discharge plays an important role in central Europe (Rhine-Meuse) where it amounts to mean annual values of 30 mm/month, which is one third of precipitation. While discharge in the Neva basin is relatively constant with a weak annual cycle, in the Rhine-Meuse and Danube basins, it is clearly related to individual precipitation events. For most basins, discharge contributes about 10%–20% to the error of the left-hand side of the water budget equation. The highest error contribution from discharge is found in the Danube basin with 40%, which is mainly due to the small standard deviation of GRACE data caused by the large basin size.

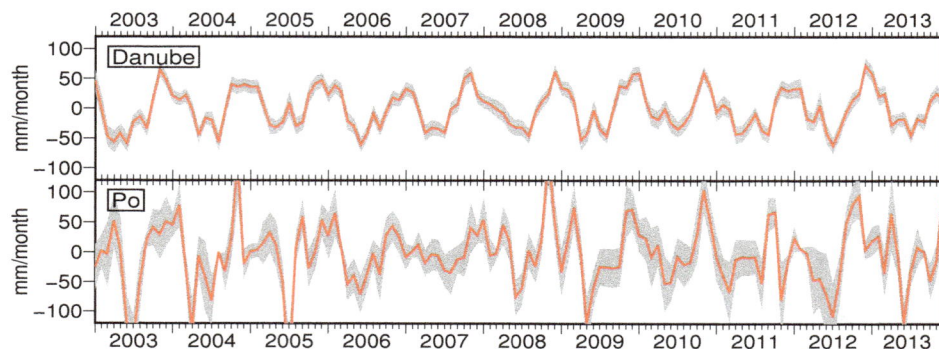

Figure 4. Total water storage (TWS) change ΔS (red) and its standard deviation (grey) for two selected river basins.

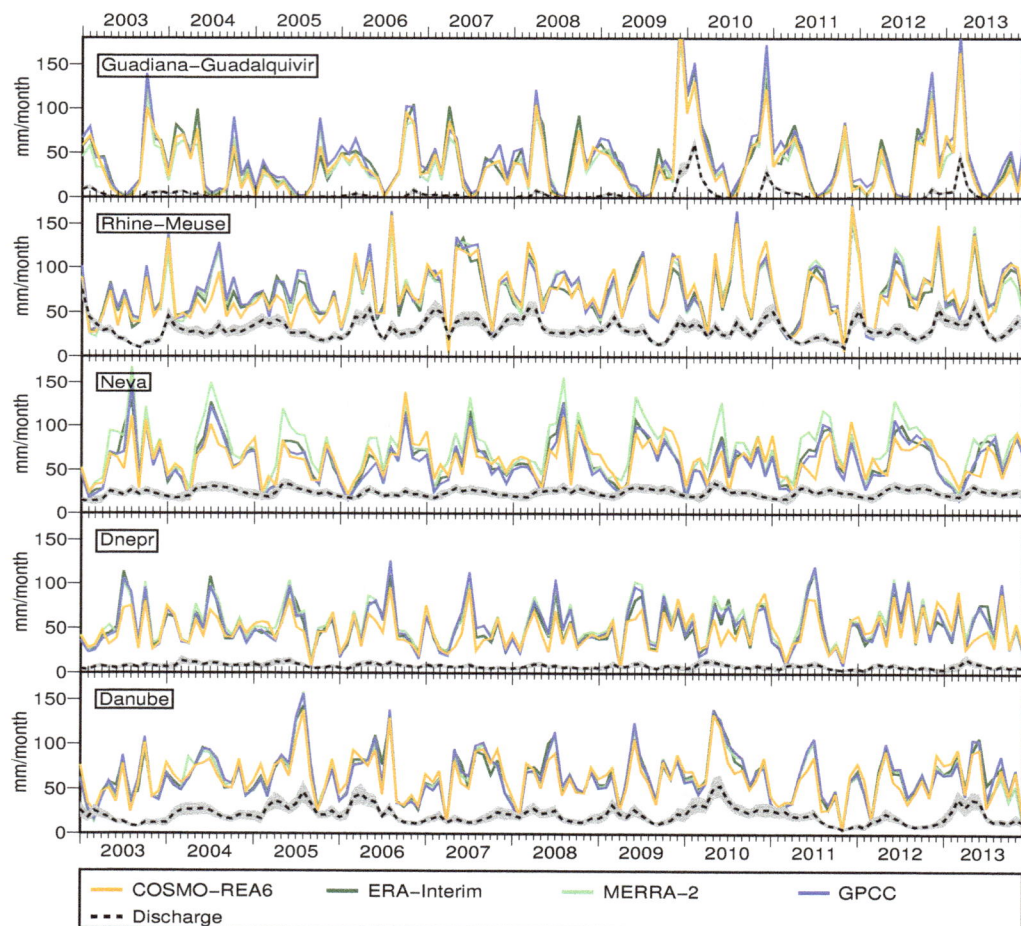

Figure 5. Monthly total precipitation P from different models and simulated discharge R (black dashed) for selected catchments of the study area.

Monthly total precipitation is highly variable for European river basins (Figure 5). The chosen catchments represent different climatic conditions over Europe, e.g., there is nearly no precipitation in summer in Guadiana-Guadalquivir; Neva has a very pronounced annual cycle; and in Rhine-Meuse, individual precipitation events are visible. All model-derived time series show similar features as the observation-based dataset GPCC. However, while in the Danube basin, peaks from all models have about the same size, we find in the Dnepr and Neva basins smaller peaks for the COSMO-REA6 model compared to the other datasets. The authors of [54] showed that COSMO-REA6 has an enhanced skill in representing individual precipitation events compared to ERA-Interim. Moreover, the authors of [9] found that COSMO-REA6 and ERA-Interim overestimate P over parts of Russia and parts of the Alps compared to GPCC. In our study, this is confirmed by particularly high P rates in the Rhone basin for COSMO-REA6 with a temporal mean of 90 mm/month and peaks of up to 180 mm/month. Furthermore, in line with [9], we found that all reanalyses underestimate P in the catchment of Garonne-Loire-Seine compared to GPCC. Besides, MERRA-2 simulates exceptional high P rates in the Neva basin where its temporal mean is more than 10 mm/month higher than the temporal mean from the other datasets.

Evapotranspiration is characterized by a distinct annual cycle over Europe with different amplitudes depending on the regional climate (Figure 6). Large differences are found between the individual E datasets with a mean spread of 17 mm/month (maximum 43 mm/month) in the Danube basin, up to a mean spread of 28 mm/month (maximum 96 mm/month) in the Neva basin, and a mean spread of 29 mm/month (maximum 68 mm/month) in the catchment of Guadiana-Guadalquivir. Interestingly, E for the catchments on the Iberian peninsula does not only show a large spread between the models, but also patterns beyond the annual cycle. Moreover, all models except ERA-Interim suggest a small, but not regularly occurring secondary peak of evapotranspiration in winter in Guadiana-Guadalquivir. One might speculate that this sub-annual variability is due to the interaction of large-scale variability with regional characteristics, e.g., effects of the North Atlantic Oscillation (NAO). In summer, evapotranspiration from ERA-Interim is twice as large as E from COSMO-REA6 in Guadiana-Guadalquivir. Furthermore, in the Guadiana-Guadalquivir basin, MERRA-2 fits best to the observation based dataset GLEAM.

In general, MERRA-2 overestimates E compared to the observation-based dataset GLEAM with the most extreme values in the Neva catchment, where the temporal mean is 20 mm/month higher than the temporal mean from the other models. In contrast, outputs from COSMO-REA6 show the smallest E in all considered basins. It is striking that for most basins, the maximum value in summer is quite the same for COSMO every year, whereas for the global models, it varies from year to year. The year 2010 stands out with high evapotranspiration values for some models, which could be explained by the heat waves over Russia and the Iberian peninsula.

The previous analyses already indicated that the performance of the models shows regional differences. This also impacts the closure of the water budget equation (Figures 7 and 8). Modeled $P - E$ is mostly within the error bars of $\Delta S + R$ even in smaller basins (Figure 7). However, MERRA-2 tends to underestimate $P - E$ especially in summer due to large E estimates. Furthermore, ERA-Interim cannot close the water budget equation on the Iberian peninsula as E is overestimated in this region. It is worth noticing that $P - E$ from COSMO-REA6 matches $\Delta S + R$ well for all selected catchments. The other models mainly differ from $\Delta S + R$ by a constant shift. Correlations between $\Delta S + R$ and $P - E$ are between 0.5 and 0.9 and mostly about 0.7 for all models (Figure 9a), which is mainly attributed to the annual cycle.

A closer look at short-term variability is taken in Figure 8. For this purpose the mean, trend and annual signals according to Equation (4) (extended by semiannual signals) were subtracted from the flux time series. De-seasoned and de-trended time series show similar patterns for both sides of the water budget equation in all basins. Short-term variability also depends on the climatic conditions of the target catchment, e.g., in the Dnepr basin de-trended and de-seasoned $P - E$ and $\Delta S + R$ have maximum values of 50 mm/month, whereas in Guadiana-Guadalquivir catchment, maximum values of 100 mm/month exist.

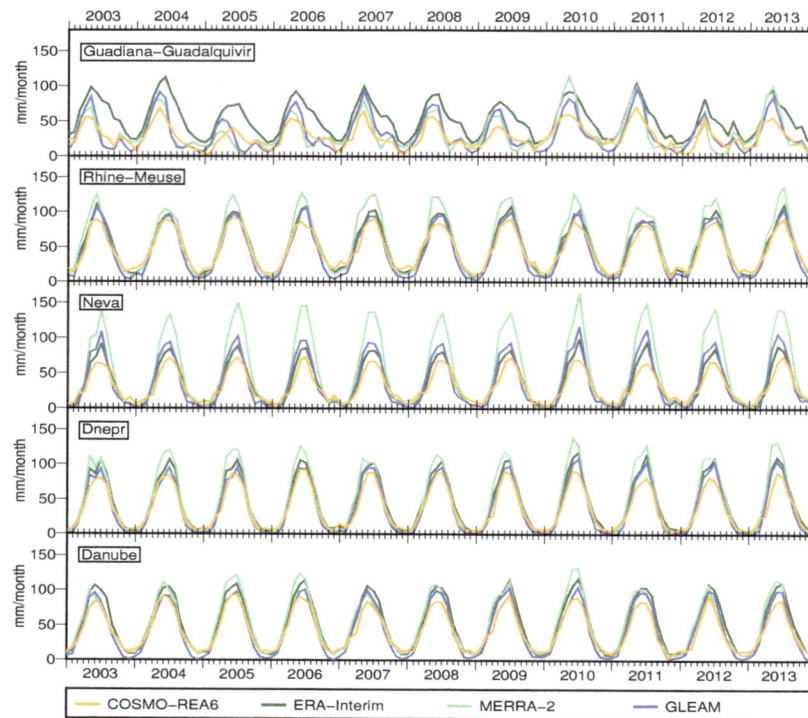

Figure 6. Monthly total evapotranspiration E from different models for selected catchments of the study area.

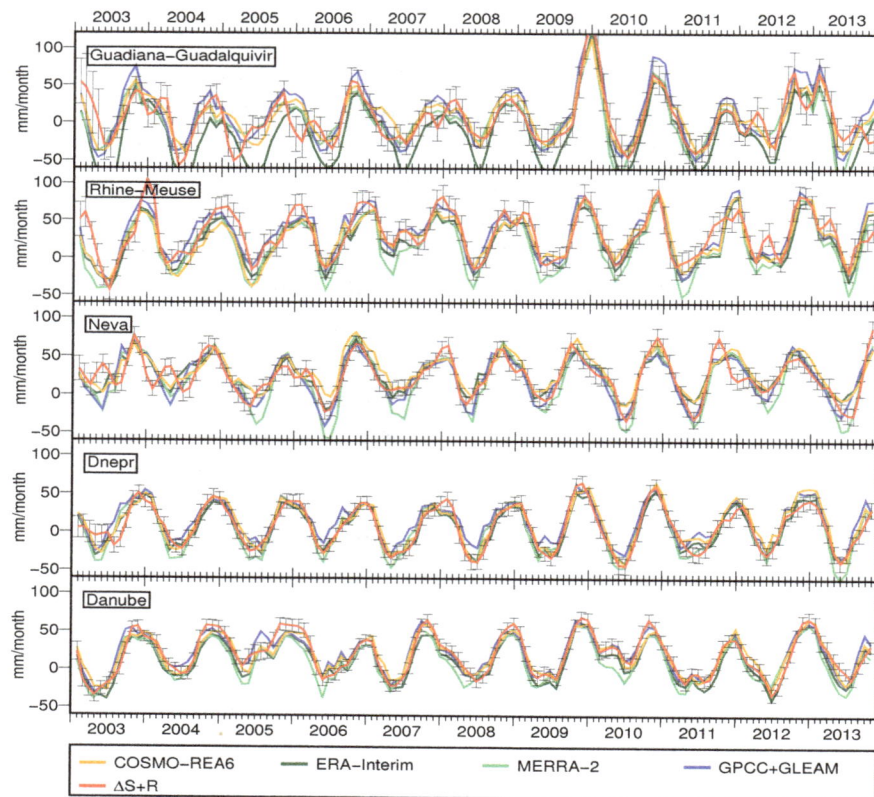

Figure 7. The closure of the water budget equation $\Delta S + R = P - E$ is assessed for selected river basins. The left side of the equation is represented by the red line and includes propagated standard deviations (black). All time series are smoothed with a three-month moving average filter to facilitate interpretation.

Best agreement between de-trended and de-seasoned $\Delta S + R$ and $P - E$ is achieved for the Danube with correlations of about 0.7 (Figure 9b). In most basins, the correlation is about 0.5, with a few exceptions like the Ebro, Neman, Don and Neva, where correlation is smaller than 0.3. Especially in smaller basins, outliers in the GRACE time series have a huge impact on correlation. Furthermore, in some basins, the agreement between $\Delta S + R$ and $P - E$ changes with time, e.g., in the Guadiana-Guadalquivir and Neva basins, models and $\Delta S + R$ disagree during the first three years of the study period, but then again match very well (Figure 8).

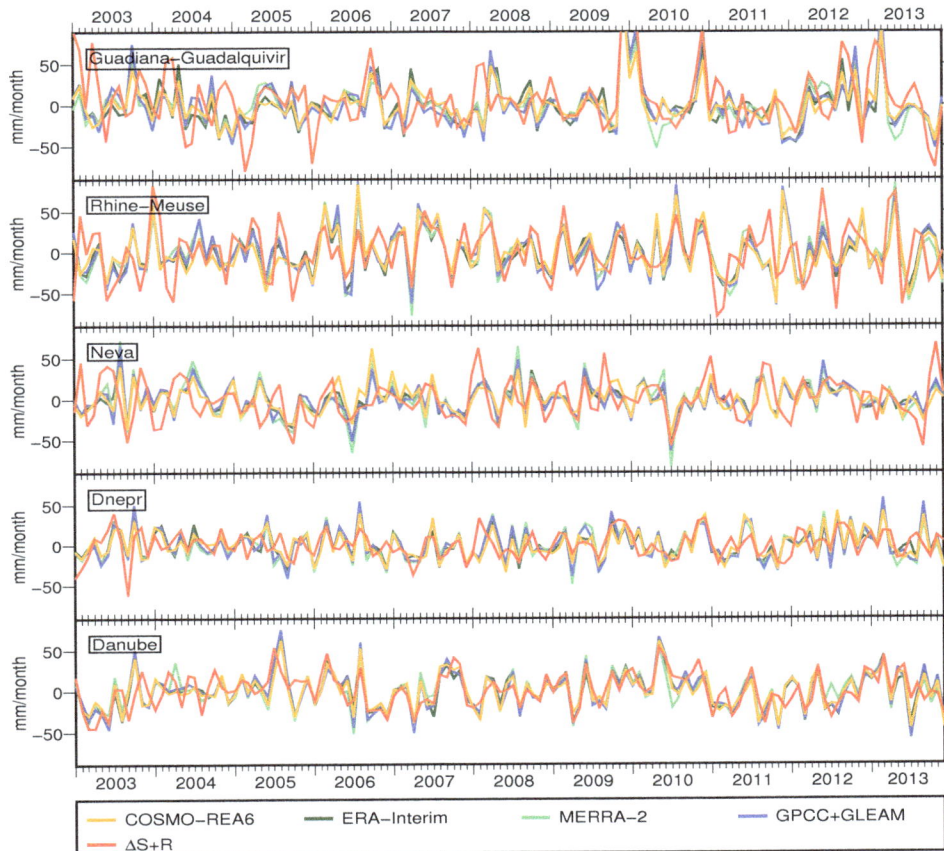

Figure 8. De-seasoned and de-trended time series of $P - E$ and $\Delta S + R$ (red) for selected river basins.

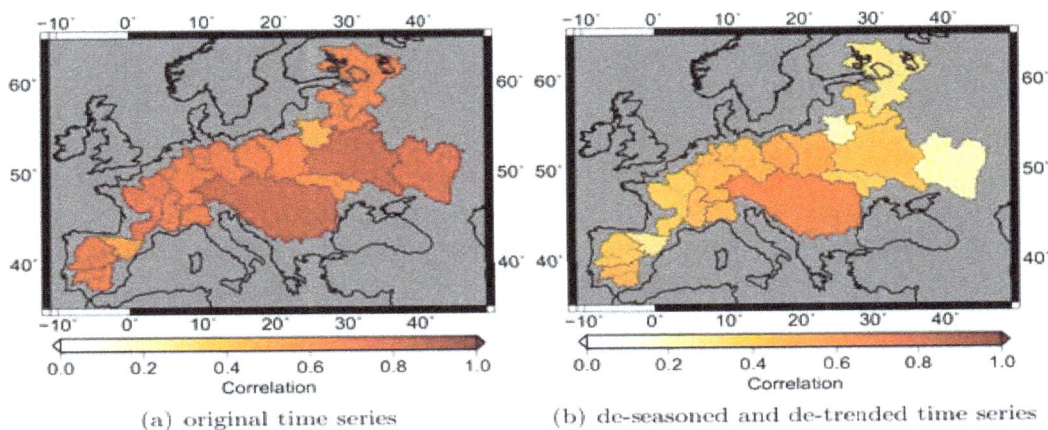

(a) original time series

(b) de-seasoned and de-trended time series

Figure 9. Correlation of $\Delta S + R$ and $P - E$ from COSMO-REA6 Reanalysis for (**a**) the original time series, as shown in Figure 7, and (**b**) de-trended and de-seasoned time series, as shown in Figure 8.

4.3. Statistics of the River Basins

After integration, biases in fluxes produce incorrect trends in storage, i.e., underestimating $P - E$ leads to a negative trend in storage, and overestimating $P - E$ leads to a positive trend in storage. Figure 10a–d shows the bias of the water budget equation for the individual $P - E$ datasets. The global models ERA-Interim and MERRA-2 tend to underestimate $P - E$ (indicated by the red color). For ERA-Interim, the largest biases (about 20 mm/month) are obtained on the Iberian peninsula, which we attribute to deficiencies in the E estimate as discussed before. In contrast, in eastern Europe, both global models perform quite well with biases smaller than 10 mm/month. Although MERRA-2 shows the largest biases in central Europe (15 mm/month for Rhine-Meuse), it improved with respect to MERRA, which was assessed in [8], and had biases of about 25 mm/month in some central European basins. Results for the basins of Po and Rhone should be interpreted with caution as the basins are too small for deriving reasonable storage changes from GRACE data, and moreover, simulated runoff is rather uncertain in these basins (Table 2).

Biases of COSMO-REA6 are smaller than 10 mm/month in all basins, and in many basins, even smaller than 5 mm/month. In central Europe, COSMO-REA6 tends to underestimate $P - E$, whereas in eastern and western Europe, it overestimates $P - E$. The relevance of the bias is illustrated by Figure 10e, which provides the standard deviation of the temporal mean of $\Delta S + R$. This value is mostly between 2 mm/month and 3 mm/month; it is smaller for large basins like Danube and Dnepr and higher for the very small basins like Po, Rhone and Ebro. We can conclude that the biases found in COSMO-REA6 are relevant only for a few basins. The observation-based datasets GPCC and GLEAM also tend to overestimate $P - E$ with large values in the catchments of Garonne-Loire-Seine (-9 mm/month), Dnepr (-7 mm/month) and Dniester-Southern Bug (-8 mm/month). Interestingly, no obvious biases are found in the eastern basins for all model-based time series of $P - E$.

A more complete picture of the closure of the water budget is obtained by analyzing the amplitude, phase and trend of $\Delta S + R$ and $P - E$ derived from the parameters estimated from Equation (4). The amplitude of $P - E$ is predominantly affected by modeled evapotranspiration, which differs largely for the individual models (Figure 6). The regional patterns and the magnitudes of the amplitudes of COSMO-REA6 fit well to the amplitudes of $\Delta S + R$ with exceptions only in a few basins, e.g., on the Iberian peninsula (Figure 11). In contrast, GPCC in combination with GLEAM and ERA-Interim overestimate the amplitude of $P - E$ with respect to GRACE by 15–20 mm/month in all basins west of the Rhine. The amplitude of $P - E$ from MERRA-2 is particularly large in the French basins, the Rhine and the eastern basins with values of about 50 mm/month. The Danube basin stands out as all $P - E$ time series and also $\Delta S + R$ agree very well regarding the amplitude with values between 31 and 35 mm/month. While COSMO-REA6 and ERA-Interim have nearly no phase shift, MERRA-2 and GPCC+GLEAM are contaminated with phase shifts of up to 10 days.

Negative trends in fluxes imply increasingly fast drying; positive trends mean that a region experiences increasingly fast wetting. Within the time span considered here, all models agree that eastern Europe tends to become more and more dry and central and northwestern Europe more and more wet (Figure 12). This conclusion is confirmed by $\Delta S + R$; however, GRACE and discharge time series indicate larger negative trends in eastern Europe than the models. The authors of [55] found similar trend patterns investigating runoff over Europe, with negative trends over eastern Europe and positive trends over northwestern Europe for the time period 1963–2000. In contrast, the authors of [3] found negative trends for changes of $P - E$ between 1976–2005 and 2070–2099 in our whole study region. The authors of [56] compared P and E for the time periods 1948–1968 and 1985–2005 and identified drying trends over parts of eastern Europe and the Iberian peninsula, but no sign of wetting trends over northwestern Europe. The largest positive trends in northwestern Europe are found by COSMO-REA6 with 1–3 mm/month/year. Over the Iberian peninsula, the results are unclear. While ERA-Interim and COSMO-REA6, reanalyses show positive trends of $P - E$ on the Iberian peninsula; GPCC + GLEAM and $\Delta S + R$ identify positive trends only in the southern catchment

Guadiana-Guadalquivir and negative trends in the northern part of the peninsula; and MERRA-2 indicates no trend at all.

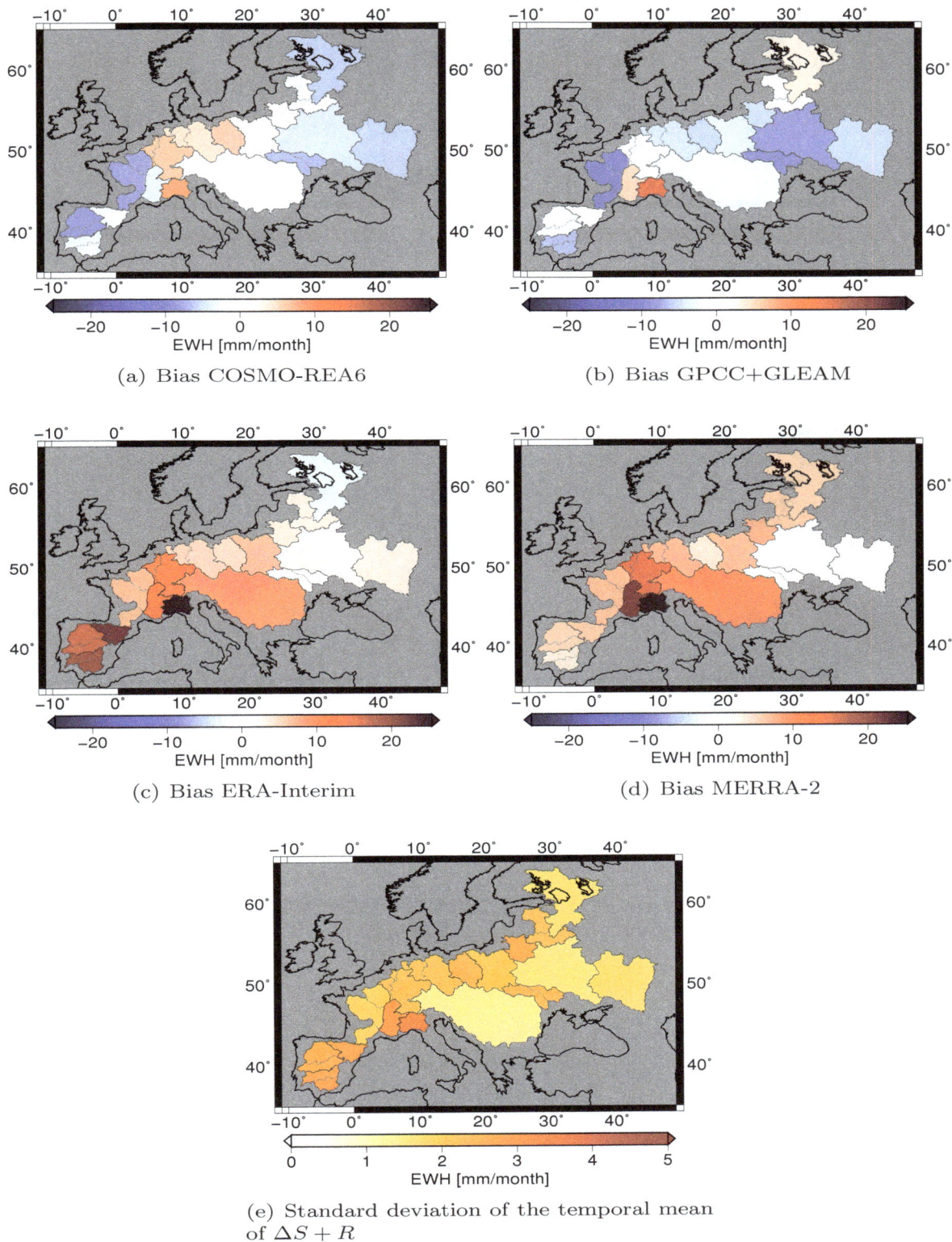

(a) Bias COSMO-REA6

(b) Bias GPCC+GLEAM

(c) Bias ERA-Interim

(d) Bias MERRA-2

(e) Standard deviation of the temporal mean of $\Delta S + R$

Figure 10. (**a–d**) Bias of the water budget equation given in equivalent water height (EWH): red color means that $P - E$ is underestimated, and blue color means that $P - E$ is overestimated. (**e**) provides as a reference the propagated error of the left side of the water budget equation.

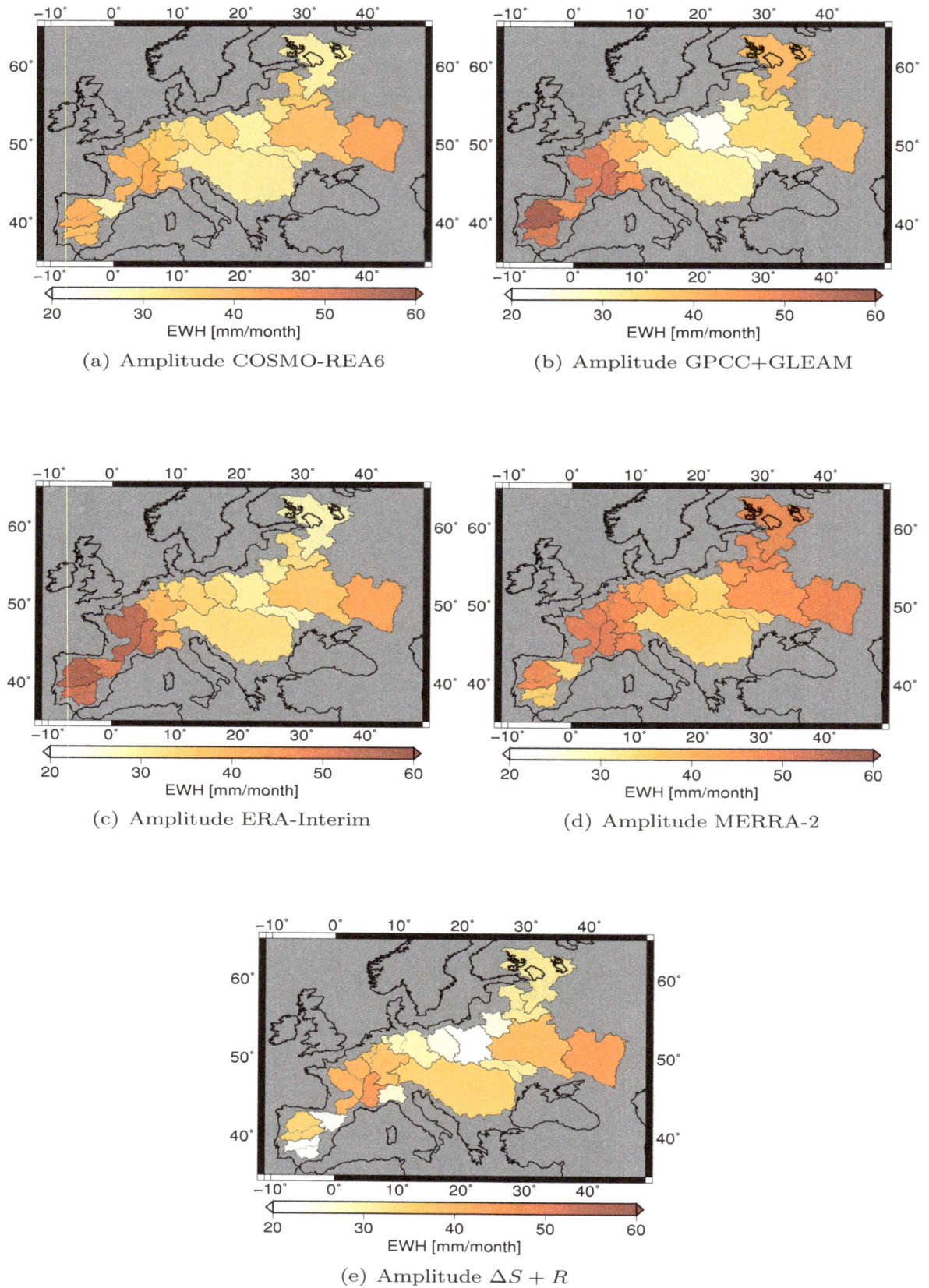

(a) Amplitude COSMO-REA6

(b) Amplitude GPCC+GLEAM

(c) Amplitude ERA-Interim

(d) Amplitude MERRA-2

(e) Amplitude $\Delta S + R$

Figure 11. Amplitudes of $P - E$ and $\Delta S + R$ given in equivalent water height (EWH).

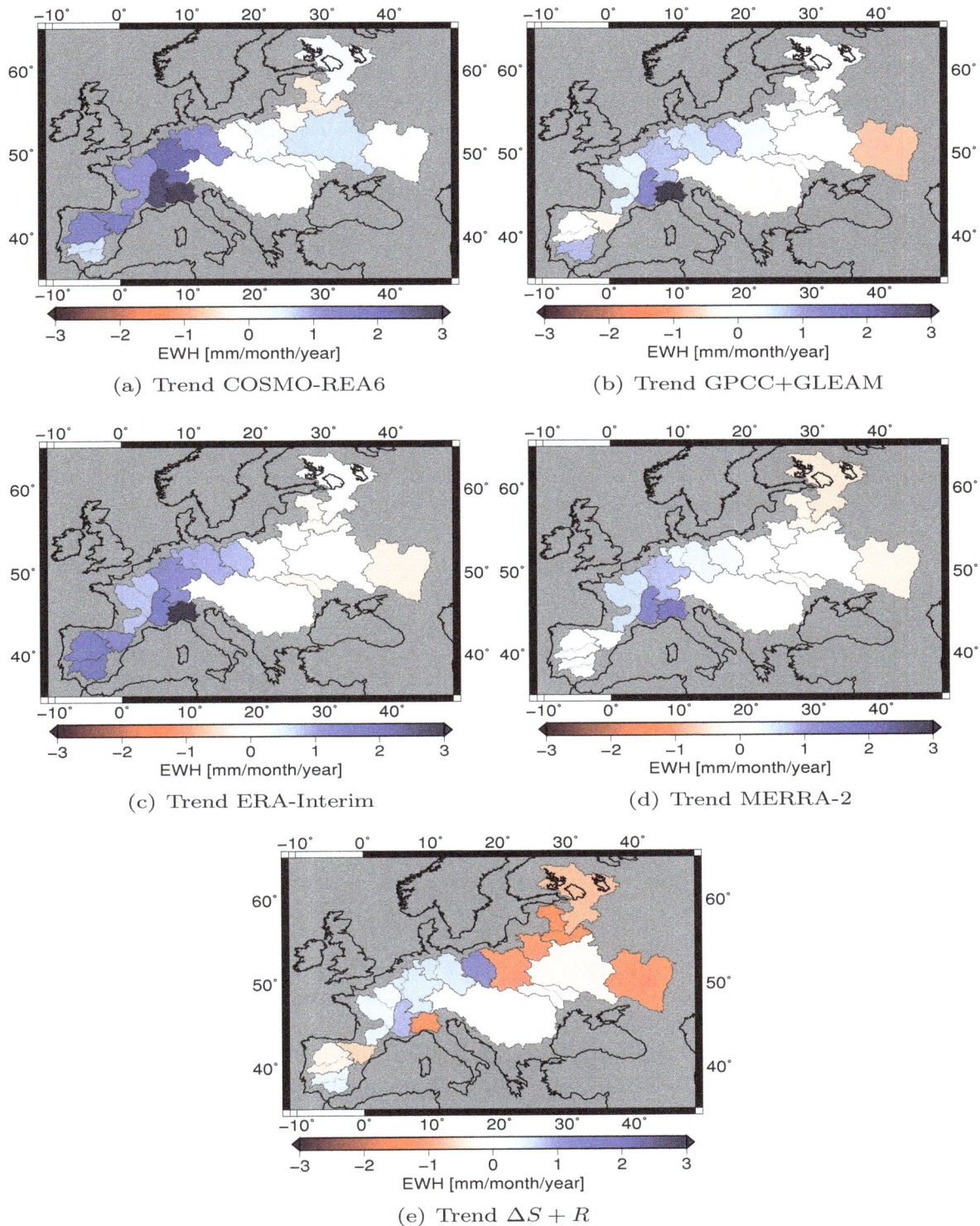

(a) Trend COSMO-REA6

(b) Trend GPCC+GLEAM

(c) Trend ERA-Interim

(d) Trend MERRA-2

(e) Trend $\Delta S + R$

Figure 12. Trend of $P - E$ and $\Delta S + R$ given in equivalent water height (EWH).

5. Discussion and Outlook

Our results suggest that the high resolution COSMO-REA6 reanalysis better closes the water budget equation than global reanalyses. This means COSMO-REA6 can be seen as an important step forward in the consistent representation of the water cycle over Europe and, thus, advances climate monitoring on regional scales. In fact, in comparison to global NWP models, COSMO-REA6 improved the representation of small-scale variability, especially with respect to the accuracy of precipitation events [9].

We investigated the closure of the water budget equation for all major European river basins, aggregated to 17 catchment areas. This allowed us to distinguish regional patterns for the performance of the individual models over Europe. Previously, most water budget studies focused on large river basins worldwide (e.g., [14,57,58]). Only a few studies investigated the closure of the water budget over different river basins in Europe [8,10]. However, [10] assessed moisture flux convergence instead of $P - E$. Due to the limited resolution of the early GRACE releases, they found very poor agreement between GRACE and ECMWF data. In [8], we focused on only four catchment areas in central Europe, which were too few for determining regional patterns. We summarize that the current study provides unique insights into the performance of different $P - E$ datasets over Europe due to (i) the extension of the study area, (ii) the long study period of 11 years and (iii) the usage of the latest GRACE release.

Discharge was modeled applying a simple calibrated rainfall-runoff model instead of using gauge observations like in most water budget studies (e.g., [8,14]). In doing so, we circumvented the problem of lacking recent discharge data; yet, this approach may have limitations, and further research is required. Modeling introduces additional uncertainty to the left-hand side of the water budget equation, which was taken into account by modeling the error of simulated discharge via ensemble runs. We obtained errors of 20–30% of the magnitude of discharge, which we believe is a rather conservative estimate. Furthermore, the NS coefficients indicate a successful calibration of the model in most river basins. However, it should be kept in mind that the computed errors do not take into account changes due to human activities (e.g., dams, redirection) between the calibration time span and the GRACE time span. Nevertheless, the largest contribution to the error of $\Delta S + R$ (about 60–80%) originates from GRACE as discharge is much smaller in most basins. Especially in the very small river basins (e.g., Rhone, Po), the GRACE data should be interpreted with caution due to their excessive noise on such small spatial scales.

For most catchments, modeled $P - E$ from COSMO-REA6 lies within the error ranges derived for $\Delta S + R$. Biases are mostly smaller than 5 mm/month. In contrast, ERA-Interim and MERRA-2 underestimate $P - E$ with particularly large biases over the Iberian peninsula for ERA-Interim and over Rhine-Meuse for MERRA-2. This confirms the results obtained in [8] for central Europe. However, it is worth noticing that the bias of MERRA-2 became smaller compared to the bias obtained for MERRA in [8]. We attribute this to changes in the data assimilation system. For hydrological studies and climate modeling, it is particularly important to reduce the bias of $P - E$ in order to avoid introducing unrealistic trends in storage.

Evapotranspiration was found to be the most uncertain component of the water budget equation over Europe. In accordance with [14], who compared E from various models to GRACE-based estimates and detected significant differences in the mean annual cycles, we found differences in the annual amplitude of up to 50 mm/month between the individual E datasets. In particular, on the Iberian peninsula and for the northeastern catchments, large uncertainties for E were determined. It is likely that the spread of the models on the Iberian peninsula arises from the differences of potential evaporation and actual evaporation in this region.

The annual amplitude of $P - E$ is mainly affected by the annual cycle of E. Compared to $\Delta S + R$, the global datasets overestimate the annual amplitude especially in western Europe. In contrast to this, the amplitude of COSMO-REA6 agrees well with the amplitude of $\Delta S + R$.

Finally, trends from $P - E$ and $\Delta S + R$ indicate that northwestern and central Europe becomes increasingly wetter, whereas eastern Europe becomes increasingly drier. However, the trend estimates are only representative for the study period of 11 years, and no reliable conclusions can be drawn for longer time spans.

Our investigations of the closure of the water budget over Europe show regional patterns that can be associated with different regional climatic conditions. Therefore, the strengths and weaknesses of the individual datasets were analyzed for regions representing these varying characteristics. All in all, the regional COSMO-REA6 allows a better modeling of the water cycle than the global reanalyses, which we attribute to its higher spatial resolution.

For future studies, the assessment of the closure of the water budget equation on a grid instead of on catchment scale may provide a more detailed picture of regional differences. In this scope, the availability of gridded runoff is critical. Besides, for actually closing the water budget equation, contributions from pumping, aquifer systems and runoff into the ocean need to be investigated.

Acknowledgments: We thank The Global Runoff Data Centre, 56068 Koblenz, Germany, for providing river discharge data and watershed boundaries. We are grateful to Laurent Longuevergne for his help and advice with the GR2M-snow model. We acknowledge funding of Anne Springer by the COAST (Studying changes of sea level and water storage for coastal regions in West-Africa using satellite and terrestrial datasets) project, supported by the Deutsche Forschungsgemeinschaft under Grant No. KU1207/20-1. Finally, the authors thank editor Evelyn Ning, Vincent Humphrey and two unknown reviewers for their helpful comments.

Author Contributions: Anika Bettge performed the computations with the GR2M-snow model, and Anne Springer performed all other computations and analyzed the data. Anne Springer and Annette Eicker designed the structure of the paper and the figures. Jürgen Kusche wrote Section 1; Annette Eicker wrote Section 2.2; Andreas Hense wrote Section 2.3.1; Anne Springer wrote all other sections. All authors contributed to the interpretation of the results. Annette Eicker and Jürgen Kusche improved the grammar and the readability of the manuscript.

References

1. Reager, J.T.; Gardner, A.S.; Famiglietti, J.S.; Wiese, D.N.; Eicker, A.; Lo, M.H. A decade of sea level rise slowed by climate-driven hydrology. *Science* **2016**, *351*, 699–703.
2. Rietbroek, R.; Brunnabend, S.E.; Kusche, J.; Schröter, J.; Dahle, C. Revisiting the contemporary sea-level budget on global and regional scales. *Proc. Natl. Acad. Sci. USA* **2016**, *113*, 1504–1509.
3. Byrne, M.P.; O'Gorman, P.A. The Response of Precipitation Minus Evapotranspiration to Climate Warming: Why the "Wet-Get-Wetter, Dry-Get-Drier" Scaling Does Not Hold over Land. *J. Clim.* **2015**, *28*, 8078–8092.
4. Tapiador, F.J.; Turk, F.; Petersen, W.; Hou, A.Y.; García-Ortega, E.; Machado, L.A.; Angelis, C.F.; Salio, P.; Kidd, C.; Huffman, G.J.; et al. Global precipitation measurement: Methods, datasets and applications. *Atmos. Res.* **2012**, *104–105*, 70–97.
5. Wang, K.; Dickinson, R.E. A review of global terrestrial evapotranspiration: Observation, modeling, climatology, and climatic variability. *Rev. Geophys.* **2012**, *50*, RG2050.
6. Swenson, S.; Wahr, J. Estimating Large-Scale Precipitation Minus Evapotranspiration from GRACE Satellite Gravity Measurements. *J. Hydrometeorol.* **2006**, *7*, 252–270.
7. Lorenz, C.; Kunstmann, H. The Hydrological Cycle in Three State-of-the-Art Reanalyses: Intercomparison and Performance Analysis. *J. Hydrometeorol.* **2012**, *13*, 1397–1420.
8. Springer, A.; Kusche, J.; Hartung, K.; Ohlwein, C.; Longuevergne, L. New Estimates of Variations in Water Flux and Storage over Europe Based on Regional (Re)Analyses and Multisensor Observations. *J. Hydrometeorol.* **2014**, *15*, 2397–2417.
9. Bollmeyer, C.; Keller, J.D.; Ohlwein, C.; Wahl, S.; Crewell, S.; Friederichs, P.; Hense, A.; Keune, J.; Kneifel, S.; Pscheidt, I.; et al. Towards a high-resolution regional reanalysis for the European CORDEX domain. *Q. J. R. Meteorol. Soc.* **2015**, *141*, 1–15.
10 Hirschi, M.; Viterbo, P.; Seneviratne, S.I. Basin-scale water-balance estimates of terrestrial water storage variations from ECMWF operational forecast analysis. *Geophys. Res. Lett.* **2006**, *33*, L21401.
11 Seitz, F.; Schmidt, M.; Shum, C. Signals of extreme weather conditions in Central Europe in GRACE 4-D hydrological mass variations. *Earth Planet. Sci. Lett.* **2008**, *268*, 165–170.
12. Sahoo, A.K.; Pan, M.; Troy, T.J.; Vinukollu, R.K.; Sheffield, J.; Wood, E.F. Reconciling the global terrestrial water budget using satellite remote sensing. *Remote Sens. Environ.* **2011**, *115*, 1850–1865.
13. Ramillien, G.; Frappart, F.; Güntner, A.; Ngo-Duc, T.; Cazenave, A.; Laval, K. Time variations of the regional evapotranspiration rate from Gravity Recovery and Climate Experiment (GRACE) satellite gravimetry. *Water Resour. Res.* **2006**, *42*, W10403.
14. Rodell, M.; McWilliams, E.B.; Famiglietti, J.S.; Beaudoing, H.K.; Nigro, J. Estimating evapotranspiration using an observation based terrestrial water budget. *Hydrol. Processes* **2011**, *25*, 4082–4092.

15. Cesanelli, A.; Guarracino, L. Estimation of regional evapotranspiration in the extended Salado Basin (Argentina) from satellite gravity measurements. *Hydrogeol. J.* **2011**, *19*, 629–639.

16. Berrisford, P.; Dee, D.; Poli, P.; Brugge, R.; Fielding, K.; Fuentes, M.; Kåallberg, P.; Kobayashi, S.; Uppala, S.; Simmons, A. ERA Report Series (Version 2.0), European Centre for Medium Range Weather Forecasts, 2011. Available online: http://www.ecmwf.int/en/elibrary/8174-era-interim-archive-version-20 (accessed on 23 February 2017).

17. McCarty, W.; Coy, L.; Gelaro, R.; Huang, A.; Merkova, D.; Smith, E.B.; Sienkiewicz, M.; Wargan, K. MERRA-2 Input Observations: Summary and Assessment 2016. Available online: https://gmao.gsfc.nasa.gov/pubs/docs/McCarty885.pdf (accessed on 22 February 2017).

18. Schneider, U.; Becker, A.; Finger, P.; Meyer-Christoffer, A.; Rudolf, B.; Ziese, M. GPCC Full Data Reanalysis Version 7.0 at 0.5°: Monthly Land-Surface Precipitation from Rain-Gauges Built on GTS-Based and Historic Data, 2015. Available online: http://doi.org/10.5676/DWD_GPCC/FD_M_V7_050 (accessed on 1 January 2017).

19. Miralles, D.G.; Holmes, T.R.H.; De Jeu, R.A.M.; Gash, J.H.; Meesters, A.G.C.A.; Dolman, A.J. Global land-surface evaporation estimated from satellite-based observations. *Hydrol. Earth Syst. Sci.* **2011**, *15*, 453–469.

20. Martens, B.; Miralles, D.G.; Lievens, H.; van der Schalie, R.; de Jeu, R.A.M.; Férnandez-Prieto, D.; Beck, H.E.; Dorigo, W.A.; Verhoest, N.E.C. GLEAM v3: Satellite-based land evaporation and root-zone soil moisture. *Geosci. Model Dev. Discuss.* **2016**, *2016*, 1–36.

21. Birkinshaw, S.J.; O'Donnell, G.M.; Moore, P.; Kilsby, C.G.; Fowler, H.J.; Berry, P.A.M. Using satellite altimetry data to augment flow estimation techniques on the Mekong River. *Hydrol. Processes* **2010**, *24*, 3811–3825.

22. Tourian, M.J.; Tarpanelli, A.; Elmi, O.; Qin, T.; Brocca, L.; Moramarco, T.; Sneeuw, N. Spatiotemporal densification of river water level time series by multimission satellite altimetry. *Water Resour. Res.* **2016**, *52*, 1140–1159.

23. Tarpanelli, A.; Brocca, L.; Lacava, T.; Melone, F.; Moramarco, T.; Faruolo, M.; Pergola, N.; Tramutoli, V. Toward the estimation of river discharge variations using MODIS data in ungauged basins. *Remote Sens. Environ.* **2013**, *136*, 47–55.

24. Sichangi, A.W.; Wang, L.; Yang, K.; Chen, D.; Wang, Z.; Li, X.; Zhou, J.; Liu, W.; Kuriaa, D. Estimating continental river basin discharges using multiple remote sensing data sets. *Remote Sens. Environ.* **2016**, *179*, 36–53.

25. Mouelhi, S. Vers une Chaîne Cohérente de Modèles Pluie-Débit aux pas de Temps Pluriannuel, Annuel, Mensuel et Journalier. Ph.D. Thesis, ENGREF (AgroParisTech), Paris, France, 2003. Available online: http://hydrologie.org/THE/MOUELHI.pdf (accessed on 22 February 2017).

26. Tapley, B.D.; Bettadpur, S.; Watkins, M.; Reigber, C. The gravity recovery and climate experiment: Mission overview and early results. *Geophys. Res. Lett.* **2004**, *31*, L09607.

27. Wahr, J.; Molenaar, M.; Bryan, F. Time variability of the Earth's gravity field: Hydrological and oceanic effects and their possible detection using GRACE. *J. Geophys. Res.* **1998**, *103*, 30205–30229,

28. Dahle, C.; Flechtner, F.; Gruber, C.; König, D.; König, R.; Michalak, G.; Neumayer, K.H. GFZ GRACE Level-2 Processing Standards Document for Level-2 Product Release 0005. Scientific Technical Report STR12/02—Data, Revised Edition, 2012. Available online: http://dx.doi.org/10.2312/GFZ.b103-1202-25 (accessed on 15 January 2013).

29. Kusche, J. Approximate decorrelation and non-isotropic smoothing of time-variable GRACE-type gravity field models. *J. Geodesy* **2007**, *81*, 733–749.

30. Rietbroek, R.; Fritsche, M.; Brunnabend, S.E.; Daras, I.; Kusche, J.; Schröter, J.; Flechtner, F.; Dietrich, R. Global surface mass from a new combination of GRACE, modelled OBP and reprocessed GPS data. *J. Geodyn.* **2012**, *59–60*, 64–71.

31. Rietbroek, R.; Brunnabend, S.E.; Kusche, J.; Schröter, J. Resolving sea level contributions by identifying fingerprints in time-variable gravity and altimetry. *J. Geodyn.* **2012**, *59–60*, 72–81.

32. Cheng, M.; Tapley, B.D.; Ries, J.C. Deceleration in the Earth's oblateness. *J. Geophys. Res.* **2013**, *118*, 740–747.

33. Geruo, A.; Wahr, J.; Zhong, S. Computations of the viscoelastic response of a 3-D compressible Earth to surface loading: An application to Glacial Isostatic Adjustment in Antarctica and Canada. *Geophys. J. Int.* **2013**, *192*, 557.

34. Mesinger, F.; DiMego, G.; Kalnay, E.; Mitchell, K.; Shafran, P.C.; Ebisuzaki, W.; Jović, D.; Woollen, J.; Rogers, E.; Berbery, E.H.; et al. North American Regional Reanalysis. *Bull. Am. Meteorol. Soc.* **2006**, *87*, 343–360.

35. Bromwich, D.H.; Wilson, A.B.; Bai, L.; Moore, G.W.K.; Bauer, P. Special Issue Article. *Q. J. R. Meteorol. Soc.* **2016**, *142*, 644–658.

36. Jermey, P.M.; Renshaw, R.J. Precipitation representation over a two-year period in regional reanalysis. *Q. J. R. Meteorol. Soc.* **2016**, *142*, 1300–1310.

37. Dahlgren, P.; Landelius, T.; Kållberg, P.; Gollvik, S. A high-resolution regional reanalysis for Europe. Part 1: Three-dimensional reanalysis with the regional HIgh-Resolution Limited-Area Model (HIRLAM). *Q. J. R. Meteorol. Soc.* **2016**, *142*, 2119–2131.

38. Baldauf, M.; Seifert, A.; Förstner, J.; Majewski, D.; Raschendorfer, M.; Reinhardt, T. Operational Convective-Scale Numerical Weather Prediction with the COSMO Model: Description and Sensitivities. *Mon. Weather Rev.* **2011**, *139*, 3887–3905.

39. Schraff, C.; Hess, R. *A Description of the Nonhydrostatic Regional COSMO-Model; Part III: Data Assimilation*; Deutscher Wetterdienst: Offenbach, Germany, 2012.

40. Wahl, S.; Bollmeyer, C.; Crewell, S.; Figura, C.; Friederichs, P.; Hense, A.; Keller, J.D.; Ohlwein, C. A novel convective-scale regional reanalyses COSMO-REA2: Improving the representation of precipitation. *Meteorol. Z.* **2017**, doi:10.1127/metz/2017/0824.

41. Dee, D.P.; Uppala, S.M.; Simmons, A.J.; Berrisford, P.; Poli, P.; Kobayashi, S.; Andrae, U.; Balmaseda, M.A.; Balsamo, G.; Bauer, P.; et al. The ERA-Interim reanalysis: Configuration and performance of the data assimilation system. *Q. J. R. Meteorol. Soc.* **2011**, *137*, 553–597.

42. Reichle, R.H.; Liu, Q. Observation-Corrected Precipitation Estimates in GEOS-5 2014. Available online: https://ntrs.nasa.gov/search.jsp?R=20150000725 (accessed on 22 February 2017).

43. Reichle, R.H.; Liu, Q.; Koster, R.D.; Draper, C.S.; Mahanama, S.P.P.; Partyka, G.S. Land Surface Precipitation in MERRA-2. *J. Clim.* **2017**, *30*, 1643–1664.

44. Schneider, U.; Becker, A.; Finger, P.; Meyer-Christoffer, A.; Ziese, M.; Rudolf, B. GPCC's new land surface precipitation climatology based on quality-controlled in situ data and its role in quantifying the global water cycle. *Theor. Appl. Climatol.* **2014**, *115*, 15–40.

45. Priestley, C.H.B.; Taylor, R.J. On the Assessment of Surface Heat Flux and Evaporation Using Large-Scale Parameters. *Mon. Weather Rev.* **1972**, *100*, 81–92.

46. Sneeuw, N.; Lorenz, C.; Devaraju, B.; Tourian, M.J.; Riegger, J.; Kunstmann, H.; Bárdossy, A. Estimating Runoff Using Hydro-Geodetic Approaches. *Surv. Geophys.* **2014**, *35*, 1333–1359.

47. Mouelhi, S.; Michel, C.; Perrin, C.; Andréassian, V. Stepwise development of a two-parameter monthly water balance model. *J. Hydrol.* **2006**, *318*, 200–214.

48. Swenson, S.; Wahr, J. Methods for inferring regional surface-mass anomalies from Gravity Recovery and Climate Experiment (GRACE) measurements of time-variable gravity. *J. Geophys. Res.* **2002**, *107*, ETG 3-1–ETG 3-13.

49. Longuevergne, L.; Scanlon, B.R.; Wilson, C.R. GRACE Hydrological estimates for small basins: Evaluating processing approaches on the High Plains Aquifer, USA. *Water Resour. Res.* **2010**, *46*, W11517.

50. Haylock, M.R.; Hofstra, N.; Klein Tank, A.M.G.; Klok, E.J.; Jones, P.D.; New, M. A European daily high-resolution gridded data set of surface temperature and precipitation for 1950–2006. *J. Geophys. Res.* **2008**, *113*, D20119.

51. Harris, I.; Jones, P.; Osborn, T.; Lister, D. Updated high-resolution grids of monthly climatic observations—The CRU TS3.10 Dataset. *Int. J. Climatol.* **2014**, *34*, 623–642.

52. Gudmundsson, L.; Seneviratne, S.I. Observation-based gridded runoff estimates for Europe (E-RUN version 1.1). *Earth Syst. Sci. Data* **2016**, *8*, 279–295.

53 Nash, J.; Sutcliffe, J. River flow forecasting through conceptual models part I—A discussion of principles. *J. Hydrol.* **1970**, *10*, 282 – 290.

54. Bach, L.; Schraff, C.; Keller, J.D.; Hense, A. Towards a probabilistic regional reanalysis system for Europe: Evaluation of precipitation from experiments. *Tellus A Dyn. Meteorol. Oceanogr.* **2016**, *68*, 32209,

55. Stahl, K.; Tallaksen, L.M.; Hannaford, J.; van Lanen, H.A.J. Filling the white space on maps of European runoff trends: Estimates from a multi-model ensemble. *Hydrol. Earth Syst. Sci.* **2012**, *16*, 2035–2047.

56.   Greve, P.; Orlowsky, B.; Mueller, B.; Sheffield, J.; Reichstein, M.; Seneviratne, S.I. Global assessment of trends in wetting and drying over land. *Nat. Geosci.* **2014**, *7*, 716–721.

57.   Syed, T.H.; Famiglietti, J.S.; Chambers, D.P. GRACE-Based Estimates of Terrestrial Freshwater Discharge from Basin to Continental Scales. *J. Hydrometeorol.* **2009**, *10*, 22–40.

58.   Fersch, B.; Kunstmann, H.; Bárdossy, A.; Devaraju, B.; Sneeuw, N. Continental-Scale Basin Water Storage Variation from Global and Dynamically Downscaled Atmospheric Water Budgets in Comparison with GRACE-Derived Observations. *J. Hydrometeorol.* **2012**, *13*, 1589–1603.

7

Surface Water Monitoring within Cambodia and the Vietnamese Mekong Delta over a Year, with Sentinel-1 SAR Observations

Binh Pham-Duc [1,2,*], **Catherine Prigent** [1] **and Filipe Aires** [1]

[1]   Laboratoire d'Etudes du Rayonnement et de la Matiére en Astrophysique et Atmosphéres, UMP 8112, l'Observatoire de Paris, 61 Avenue de l'Observatoire, 75014 Paris, France; catherine.prigent@obspm.fr (C.P.); filipe.aires@obspm.fr (F.A.)

[2]   Space and Aeronautics Department, University of Science and Technology of Hanoi, 18 Hoang Quoc Viet, Cau Giay, 10000 Hanoi, Vietnam

[*]   Correspondence: pham.binh@obspm.fr

Academic Editor: Frédéric Frappart

Abstract: This study presents a methodology to detect and monitor surface water with Sentinel-1 Synthetic Aperture Radar (SAR) data within Cambodia and the Vietnamese Mekong Delta. It is based on a neural network classification trained on Landsat-8 optical data. Sensitivity tests are carried out to optimize the performance of the classification and assess the retrieval accuracy. Predicted SAR surface water maps are compared to reference Landsat-8 surface water maps, showing a true positive water detection of ∼90% at 30 m spatial resolution. Predicted SAR surface water maps are also compared to floodability maps derived from high spatial resolution topography data. Results show high consistency between the two independent maps with 98% of SAR-derived surface water located in areas with a high probability of inundation. Finally, all available Sentinel-1 SAR observations over the Mekong Delta in 2015 are processed and the derived surface water maps are compared to corresponding MODIS/Terra-derived surface water maps at 500 m spatial resolution. Temporal correlation between these two products is very high (99%) with very close water surface extents during the dry season when cloud contamination is low. This study highlights the applicability of the Sentinel-1 SAR data for surface water monitoring, especially in a tropical region where cloud cover can be very high during the rainy seasons.

Keywords: SAR; Sentinel-1; surface water monitoring; neural network; Mekong Delta; Landsat-8; MODIS

---

1. Introduction

Studying the spatial and temporal distribution of surface water resources is critical, especially in highly populated areas and in regions under climate change pressure. With an increased number of Earth-observation satellites providing a large diversity of remote sensing data, there is now the potential to monitor the surface water at regional to global scale. However, mapping surface water is still challenging. It is difficult to provide products with the accuracy required for a large range of applications (e.g., agriculture, disaster management, and hydrology).

Several methods have already been proposed to detect and monitor surface water with visible and Near-Infrared (NIR) images. Ref. [1] used positive values of the Normalized Difference Water Index (NDWI) to classify water bodies. Ref. [2] applied a threshold on NIR reflectances of the NOAA/AVHRR satellite to delineate lakes. Ref. [3] detected surface water by identifying the positive values of the Modification of Normalized Difference Water Index (MNDWI). Ref. [4] combined NIR data and the Normalized Difference Vegetation Index (NDVI) to detect surface water bodies. However, cloud contamination is a stringent constraint for these methods, limiting their application to cloud-free conditions which is very restrictive in some regions (e.g., in the Tropics). Vegetation can also mask the surface water partly or totally. This makes the water detection difficult or impossible under canopy. In addition, the NIR reflectance over highly turbid water can be higher than the red reflectance, introducing confusions in the indices used for the water detection.

Synthetic Aperture Radar (SAR) have become an important source of data to detect flood or monitor surface water as they allow observations regardless of the cloud cover, day and night, with spatial resolution comparable to visible and near-infrared satellite images [5]. SAR instruments have been available on many sensors and platforms (Envisat ASAR, PALSAR, or RADARSAT, for example) providing observations for different areas all over the globe (but normally with a limited number of images available per year in some regions). Flood detection using different SAR observations has been studied by many authors, showcasing the advantages of SAR instruments compared to optical instruments in monitoring floods. Ref. [6] used a single decision tree classifier on two sets of JERS-1 SAR data to classify surface water within the states of North Carolina and South Carolina into five land cover types (water, marsh, flooded forest, field, and non-flooded forest). Although the classifier was simple, they reported an overall classification accuracy of nearly 90%. Ref. [7] showed the potential of the COSMO-SkyMed data for flood detection by showing case studies in several locations all over the globe (e.g., Tarano River overflow, Italy, April 2009; Pakistan inundation, July–September 2010; Thailand flood, October 2010; and Australia flood, January 2011). COSMO-SkyMed instruments provided very high resolution X-band SAR images, but covered limited areas (the highest spatial resolution is $\sim$1 m for an observation area of 10 km $\times$ 10 km). X-band data from TerraSAR-X instrument were also reported suitable for flood mapping under forest canopy in the temperate forest zone in Estonia [8]. Ref. [9] compared four flood detection approaches over five areas (Vietnam, the Netherlands, Mali, Germany, and China) using SAR data from the TanDEM-X mission. Although these four approaches were designed according to different requirements, their performances were satisfactory over the studied areas (17 out of 20 water masks reaching an overall accuracy larger than 90%). Other studies using SAR data for water monitoring locally and regionally under different environments can be listed, such as [10–12]. Mapping water bodies at global scale using SAR data was limited due to the lack of global observations, and the fact that SAR data are not easy to access freely. Ref. [13] used multi-year (2005–2012) Envisat ASAR observations to create, for the first time, a global potential water body map at a spatial resolution of 150 m. Errors concentrated along shorelines and coastline, but this global water map has an accuracy of $\sim$80% compared to the reference data.

The Mekong Delta in Southeast Asia (one of the largest deltas in the world) is a vast triangular plain of approximately 55,000 km^2, most of it lower than 5 m above sea level. The seasonal variation in water level results in rich and extensive wetlands. For instance, the Mekong Delta region covers only 12% of Vietnam but produces $\sim$50% of the annual rice (with two or three harvests per year depending on the provinces), represents $\sim$50% of the fisheries, and $\sim$70% of the fruit production. In the Delta, the dry season extends from November to April and the rainy season from May to October. Many researches have been carried out to monitor the surface water in the Delta, using both optical and active microwave satellite images. Ref. [14] produced a monthly mean climatology of the water extent from 2000 to 2004 with a spatial resolution of 500 m, using visible and NIR MODIS/Terra data. However, with 85% to 95% cloud cover during the wet season over the Mekong Delta [15], remote sensing methods derived from visible and NIR images present some limitations. Different SAR observations have also been exploited to study floods and wetlands over the Delta. Ref. [16] mapped flood occurrence for the year 1996 over the Delta using five ERS-2 observations. Ref. [17] used 60 Envisat ASAR observations during the years 2007–2011 to study the flood regime in the Delta. Thanks to the launch of the Sentinel-1A &B satellites, as well as the free data policy of the European Space Agency (ESA), Sentinel-1 SAR observations are now regularly and freely accessible for scientific and educational purposes, over large parts of the globe. Similar to previous SAR instruments, Sentinel-1 instruments show strong potential for detecting open water bodies at high spatial resolution [18,19].

With the advantage of higher temporal resolution than previous SAR instruments, Sentinel-1 has the ability to monitor the seasonal cycle of water extent every six days over Europe and the boreal region, and with slightly reduced temporal sampling elsewhere. In this study, we propose a methodology using Sentinel-1A SAR observation for monitoring water surface extent within Cambodia and the Mekong Delta for the year 2015. It is based on a Neural Network (NN) algorithm, trained on visible Landsat-8 images (30 m spatial resolution). At the time of this study, the temporal resolution of Sentinel-1 over the Delta was 12 days: it reduced to 6 days after the launch of the Sentinel-1B in April 2016.

The Sentinel-1 SAR data and the ancillary observations are described in Section 2, including the pre-processing steps. Section 3 presents the NN methodology, along with sensitivity tests. Results and comparisons with other products are provided and discussed in Section 4. Section 5 concludes this study.

2. Sentinel-1 SAR Data and the Ancillary Datasets

2.1. Sentinel-1 SAR Data

Sentinel-1 is a satellite project funded by the European Union and carried out by the European Space Agency. It is a two satellite constellation working at C-band (5.405 GHz). The major objective of the satellites is the observations and monitoring of land and ocean surfaces day and night, under all weather conditions [20]. The satellite operates in four exclusive imaging modes with different spatial resolutions (the highest being 5 m) and swaths (up to 400 km). The first Sentinel-1A satellite of the pair was launched on 3 April 2014, while the second Sentinel-1B satellite was launched on 22 April 2016. The Sentinel-1 satellites fly along a sun-synchronous, near-polar circular orbit at an altitude of ∼693 km. Incidence angle varies between 29° and 46°. The two satellites provide a re-visiting time of 6 days (it was 12 days before the launch of the Sentinel-1B satellite). Sentinel-1 satellites have dual polarization capabilities (HH, VV, HH + HV and VV + VH), giving final users the ability to access a large variety of applications, including the monitoring of surface water. SAR images from Sentinel-1 satellites are freely downloaded from the sentinel scientific data hub [21].

In this project, 20 m resolution (10 m pixel spacing) Level-1 Ground Range Detected (GRD) Sentinel-1 images are used, from the Interferometric WideSwath (IW) mode. These images have been detected and projected to ground range using an Earth ellipsoid model provided by ESA. Over the Mekong Delta, there are two polarizations available: the VH and VV polarizations. Some pre-processing steps have to be carried out using the free Sentinel Application Platform (SNAP) software developed by ESA, before moving to the analysis steps (see Figure 1). These pre-processing steps are described in the "SAR Basics with the Sentinel-1 Toolbox in SNAP tutorial" [22].

First, multi-looking processing is applied to each single Sentinel-1 image (both polarizations) to convert to 30 m spatial resolution (to match with Landsat-8 images). Applying multi-looking at the beginning of the chain reduces the processing time for the next steps since the size of the image is several times smaller than the original one. Second, the image is calibrated to convert values of the raw image from digital number to radar backscatter coefficient (σ^0). Third, the Refined Lee filter is applied to reduce the speckle noise and to smooth the radar backscatter coefficient data because this filter maintains details of the standing water boundary [23]. Other filters (Lee, Lee Sigma or Median, for example) were tested, and results showed little differences in terms of water detection. Next, the "terrain correction" tool is used to compensate for distortions in the SAR images, so that the geometric presentation of the image will be as close as possible to the real world. At the end of this step, the image is also re-projected from the satellite projection to the Earth geographic projection, and is ready for applications. To fully cover Cambodia and the Vietnamese Mekong Delta, at least five Sentinel-1 SAR images are needed. Figure 2 (top) provides examples of the SAR backscatter coefficients for VH (a) and VV (b) polarizations, along with the incidence angle (c), over the Tonle Sap Lake, on 17 December 2015.

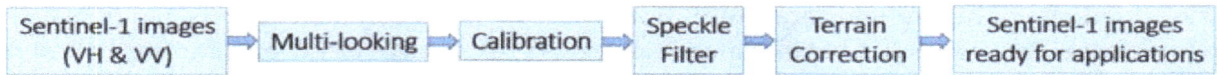

Figure 1. Pre-processing steps for Sentinel-1 Synthetic Aperture Radar (SAR) images.

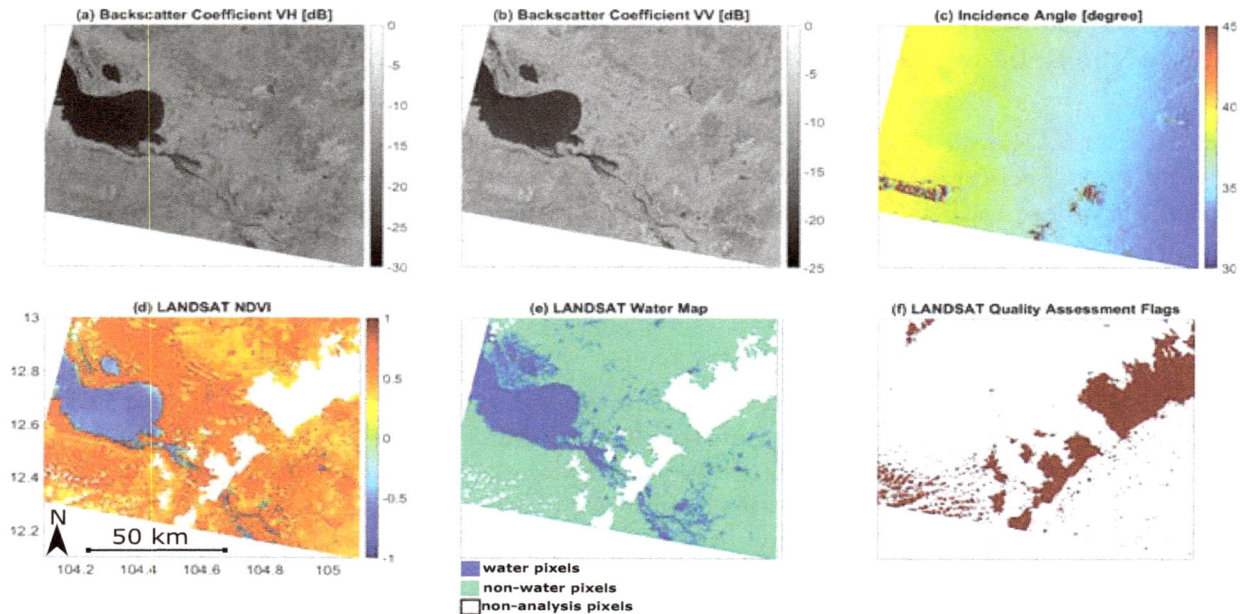

Figure 2. Examples of satellite observations from Sentinel-1 (top) and from Landsat-8 (bottom), over the lower part of the Tonle Sap Lake (Cambodia) after the pre-processing steps: (**a**) SAR backscatter coefficient at VH polarization; (**b**) SAR backscatter coefficient at VV polarization; (**c**) SAR incidence angle; (**d**) The Normalized Difference Vegetation Index (NDVI) from Landsat-8; (**e**) Surface water estimated from Landsat-8; and (**f**) Landsat-8 quality flags. The white areas are cloud-covered pixels detected by the Landsat quality flags, and have been removed. Both Sentinel-1 and Landsat-8 images were taken on 17 December 2015.

2.2. Ancillary Datasets

2.2.1. Inundation Maps Derived from Landsat-8 Data

Landsat-8 satellite collects visible and shortwave images (30 m spatial resolution). NIR wavelength reflects less solar radiation than the red wavelength over water bodies [1,3], and surface water maps can be derived from the NDVI maps (water pixels and non-water pixels correspond to negative and positive values of NDVI, respectively) [24,25]. Other indices have been used to detect water, but the NDVI is effective when properly corrected from the atmospheric contamination. In this study, official and reliable atmospheric corrected Landsat-8 NDVI images are ordered directly from the U.S. Geological Survey (USGS) website (https://espa.cr.usgs.gov/index/). To limit cloud effects, only images with less than 10% of cloud contamination are used. The selected images are further filtered using the Landsat-8 quality assessment to remove pixels that might be affected by instrument artifacts or subject to cloud contamination. Figure 2 (bottom) shows the NDVI from Landsat-8 (d), the resulting surface water map based on negative NDVI values (e), and the quality flag (f), for the same regions and the same day (17 December 2015) as previously presented. Over the Lower Mekong Delta (lower than latitude number 15), there are ~250 Landsat-8 images available between January 2015 and January 2016. However, there is only ~10% (27 images) with less than 10% cloud contamination. Among the remaining images, only 1/3 was selected for this study since they were observed with a time difference of less than 3 days from a Sentinel-1 image.

2.2 2. Inundation Maps Derived from MODIS/Terra Data

In this study, the surface reflectance 8-Day L3 Global 500 m products from MODIS/Terra (MOD09A1) are used to create surface water maps, mainly based on values of the Enhanced Vegetation Index (EVI), the Land Surface Water Index (LSWI), and the difference between EVI and LSWI by a methodology described in [14]. MODIS surface water maps (500 m spatial resolution) over the Mekong Delta will be used to compare to the corresponding surface water maps derived from SAR Sentinel-1 observations for 2015. MODIS/Terra data can be downloaded from http://reverb.echo.nasa.gov/reverb/.

All Sentinel-1, Landsat-8 and MODIS/Terra observations used in this study are listed in Tables 1 and 2. Sentinel-1 and Landsat-8 training observations are used to train the NN (Section 3.2). Sentinel-1 and Landsat-8 test observations are used to test, optimize, and evaluate the performance of the NN (Sections 3.3 and 4.1). NN evaluation is also based on comparisons with MODIS surface water estimates (Section 4.3).

Table 1. List of 9 Sentinel-1 and corresponding Landsat-8 training (top) and test (bottom) observations used in this study over Cambodia and the Vietnamese Mekong Delta. Maximum gap between Sentinel-1 and Landsat-8 observations is only 3 days. The cloud cover percentage is indicated for each Landsat-8 observation.

Image No	Sentinel-1	Landsat-8	Clouds
	**Sentinel-1 and Landsat-8 Training Observations**		
1	16 April 2015	14 April 2015	6.29%
2	21 April 2015	21 April 2015	0.05%
3	19 August 2015	18 August 2015	7.94%
4	17 December 2015	17 December 2015	4.84%
5	29 March 2016	31 March 2016	6.22%
6	9 June 2016	10 June 2016	3.94%

Image No	Sentinel-1	Landsat-8	Clouds
1	5 January 2016	2 January 2016	0.16%
2	3 February 2016	3 February 2016	7.5%
3	22 February 2016	19 February 2016	0.29%

Table 2. List of 20 Sentinel-1 and corresponding MODIS/Terra observations used in this study over Cambodia and the Vietnamese Mekong Delta.

Image No	Date	Image No	Date
1	10 January 2015	11	14 August 2015
2	3 February 2015	12	26 August 2015
3	15 February 2015	13	7 September 2015
4	11 March 2015	14	19 September 2015
5	4 April 2015	15	1 October 2015
6	28 April 2015	16	13 October 2015
7	15 June 2015	17	25 October 2015
8	27 June 2015	18	6 November 2015
9	9 July 2015	19	30 November 2015
10	21 July 2015	20	24 December 2015

3. Methodology

3.1. Surface Water Information from the Sentinel-1 SAR Images

Flat water surfaces act like mirrors and reflect almost all incoming energy in the specular direction, thus providing very low backscatter. With this physical principle, detection of surface water is often based, at least partly, on the application of a threshold on the SAR backscatter coefficient, with the low backscatter values attributed to water bodies [6,7,16,17]. However, SAR backscatter coefficients over water surfaces are also affected by several mechanisms related to the interaction of the signal with vegetation or with possible surface roughness. The backscattered signals over flooded vegetation in wetlands can be enhanced due to the double-bounce scattering mechanism [26–28]. On the other side, the backscatter coefficients can be affected by vegetation canopy (e.g., rice) above the water surfaces due to volume scattering from the plant components (stems or leaves) [29]. The backscatter coefficients (especially the VV polarization) can also be influenced by the wind-induced surface roughness over open water [17,30]. Finally, there might be ambiguities between surface water and other very flat surfaces (such as arid regions), that could provide very similar backscatter signatures [31].

Based on a reference water mask derived from Landsat-8 NDVI, Figure 3 presents the histograms of the backscatter coefficients for VH and VV polarizations, separately for water and non-water pixels over the incidence angle range of 30°–45° for the area shown in Figure 2. For both polarizations, the water and non-water histograms are rather well separated, with thresholds of −22 dB and −15 dB for the VH and VV polarizations, respectively. Using these thresholds, the surface water has been classified separately for each polarization. The classification derived from the VH polarized image had a stronger spatial linear correlation with the reference water mask than the one derived from the VV polarized image (72% compared to 62%), confirming a higher sensitivity of the VH polarization to the presence of surface water [19]. Using both polarizations for the classification increased the correlation (76%), confirming that the two polarizations carry different information and that using both of them increases the retrieval accuracy. These findings confirmed the study by [32] where water detection with VV polarization was further refined using multiple-polarization.

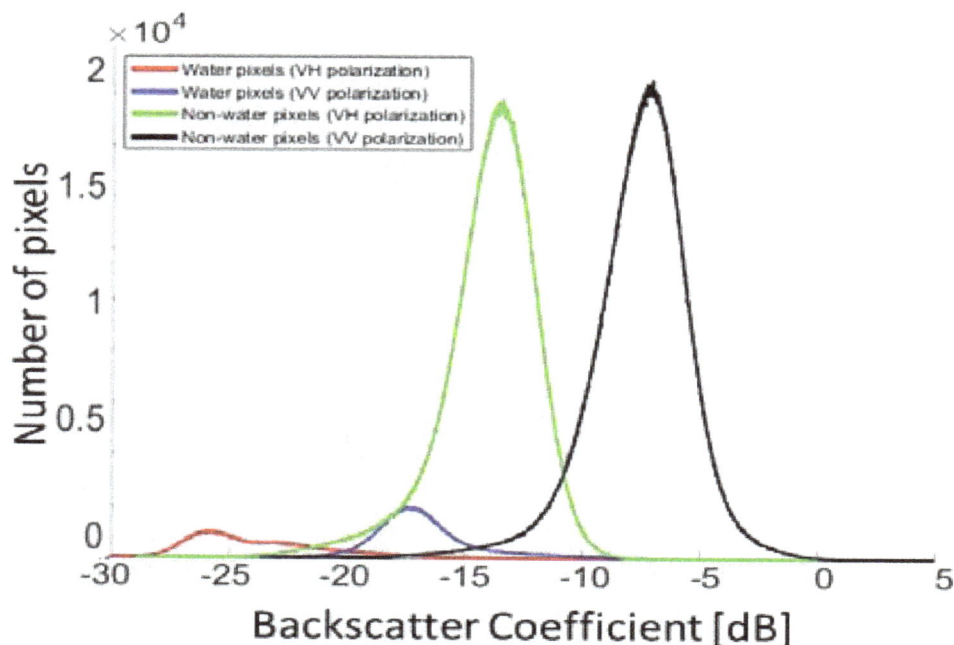

Figure 3. For surface water delineated with Landsat-8, histograms of the water and non-water pixels for the SAR backscatter coefficients in VH and VV polarizations for the area shown in Figure 2 (over the incidence angle range of 30° to 45°).

The effect of the backscatter incidence angle is also tested here. For a collection of pixels located over water (rivers, reservoirs, or lakes), the backscatter coefficient is plotted as a function of the incidence angle between 30° and 45° (Figure 4). Similar negative correlations between incidence angle and backscatter coefficients can also be found in [13] with ASAR data over water bodies (from ~-5 dB at 20° to ~-20 dB at 45° of incidence angle).

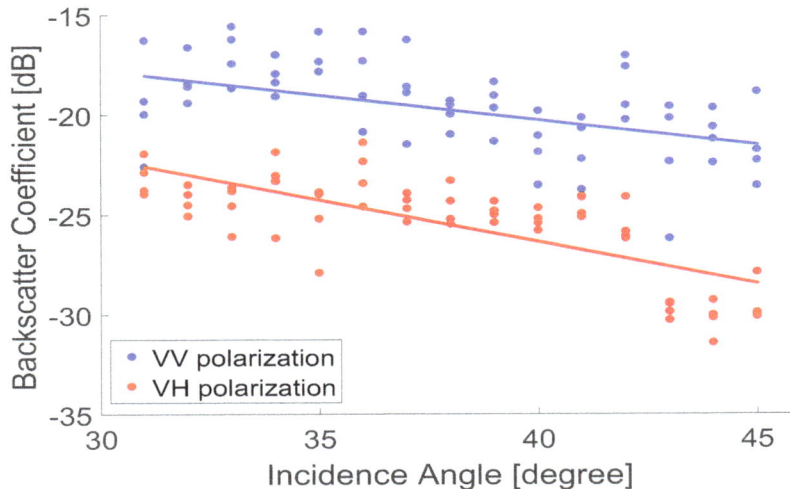

Figure 4. The SAR backscatter coefficients (VH and VV polarizations) from the Sentinel-1 as a function of the incidence angle over water bodies. The linear regression lines are also plotted.

As a conclusion, the SAR backscatter coefficients (VH and VV polarizations) are both sensitive to the presence of water, but with slightly different sensitivities. The effect of the incidence angle, although rather limited within the 29°–46° range of Sentinel-1 SAR, has to be accounted for if a high detection accuracy is required. Simple tests on thresholding techniques illustrated the limitations of these approaches and here we suggest developing a new scheme to delineate the surface water based on Neural Networks.

The temporal dynamics of the backscatter coefficients can also be a source of information and can help disentangle the influence of the other surface parameters [13]. However, this temporal information will not be investigated here.

3.2. A Neural Network-Based Classification

Here, we propose training a NN to produce surface water maps from SAR images, over the Mekong Delta. In the remote sensing field, NNs are often used as a regression tool to estimate a quantity. For each pixel, NN input satellite observations are represented by a vector x, and the network outputs (i.e., the retrieval) is represented by a vector y. However, NNs can also be used as classifiers. In this case, when trained with binary output values ($y = 0$ for non-water, 1 for water surfaces), the NN becomes a statistical model for the conditional probability $y = P(surface = water/x)$, i.e., the probability of the surface being covered by water knowing the satellite observations x. The NN output can then directly be used as an index for water presence probability, but a threshold can also be applied to classify the state as being covered by water or not. The threshold needs to be optimized in order to satisfy some quality criteria, such as overall accuracy or false alarm rates.

The NN classifier needs to be trained in order to perform an optimal discrimination between water and non-water states. A supervised learning is chosen: the NN will be designed to reproduce an already existing classification. A dataset including a collection of SAR information x and associated surface water state y is first built. Part of it is then used during the training stage in order to determine the optimal parameters of the NN model. The reference dataset in the selected area is provided here by a Landsat-8 surface water map (NN outputs), in spatial and temporal coincidence with the Sentinel-1 SAR data (NN inputs). A maximum time difference of 3 days is tolerated, as the two satellites do not fly in phase. Six Landsat-8 surface water maps are selected, along with the corresponding Sentinel-1 SAR observations (see Table 1 for more details on the training dataset). The selection process for the Landsat-8 images has been described in Section 2.2.1. The images cover parts of the lower Mekong Delta in Vietnam and Cambodia. For each image in the training dataset, the number of non-water pixels is much higher than the number of water pixels. To avoid giving too much weight to the non-water pixels, an equalization of the training dataset is performed: an equal number of non-water and water pixels is selected in the training dataset. For this purpose, non-water pixels are selected randomly in the images, to match the number of water pixels. The total number of training samples is ~10 million pixels, half water pixels, half non-water pixels. It takes ~5 h to train the NN (with the use of a personal computer), but when the training is completed, a surface water map can be produced quickly (after ~3–4 min) from any new set of satellite inputs x. A test dataset is chosen to measure the performance of the NN retrieval scheme with data not used in the training process. The NN methodology is summarized in Figure 5.

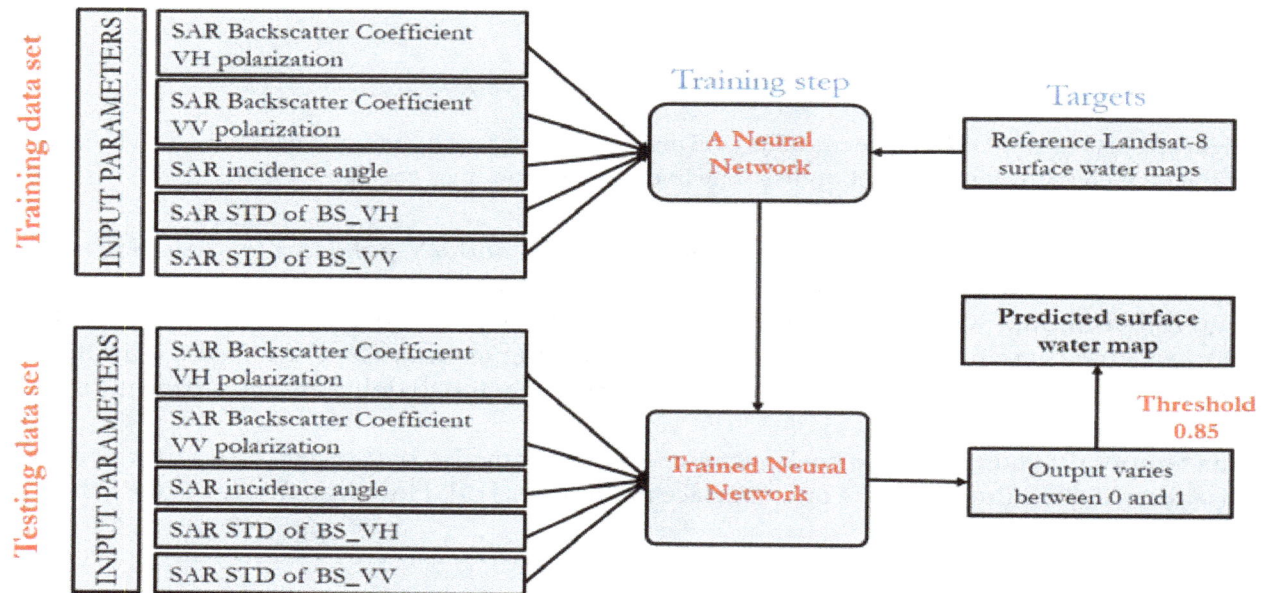

Figure 5. The block diagram of the proposed Neural Network (NN) algorithm.

Several tests were necessary to determine the optimum inputs to the NN, in addition to the obvious ones, i.e., the backscatter coefficients for both polarizations. To limit ambiguities between flat arid surfaces and surface water, and to better capture small rivers, the spatial homogeneity of the backscatter coefficients appeared to be a relevant parameter. The standard deviation of the backscatter coefficients are computed locally over 100 m × 100 m boxes. As a result, the NN uses a maximum of five different inputs x:

- SAR backscatter coefficient VH polarization (BS_VH);
- SAR backscatter coefficient VV polarization (BS_VV);
- SAR incidence angle;
- SAR standard deviation of backscatter coefficient VH over 100 m × 100 m (STD_VH);
- SAR standard deviation of backscatter coefficient VV over 100 m × 100 m (STD_VV);

Figure 6 presents an example of the set of five input images and the target surface water map used to train the NN. Missing areas in the maps correspond to Landsat-8 low quality pixels and are excluded from the training. The NN model is asked to find a relationship between these five input parameters and the corresponding water and non-water state.

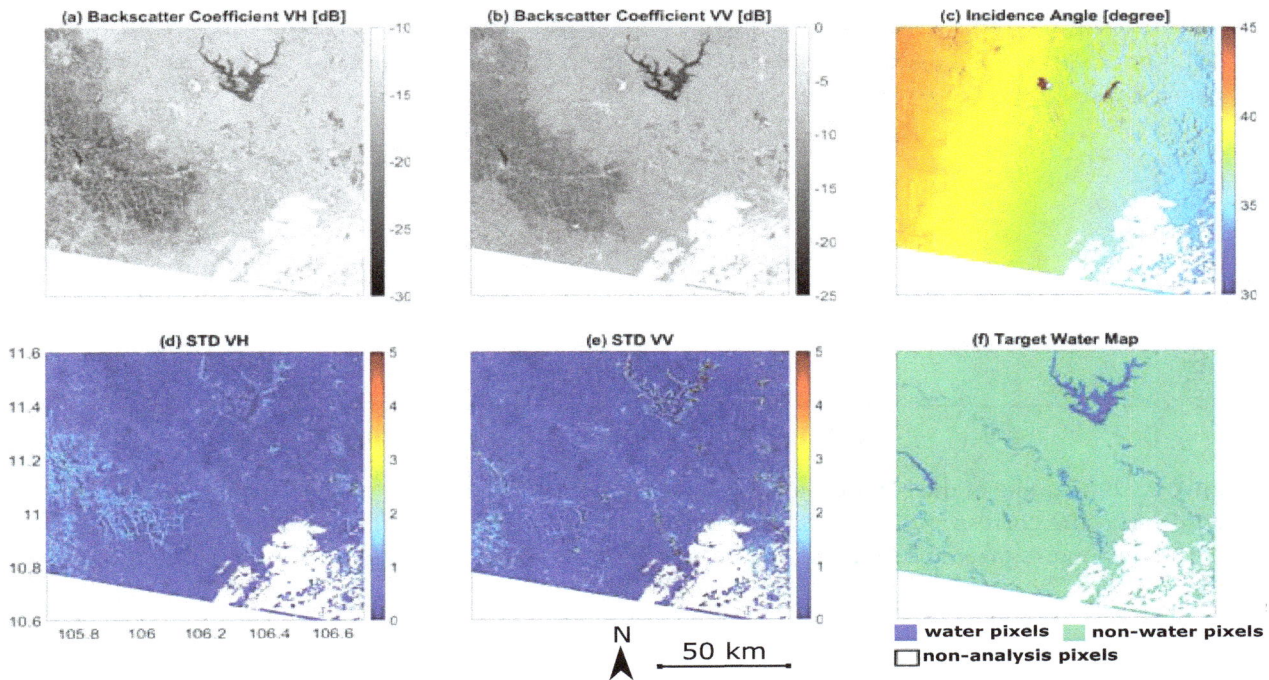

Figure 6. Examples of the five inputs and the target for the NN. (**a**) SAR backscatter coefficient VH polarization; (**b**) SAR backscatter coefficient VV polarization; (**c**) SAR incidence angle; (**d**) SAR standard deviation of backscatter coefficient VH polarization; (**e**) SAR standard deviation of backscatter coefficient VV polarization; and (**f**) Target surface water map based on NDVI from Landsat-8. The white areas are cloud-covered pixels detected by the Landsat quality flags, and they have been removed. Sentinel-1 and Landsat-8 images were acquired on 16 and 14 April 2015, respectively.

3.3. NN Sensitivity Tests

In this section, we use a test dataset of three SAR Sentinel-1 images and three corresponding Landsat-8 reference surface water maps to make several sensitivity tests in order to optimize the performance of the NN classification (see details of the test data sets in Table 1). Three different sensitivity tests were carried out: (1) selecting the best threshold of the NN output to classify land/water surface; (2) understanding the effect of the equalization of the water and non-water pixels in the NN training dataset; (3) finding the most important satellite NN inputs. The NN performances have been evaluated based on: spatial correlation between the SAR and Landsat-8 surface water maps, overall accuracy of the NN, as well as higher values of true positive (TP) and true negative (TN) percentages. True positive value indicates the NN ability to correctly detect water pixels, while true negative value illustrates its ability to correctly detect non-water pixels (compared to the Landsat-8 surface water maps).

3.3.1. Selection of an Optimized Threshold for the NN Output

The first test is conducted to optimize the output threshold to distinguish water from non-water pixels. Figure 7 shows the histogram of the output of the NN, separating the water and non-water pixels according to the related Landsat-8 surface water map. The histograms of the water and non-water clusters intersect around 0.9, meaning that the optimal threshold to separate water from non-water pixels is close to this number. Different thresholds on the NN output values were tested (0.80, 0.85, and

0.90): for each one, the confusion matrix and the overall accuracy are calculated, with the corresponding Landsat-8 images as references. The overall accuracy and the spatial correlation increase from 98% to 99% when the threshold increases from 0.80 to 0.90 (Table 3), but the true positive pixel detection decreases from 92% (with threshold 0.80) to 89% (with threshold 0.90) and the false negative pixel detection increases from 8% to 11%. A threshold of 0.85 is selected here because of its good water detection performance and because it results in the predicted water surface closest to the reference map: 4430 km^2 from the Landsat-8 versus 4420 km^2 from the SAR results, i.e., a limited overestimation of 0.4% as compared to the reference map.

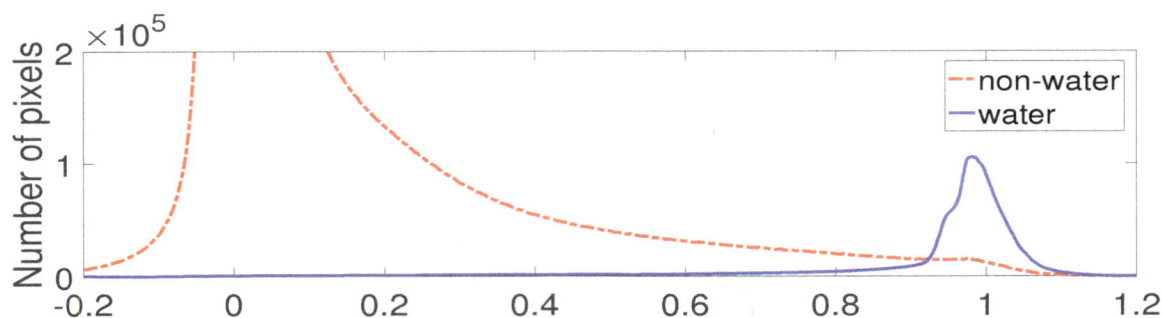

Figure 7. Histograms of the NN outputs, for water (blue) and non-water (dashed red) pixels separately, according to the corresponding Landsat-8 surface water maps. The NN uses the five initial inputs and the training dataset is equalized. The y axis range is selected to illustrate the peak of the water histogram.

Table 3. Confusion matrix of the NN classification for different thresholds. The NN uses the five initial inputs and the training dataset is equalized.

	Output Threshold: 0.80			
	Non-Water(0) (Predicted)	Water(1) (Predicted)	Overall Accuracy	Spatial Correlation
Non-water(0) (Actual)	99.3%	0.7%	98%	91%
Water(1) (Actual)	8%	92%		
	Output Threshold: 0.85			
	Non-Water(0) (Predicted)	Water(1) (Predicted)	Overall Accuracy	Spatial Correlation
Non-water(0) (Actual)	99.5%	0.5%	99%	92%
Water(1) (Actual)	9%	91%		
	Output Threshold: 0.90			
	Non-Water(0) (Predicted)	Water(1) (Predicted)	Overall Accuracy	Spatial Correlation
Non-water(0) (Actual)	99.6%	0.4%	99%	91%
Water(1) (Actual)	11%	89%		

3.3.2. Equalization of Water and Non-Water Pixel Number

For this test, instead of using an equal number of water and non-water pixels in the training dataset, 10% of each Sentinel-1 image is selected randomly to train the neural network, meaning that the number of non-water pixels is several times higher (10–15 times depending on each image in the training dataset) than the number of water pixels (as seen in Figure 7). The intersection between histograms of the NN outputs for water pixels (blue) and non-water pixels (red) moves to 0.5 (see the histogram in Figure 8), meaning that the value 0.5 should be selected to separate water from non-water clusters. As shown in Table 4, the resulting NN is very efficient at detecting non-water pixels with a true negative detection of 99.7%, but it misses 14% of the actual water pixels (86% of true positive detection only, compared to 91% with the equalized training dataset—Table 3). The true positive detection of water pixels decreases because in the training database, the non-water pixels are more numerous and as such have more weight in the retrieval than the water pixels. As a consequence, the NN is more effective at detecting non-water pixels, and less effective at detecting water pixels. It is concluded that the use of an equalized training data set is very important in this classification framework.

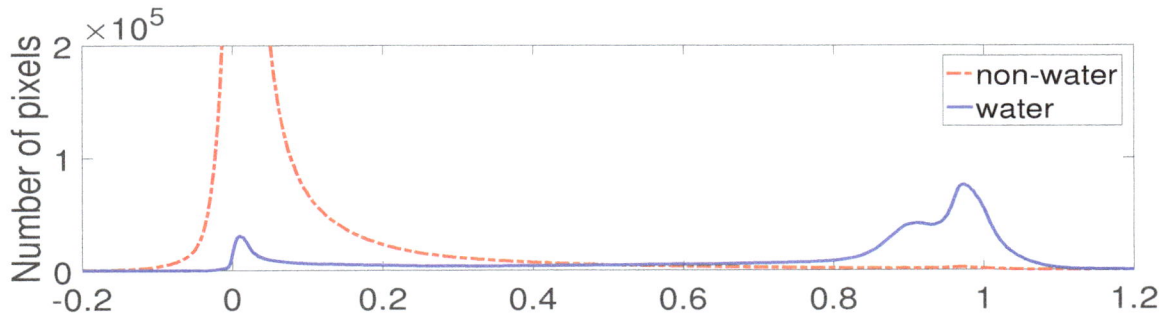

Figure 8. Histograms of the NN outputs, for water (blue) and non-water (dashed red) pixels separately, according to the corresponding Landsat-8 surface water maps. The NN uses the five initial inputs but the training dataset is not equalized. The y axis range is selected to illustrate the peak of the water histogram.

Table 4. Confusion matrix of the NN without equalization of the training dataset, for an optimum threshold of 0.5 on the NN outputs.

	Non-water(0) (Predicted)	Water(1) (Predicted)
Non-water(0) (Actual)	99.7%	0.3%
Water(1) (Actual)	14%	86%

3.3.3. Analyzing the Weight of Each NN Satellite Input

To identify the most relevant inputs for the NN classification of the water surface, 15 NNs are trained based on all 15 different combinations of five input parameters, and their performances are evaluated following various criteria. Table 5 presents the best results with one to five inputs and illustrates how the overall accuracy of the NN classification increases when the number of satellite inputs increases, as compared to the reference Landsat-8 dataset. The NN trained with only the VH backscatter coefficient has a spatial correlation of 78% and a true positive accuracy (correctly detecting water pixels) of 77% compared to the reference data. The spatial correlation increases to 79%, and the true positive accuracy rises to 85% when the standard deviation of the VV backscatter coefficient is added as an input to the NN. The VV backscatter coefficient helps to increase the performance of the NN since both spatial correlation and true positive accuracy increase to 87% and 90%, respectively. The standard deviation of the VH backscatter coefficient does not significantly improve the accuracy of the NN classification. This is due to the strong linear correlation (88%) between the spatial standard deviations of the VH and the VV backscatter coefficients (the other linear correlations among the five input parameters of the NN are provided in Table 6). Similar to the standard deviation of the VH backscatter coefficient, the incidence angle does not have a strong impact on the performance of the NN since its accuracy remains nearly the same after adding the incidence angle as a new input. The input parameters of the NN classification are listed below, from the most important to the least important one in the NN processing:

- Backscatter coefficient VH polarization (BS_VH)
- Standard deviation of backscatter coefficient VV polarization (STD_VV)
- Backscatter coefficient VV polarization (BS_VV)
- Incidence angle
- Standard deviation of backscatter coefficient VH polarization (STD_VH)

Table 5. The NN classification performances when adding input parameters, one at a time.

One Input: BS_VH			
	Non-Water(0) (Predicted)	Water(1) (Predicted)	Spatial Correlation
Non-water(0) (Actual)	98%	2%	78%
Water(1) (Actual)	23%	77%	

Two Inputs: BS_VH + STD_VV			
	Non-Water(0) (Predicted)	Water(1) (Predicted)	Spatial Correlation
Non-water(0) (Actual)	98%	2%	79%
Water(1) (Actual)	15%	85%	

Three Inputs: BS_VH + STD_VV + BS_VV			
	Non-Water(0) (Predicted)	Water(1) (Predicted)	Spatial Correlation
Non-water(0) (Actual)	99%	1%	87%
Water(1) (Actual)	10%	90%	

Four Inputs: BS_VH + STD_VV + BS_VV + Angle			
	Non-Water(0) (Predicted)	Water(1) (Predicted)	Spatial Correlation
Non-water(0) (Actual)	99.5%	0.5%	91%
Water(1) (Actual)	10%	90%	

Five Inputs: BS_VH + STD_VV + BS_VV + Angle + STD_VH			
	Non-Water(0) (Predicted)	Water(1) (Predicted)	Spatial Correlation
Non-Water(0) (Actual)	99.5%	0.5%	92%
Water(1) (Actual)	9%	91%	

Table 6. Linear correlations among the five potential NN inputs.

	BS_VH	BS_VV	STD_VH	STD_VV	ANGLE
BS_VH	100%				
BS_VV	84%	100%			
STD_VH	24%	20%	100%		
STD_VV	21%	21%	88%	100%	
ANGLE	25%	22%	11%	6%	100%

To conclude, the water detection ability of the proposed NN increased when the input parameters are carefully selected and when an optimal output threshold is selected. An equal number of water and non-water pixels should be used in the training dataset to ensure that the NN performs equally well in classifying water and non-water clusters. The STD_VH provides limited additional information to the NN due to its strong linear correlations with the other NN inputs. The incidence angle also plays a limited role in the NN performance. This is partly due to the rather narrow range of incidence angle, from $29°$ to $46°$.

4. Results and Comparisons with Other Surface Water Products

The following results and comparisons involve the optimized version of the NN classification with five input parameters (an equalization of water and non-water pixels, and the output threshold is 0.85). In Section 4.1, the SAR-predicted surface water maps are calculated for two test areas in the Mekong Delta, and compared to Landsat-8 surface water maps over the Tonle Sap Lake in Cambodia and over the Mekong river in Vietnam (see test dataset in Table 1). Other regions were tested but the results are not shown here. Due to the lack of in-situ local inundation maps at the time of this study, we do not have a reference dataset to confirm the accuracy of the Landsat-8 based maps. Therefore, an inter-comparison between Sentinel-1 estimate and other existing estimates is the only way to evaluate the new wetland product based on SAR Sentinel-1 data. First, the results are evaluated with respect to the floodability map derived mainly from the HydroSHEDS topography dataset [33], developed by [34] (Section 4.2). Second, time series of the SAR-derived surface water over the Mekong Delta is compared to the MODIS/Terra-derived inundation maps based on the methodology described by [14], for 2015 (Section 4.3)

4.1. Evaluation of the SAR NN Classification Method

Figure 9 shows the results of the classification applied over the Tonle Sap Lake in Cambodia (top) and over the Mekong river in Vietnam (bottom), in February 2016. Figure 9a,d show the SAR-predicted surface water maps, Figure 9b,e present the reference Landsat-8 surface water maps, whereas the differences between these two surface water maps are shown in Figure 9c,f.

Over the Tonle Sape Lake, both Sentinel and Landsat images were acquired on the same day (3 February 2016). The spatial correlation between the two surface water maps is 94%. The confusion matrix for this area is given in Table 7 (left). Overall accuracy of the classification is 99%, with a true positive water detection of 93.5%, and a false negative percentage of 6.5%. The classification correctly detects more than 99.6% of non-water pixels compared to the reference map. The classification slightly underestimates the surface water coverage by ~2.5%. This is 961 km^2 compared to the reference surface water map derived from the Landsat-8 images of 986 km^2.

The second case study is carried out over the Mekong river and its surrounding areas (latitude range [$10.8°$N–$11.8°$N] and longitude range [$104.6°$E–$105.6°$E]). The optical Landsat-8 images were taken on 19 February 2016 and the SAR Sentinel-1 images were taken 3 days later, on 22 February 2016. These Sentinel and Landsat images were not acquired on the same day, but within 3 days in the middle of the dry season when land surfaces in this area are not expected to change much. Similar to the first case study, the classification works well, even though the environment here is rather complex, with rivers and vegetated wetlands. The overall accuracy is 98.8%, with a spatial correlation of nearly 82%

with the Landsat-8 reference surface water map. Confusion matrix for this area is shown in Table 7 (right) where the true positive percentage is 85.7%, the false negative percentage 14.3%, and 99.2% of non-water pixels are classified correctly. The total surface water area derived from Landsat data is 325 km^2, and it is 355 km^2 predicted from the NN.

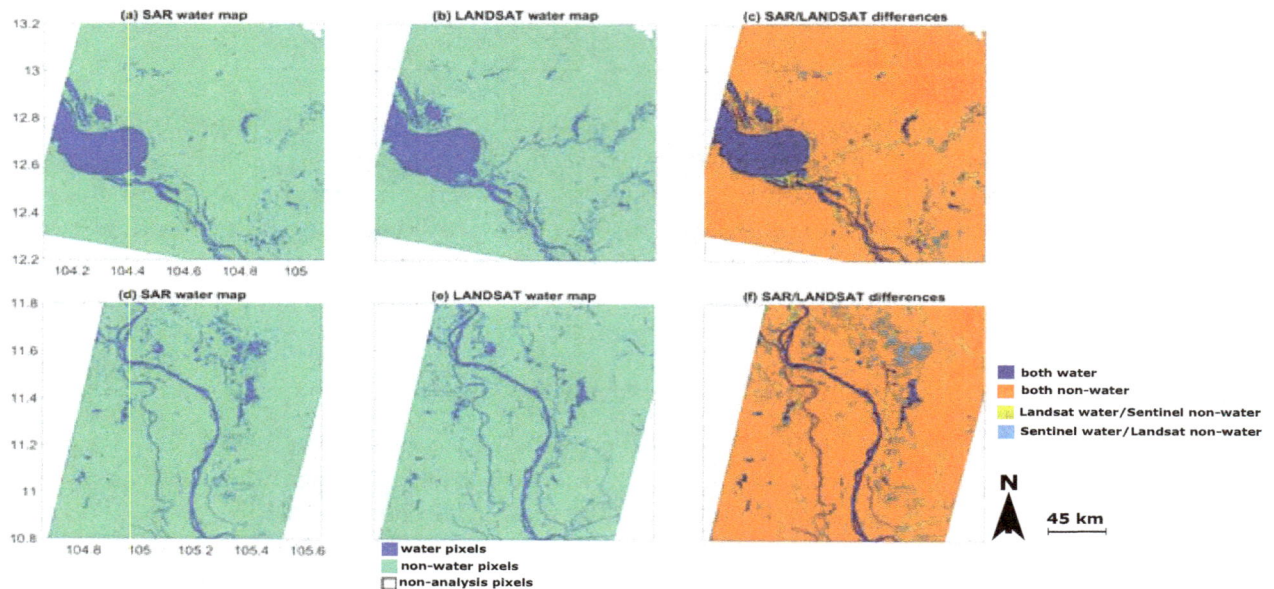

Figure 9. (**a**,**d**) SAR surface water maps; (**b**,**e**) Landsat-8 surface water maps; and (**c**,**f**) their differences; over the Tonle Sap Lake (left), and over the Mekong river (right), for February 2016. Blue color presents water pixels while orange color presents non-water pixels detected by both Sentinel and Landsat, green color is Landsat water/Sentinel non-water pixels, and light blue color is Sentinel water/Landsat non-water pixels.

Table 7. Confusion matrices (in numeric and percentage forms) of the SAR-predicted surface water maps and the Landsat-8 reference surface water maps, over the Tonle Sap Lake (**Left**) and over the Mekong River (**Right**).

	Tonle Sap Lake			Mekong River	
	Non-water(0) (Predicted)	Water(1) (Predicted)		Non-water(0) (Predicted)	Water(1) (Predicted)
Non-water(0) (Actual)	11,641,078 (99.6%)	44,493 (0.4%)	**Non-water**(0) (Actual)	10,983,583 (99.2%)	85,096 (0.8%)
Water(1) (Actual)	71,884 (6.5%)	1,023,457 (93.5%)	**Water**(1) (Actual)	51,611 (14.3%)	309,982 (85.7%)

The same results are found when applying the NN classification to other areas. To conclude this comparison, the proposed NN methodology correctly detected ~90% of the water pixels observed by Landsat-8, with a spatial correlation of ~90%. The NN works better over open water bodies than over other heterogeneous environments. For instance, the NN has difficulties detecting small river branches (Southeast of the Tonle Sap Lake in Figure 9—top panel) although they are clearly detected with Landsat-8 images. The NN can provide water maps with high accuracy compared to the reference Landsat-8 water maps; there are differences between them. Errors could come from the following factors:

- The SAR responses can be affected by complex interactions with the terrain and the vegetation, especially along small river banks. It can be difficult to account for this local complexity in the methodology.
- In the SAR water detection method, as in any other classifications method scheme, different parameters were selected to optimize the overall performance of the method, but local ambiguities can still exist.
- Sentinel-1 and Landsat-8 data are not always acquired on the same day.
- Using Landsat-8 quality flags, we can remove cloud-covered pixels, but we cannot detect cloud-shadow pixels causing ambiguities in the NN training dataset.
- Reference surface water maps derived from negative NDVI values on the Landsat-8 images are not always perfect. Water under vegetation can be difficult to detect with Landsat-8 observations. The NDVI values can also be impacted for highly turbid waters where the NIR reflectance can be higher than the red reflectance.

4.2. Evaluation Using a Topography-Based Floodability Index

A global floodability index based on topography has been developed by [34]. It uses mainly the Hydrological data and maps based on SHuttle Elevation Derivatives at multiple Scales (HydroSHEDS) dataset [33] that has been derived from elevation measured by the Shuttle Radar Topography Mission (SRTM) satellite. This floodability index provides a static map of an estimate of the probability for a pixel to be inundated (between 0% and 100%) at the spatial resolution of 90 m, based only on topography information (such as slope in the pixel, distance to the closest river, difference of elevation with the closest river). Figure 10a presents this floodability index map over the whole Mekong Delta. As expected, all rivers and lakes in this area have a very high probability of being inundated (over 80%). Since this index is based only on topography, its reliability is higher for natural environments and it can be less precise over regions with strong anthropic impact such as irrigated areas. The floodability data is upscaled from 90 m to 30 m spatial resolution to compare with predicted SAR surface water maps over the Tonle Sap Lake and the Vietnamese Mekong Delta. Each floodability pixel is divided into a 3×3 matrix with the same value, and projected onto the Sentinel-1 grid. By comparing these two products, we can see where and how Sentinel-1 water pixels are located with respect to the floodability index, and test the consistency between two independent products. Figure 10b–e show floodability maps at 30 m spatial resolution and predicted Sentinel-1 water maps, over four different areas in the Mekong Delta. SAR surface water areas are generally located in areas with high predicted inundation probabilities, as expected (see Table 8). A total of 98% of the SAR surface water pixels are located in areas where the floodability index is greater than 60%, while only 2% of the SAR surface water pixels are located in areas with a lower floodability index ($\leq$60%). As mentioned earlier, the floodability index only relies upon topography information, and it can be less precise over regions with strong anthropic activities, such as irrigation. There are many rice paddies in the Lower Mekong Delta, and these irrigated fields can be missed by the floodability index, contributing to the 15% errors of SAR water pixels located in areas with a floodability index less than 80%. In the future, in complex-topography environments where SAR only data could not provide the required accuracy for the water classification (the Red River Delta in the North of Vietnam, for example), the floodability index information could be added as another input to the NN to improve the classification performance.

Table 8. Performance of the SAR surface water classification for different ranges of floodability index.

Floodability Index	$\leq$40	40–60	60–80	$\geq$80
Percentage of surface water pixels detected by the NN classification	1%	1%	13%	85%

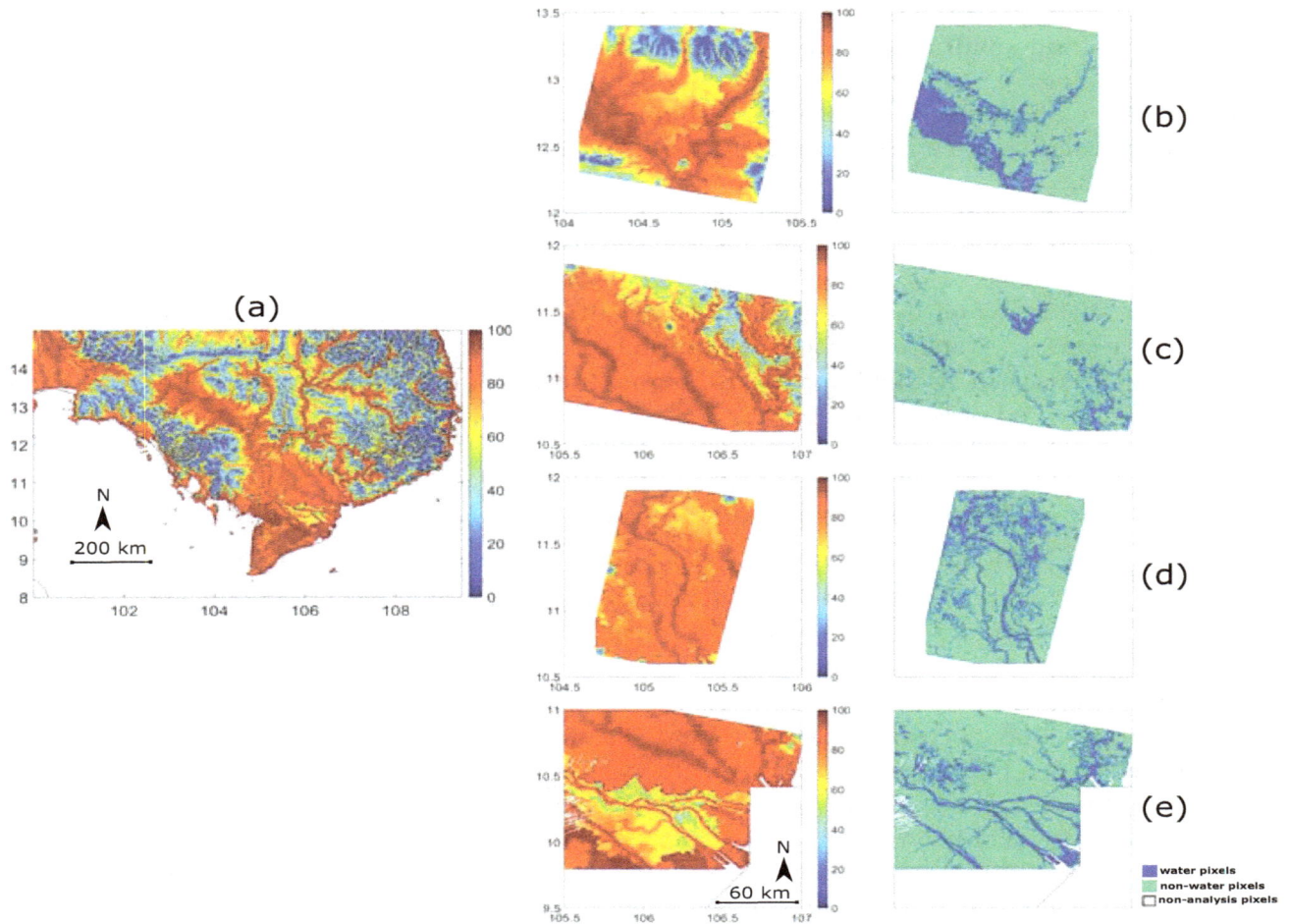

Figure 10. (a) Topography-based floodability index map over the Mekong Delta from [34]. (b–e) Comparisons of floodability index maps and SAR-predicted surface water maps for four areas over Cambodia and the Vietnamese Mekong Delta.

4.3. Comparisons with MODIS/Terra-Derived Inundation Maps

In this section, the 30 m SAR surface water maps are compared to the 500 m MODIS/Terra-derived inundation maps, for a region in the Mekong Delta. One year (2015) of SAR Sentinel-1 and MODIS/Terra data are extracted, over the same region (latitude [9.8°N–11.3°N]; longitude [104.75°E–107°E]). The MODIS inundation maps are derived from the method described by [14]. We re-produced their methodology to calculate inundation maps with three different states of non-water, water, and mixed pixels, respectively. The total MODIS surface water is the sum of the water pixels (100% area is inundated) and mixed pixels (part of these pixels is inundated). For a mixed pixel, we tested two hypothesis: 25% or 50% of the pixel is inundated.

Twenty Sentinel-1 SAR observations are available over the selected region for the year 2015 (less than two images per month—see Table 2). The surface water extent calculated from the SAR and MODIS data are presented in Figure 11. With the first assumption (25% of a mixed MODIS pixel is covered by water), the two surface water extents have very similar seasonal cycles and amplitudes, with a correlation of ~99% (Figure 11-bottom). For the second assumption (the surface water extent of a mixed pixel is increased to 50%), the difference in surface water areas increases (without significant changes in the seasonal cycle with still high correlation with the SAR surface water time series). With both hypotheses, the SAR and MODIS surface water extents reach their maximum at the same time (around 20 October 2015). Total inundated areas derived from SAR and MODIS are very close during the dry season (January to July). The cloud contamination of the MODIS estimate is low during that season. During the rainy season, more cloud contamination is expected in the MODIS estimates, and the discrepancies between the two surface water extents increase. The SAR-derived surface water estimate is expected to be more reliable due to its insensitivity to the cloud cover, but at this stage there is no convincing dataset at this spatial resolution to confirm it, as mentioned before.

To evaluate the consistency of the spatial structure between the SAR-derived and the MODIS-derived surface water maps, 10 SAR Sentinel-1 images were downloaded to cover the whole Mekong Delta and the Tonle Sap Lake (five images in May and five images in October 2015). For comparison purposes and to calculate the spatial correlation, the SAR surface water maps are aggregated from the 30 m resolution to the 500 m resolution of the MODIS-derived inundation maps (see Figure 12a,c). As a consequence, Sentinel-1-derived inundation maps are not binary (0 for non-water pixels or 1 for water pixels), but they are converted into a percentage of surface water at 500 m spatial resolution. For the dry season (Figure 12a,b—May 2015), the spatial correlation between the two surface water maps is 68%. A total of 4% of the area is inundated for the SAR estimation, while it is 5% for the MODIS estimates. For the rainy season (October 2015), the spatial correlation of the two maps increases to 78%, with 8% inundated area with the SAR and 11% with MODIS. For these calculations, we used the hypothesis of 25% inundation of the MODIS mixed pixels. Although SAR-derived and MODIS-derived water maps have a very similar seasonal cycle and similar spatial distribution of the water bodies, confirming the wetland seasonal cycle over this region, there are differences in the total surface of inundated areas. It comes mainly from the difference of spatial resolution between the two satellites. First, MODIS sensors cannot detect very small surface water fractions due to their spatial resolution. Second, the MODIS mixed pixels include water surfaces, vegetation surface and bare soil, and the percentage of each surface type within the pixel is not quantified.

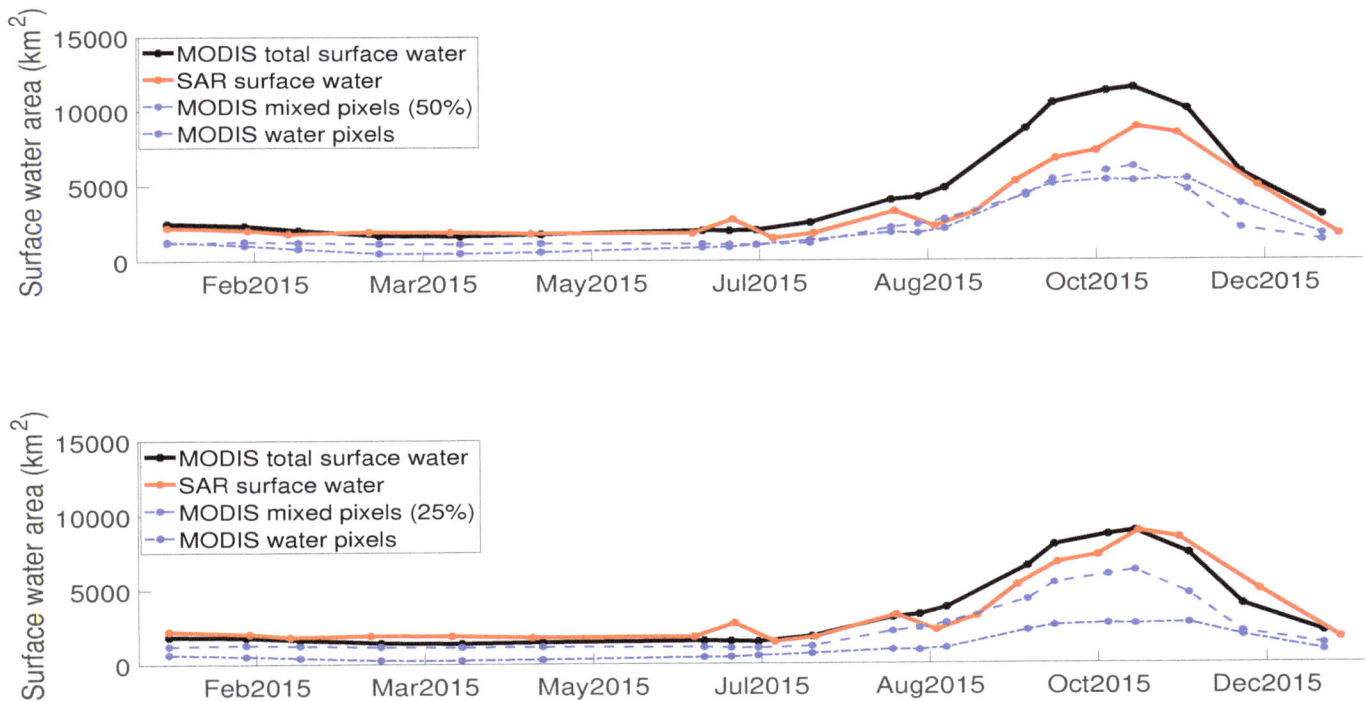

Figure 11. Time series of the surface water detected by SAR (red) and MODIS data (black), over the Mekong Delta (Latitude [9.8°N–11.3°N]; Longitude [104.75°E–107°E]), for 2015. Two hypotheses are tested for the MODIS mixed pixels: 50% inundated (**top Panel**), and 25% inundated (**bottom Panel**).

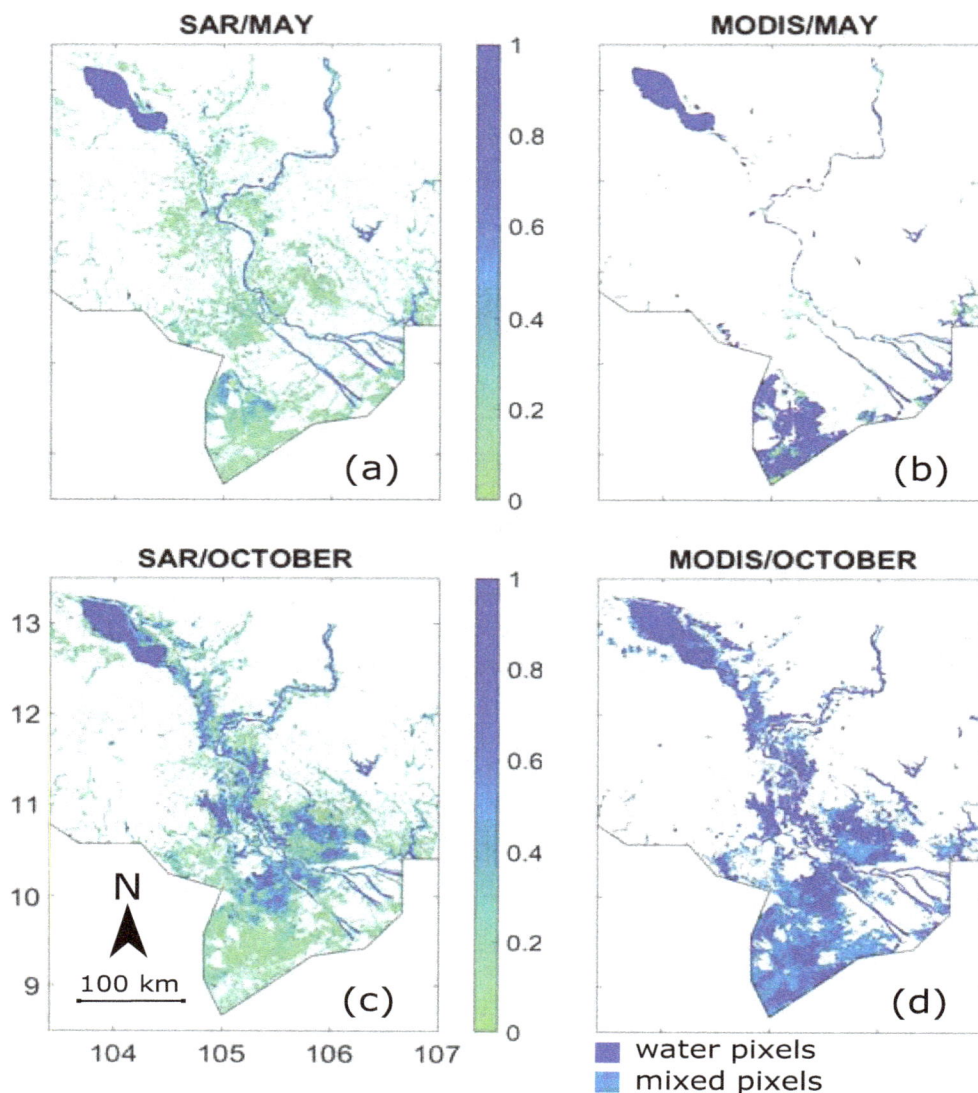

Figure 12. (**a,c**) SAR and (**b,d**) MODIS surface water maps at 500 m resolution over the Mekong Delta in May (**a,b**) and October (**c,d**) 2015.

5. Conclusions and Perspectives

This study presents a methodology to monitor and quantify surface water under all weather conditions within Cambodia and the Mekong Delta in Vietnam, using high quality Sentinel-1 SAR observations, freely available online. The methodology is based on a neural network classification trained with optical Landsat-8 images at 30 m spatial resolution. The information content of each satellite input is analyzed and the inputs are selected to optimize the performance of the classification. This method allows the detection of surface water with good accuracy when compared to visible and NIR data under clear sky conditions, as well as when compared to a floodability map derived from topography data. Surface water maps derived from the proposed NN show a spatial correlation of ∼90% when compared to Landsat-8 water maps, with a true positive water detection of ∼90%. Compared to MODIS/Terra water maps over the Delta in 2015, our products share the same wetland seasonal cycle and dynamics, with a temporal correlation of ∼99%.

In the future, we will first apply the method to other areas under similar environments in southeast Asia and in other parts of the globe, and second we will test it in more vegetated environments. The final goal is to develop a general method capable of performing at the global scale to exploit the full spatial coverage of the Sentinel-1 mission. For this purpose, several approaches will be tested to improve the retrieval scheme. First, the introduction of a priori information from a topography-based floodability index will increase information on flooding and reduce ambiguities in the SAR signal with other surface parameters. Second, with the launch of the optical Sentinel-2 satellite, Sentinel-2 observations could be used to replace Landsat-8 data, and to train the SAR surface water classification under clear sky conditions. The classification could then be extended to the cloudy areas using the SAR data. Third, the temporal information in the SAR backscatter could also be exploited (i.e., minimum or standard deviation of the time series) as this information has been shown to improve the detection of floods [13]. Finally, the high-resolution inundation extent retrieval maps could be post-processed in order to reduce the inherent noise in such high-spatial retrievals. We plan to test random walk techniques for that purpose.

Acknowledgments: We would like to thank Toshihiro Sakamoto for proving the methodology to derive the MODIS 500 m surface water maps. We are thankful to Simon Munier for interesting discussions and for sharing some of his tools in the framework of his ESA-funded postdoc (CCI Living Planet Fellowship AO/1-7829/14/I-MB). We also thank Thuy Le Toan, Stéphane Jacquemoud, Thanh Ngo Duc, and Nicolas Delbart for interesting discussions and suggestions for this work. This study was financially supported by a PhD fellowship from the Vietnam International Education Development (911 project). We would like to thank three anonymous reviewers for their helpful comments and suggestions that helped to improve the quality of this manuscript.

Author Contributions: Catherine Prigent, Filipe Aires and Binh Pham-Duc conceived and designed the experiments; Binh Pham-Duc performed the experiments; Catherine Prigent, Binh Pham-Duc and Filipe Aires analyzed the data; Filipe Aires, Catherine Prigent and Binh Pham-Duc contributed reagents/materials/analysis tools; Binh Pham-Duc, Catherine Prigent and Filipe Aires wrote the paper. All authors contributed to the discussion of results and the preparation of the manuscript.

References

1. McFeeters, S.K. The use of the Normalized Difference Water Index (NDWI) in the delineation of open water features. *Int. J. Remote Sens.* **1996**, *17*, 1425–1432.

2. Bryant, R.G.; Rainey, M.P. Investigation of flood inundation on playas within the Zone of Chotts, using a time-series of AVHRR. *Remote Sens. Environ.* **2002**, *82*, 360–375.

3. Xu, H. Modification of normalised difference water index (NDWI) to enhance open water features in remotely sensed imagery. *Int. J. Remote Sens.* **2006**, *27*, 3025–3033.

4. Cretaux, J.F.; Berge-Nguyen, M.; Leblanc, M.; Abarca Del Rio, R.; Delclaux, F.; Mognard, N.; Lion, C.; Pandey, R.K.; Tweed, S.; Calmant, S.; et al. Flood mapping inferred from remote sensing data. *Int. Water Technol. J.* **2011**, *1*, 48–62.

5. Brisco, B.; Touzi, R.; Sanden, J.J.V.D.; Charbonneau, F.; Pultz, T.J.; D'Iorio, M. Water resource applications with RADARSAT-2: A preview. *Int. J. Digit. Earth* **2008**, *1*, 130–147.

6. Wang, Y. Seasonal change in the extent of inundation on floodplains detected by JERS-1 Synthetic Aperture Radar data. *Int. J. Remote Sens.* **2004**, *25*, 2497–2508.

7. Pierdicca, N.; Pulvirenti, L.; Chini, M.; Guerriero, L.; Candela, L. Observing floods from space: Experience gained from COSMO-SkyMed observations. *Acta Astronaut.* **2013**, *84*, 122–133.

8. Voormansik, K.; Praks, J.; Antropov, O.; Jagomagi, J.; Zalite, K. Flood Mapping With TerraSAR-X in Forested Regions in Estonia. *IEEE J. Sel. Top. Appl. Earth Obs. Remote Sens.* **2014**, *7*, 562–577.

9. Martinis, S.; Kuenzer, C.; Wendleder, A.; Huth, J.; Twele, A.; Roth, A.; Dech, S. Comparing four operational SAR-based water and flood detection approaches. *Int. J. Remote Sens.* **2015**, *36*, 3519–3543.

10. Bartsch, A.; Pathe, C.; Wagner, W.; Scipal, K. Detection of permanent open water surfaces in central Siberia with ENVISAT ASAR wide swath data with special emphasis on the estimation of methane fluxes from tundra wetlands. *Hydrol. Res.* **2008**, *39*, 89–100.

11. Brisco, B.; Short, N.; van der Sanden, J.; Landry, R.; Raymond, D. A semi-automated tool for surface water mapping with RADARSAT-1. *Can. J. Remote Sens.* **2009**, *35*, 336–344.

12. Reschke, J.; Bartsch, A.; Schlaffer, S.; Schepaschenko, D. Capability of C-Band SAR for Operational Wetland Monitoring at High Latitudes. *Remote Sens.* **2012**, *4*, 2923–2943.

13. Santoro, M.; Wegmuller, U.; Lamarche, C.; Bontemps, S.; Defourny, P.; Arino, O. Strengths and weaknesses of multi-year Envisat ASAR backscatter measurements to map permanent open water bodies at global scale. *Remote Sens. Environ.* **2015**, *171*, 185–201.

14. Sakamoto, T.; Van Nguyen, N.; Kotera, A.; Ohno, H.; Ishitsuka, N.; Yokozawa, M. Detecting temporal changes in the extent of annual flooding within the Cambodia and the Vietnamese Mekong Delta from MODIS time-series imagery. *Remote Sens. Environ.* **2007**, *109*, 295–313.

15. Leinenkugel, P.; Kuenzer, C.; Dech, S. Comparison and optimisation of MODIS cloud mask products for South East Asia. *Int. J. Remote Sens.* **2012**, *34*, 2730–2748.

16. Nguyen, L.; Bui, T. Flood Monitoring of Mekong River Delta, Vietnam using ERS SAR Data. In Proceedings of the 22nd Asian Conference on Remote Sensing, Singapore, 5–9 November 2001. Available online: http://www.crisp.nus.edu.sg/~acrs2001/pdf/147nguye.pdf (accessed on 22 May 2017).

17. Kuenzer, C.; Guo, H.; Huth, J.; Leinenkugel, P.; Li, X.; Dech, S. Flood Mapping and Flood Dynamics of the Mekong Delta: ENVISAT-ASAR-WSM Based Time Series Analyses. *Remote Sens.* **2013**, *5*, 687–715.

18. Amitrano, D.; Martino, G.D.; Iodice, A.; Mitidieri, F.; Papa, M.N.; Riccio, D.; Ruello, G. Sentinel-1 for Monitoring Reservoirs: A Performance Analysis. *Remote Sens.* **2014**, *6*, 10676–10693.

19. Santoro, M.; Wegmuller, U.; Wiesmann, A.; Lamarche, C.; Bontemps, S.; Defourny, P.; Arino, O. Assessing Envisat ASAR and Sentinel-1 multi-temporal observations to map open water bodies. In Proceedings of the 2015 IEEE 5th Asia-Pacific Conference on Synthetic Aperture Radar (APSAR), Marina Bay Sands, Singapore, 1–4 September 2015; pp. 614–619.

20. ESA. Sentinel-1 Technical Guides. Available online: https://sentinel.esa.int/web/sentinel/technical-guides/sentinel-1-sar (accessed on 22 May 2017).

21. Sentinel Scientific Data Hub. Available online: https://scihub.copernicus.eu/ (accessed on 22 May 2017).

22. SAR Basics with the Sentinel-1 Toolbox in SNAP Tutorial. Available online: http://step.esa.int/main/doc/tutorials/ (accessed on 22 May 2017).

23. Liu, C. Analysis of Sentinel-1 SAR Data for Mapping Standing Water in the Twente Region. Master Thesis on Science in Geo-information Science and Earth Observation, University of Twente, Twente, The Netherlands, February 2016. Available online: http://www.itc.nl/library/papers_2016/msc/wrem/cliu.pdf (accessed on 22 May 2017).

24. Rouse, J.W., Jr.; Haas, R.H.; Schell, J.A.; Deering, D.W. Monitoring Vegetation Systems in the Great Plains with Erts. *NASA Spec. Publ.* **1974**, *351*, 309.

25. Rokni, K.; Ahmad, A.; Selamat, A.; Hazini, S. Water Feature Extraction and Change Detection Using Multitemporal Landsat Imagery. *Remote Sens.* **2014**, *6*, 4173–4189.

26. Hess, L.L.; Melack, J.M.; Simonett, D.S. Radar detection of flooding beneath the forest canopy: A review. *Int. J. Remote Sens.* **1990**, *11*, 1313–1325.

27. Kasischke, E.S.; Bourgeau-Chavez, L.L. Monitoring South Florida Wetlands Using ERS-1 SAR Imagery. *Photogramm. Eng. Remote Sens.* **1997**, *63*, 281–291.

28. Pope, K.O.; Rejmankova, E.; Paris, J.F.; Woodruff, R. Detecting seasonal flooding cycles in marshes of the Yucatan Peninsula with SIR-C polarimetric radar imagery. *Remote Sens. Environ.* **1997**, *59*, 157–166.

29. Liu, Y.; Chen, K.S.; Xu, P.; Li, Z.L. Modeling and Characteristics of Microwave Backscattering From Rice Canopy Over Growth Stages. *IEEE Trans. Geosci. Remote Sens.* **2016**, *54*, 6757–6770.

30. Gstaiger, V.; Huth, J.; Gebhardt, S.; Wehrmann, T.; Kuenzer, C. Multi-sensoral and automated derivation of inundated areas using TerraSAR-X and ENVISAT ASAR data. *Int. J. Remote Sens.* **2012**, *33*, 7291–7304.

31. Prigent, C.; Aires, F.; Jimenez, C.; Papa, F.; Roger, J. Multiangle Backscattering Observations of Continental Surfaces in Ku-Band (13 GHz) From Satellites: Understanding the Signals, Particularly in Arid Regions. *IEEE Trans. Geosci. Remote Sens.* **2015**, *53*, 1364–1373.

32. Henry, J.B.; Chastanet, P.; Fellah, K.; Desnos, Y.L. Envisat multi-polarized ASAR data for flood mapping. *Int. J. Remote Sens.* **2006**, *27*, 1921–1929.

33. Lehner, B.; Verdin, K.; Jarvis, A. HydroSHEDS Technical Documentation, Version 1.0; World Wildlife Fund US: Washington, DC, USA. Available online: https://hydrosheds.cr.usgs.gov/webappcontent/HydroSHEDS_TechDoc_v10.pdf (accessed on 22 May 2017).

34. Aires, F.; Miolane, L.; Prigent, C.; Pham-Duc, B.; Fluet-Chouinard, E.; Lerner, B.; Papa, F. A Global Dynamic Long-Term Inundation Extent Dataset at High Spatial Resolution Derived through Downscaling of Satellite Observations. *J. Hydrometeorol.* **2017**, doi:10.1175/JHM-D-16-0155.1.

8

The Performance and Potentials of the CryoSat-2 SAR and SARIn Modes for Lake Level Estimation

Karina Nielsen *, Lars Stenseng, Ole Baltazar Andersen and Per Knudsen

Department of Geodesy, DTU Space, National Space Institute, 2800 Kgs. Lyngby, Denmark;
stenseng@space.dtu.dk (L.S.); oa@space.dtu.dk (O.B.A.); pk@space.dtu.dk (P.K.)
* Correspondence: karni@space.dtu.dk

Academic Editors: Frédéric Frappart and Luc Bourrel

Abstract: Over the last few decades, satellite altimetry has proven to be valuable for monitoring lake levels. With the new generation of altimetry missions, CryoSat-2 and Sentinel-3, which operate in Synthetic Aperture Radar (SAR) and SAR Interferometric (SARIn) modes, the footprint size is reduced to approximately 300 m in the along-track direction. Here, the performance of these new modes is investigated in terms of uncertainty of the estimated water level from CryoSat-2 data and the agreement with in situ data. The data quality is compared to conventional low resolution mode (LRM) altimetry products from Envisat, and the performance as a function of the lake area is tested. Based on a sample of 145 lakes with areas ranging from a few to several thousand km^2, the CryoSat-2 results show an overall superior performance. For lakes with an area below 100 km^2, the uncertainty of the lake levels is only half of that of the Envisat results. Generally, the CryoSat-2 lake levels also show a better agreement with the in situ data. The lower uncertainty of the CryoSat-2 results entails a more detailed description of water level variations.

Keywords: satellite altimetry; CryoSat-2; water level; lakes

1. Introduction

Satellite altimetry has played an increasingly important role in lake level estimation over the past 20 years, where the number of gauges has been declining. The measuring technique provides almost global data sets, which makes it possible to study continental surface hydrology at all scales, independent of borders and national policies. The spatial and temporal coverage varies between missions. The TOPEX/Poseidon and the Jason 1–3 satellites were/are operating in a 10-day repeat cycle, while the European Remote Sensing (ERS) 1 and 2, Envisat, and Saral/Altika satellites were operating in a 35-day repeat cycle. Many of these conventional missions, with a footprint diameter of several kilometers, were originally intended for ocean applications. However, the use of satellite altimetry for inland water applications has evolved into a separate field of research. Some of the first results were obtained by [1], who estimated water level time series of lakes and reservoirs with the TOPEX/Poseidon satellite, and thereby demonstrating a successful use of satellite altimetry for hydrology applications. Since then, numerous studies have estimated not just the water levels of lakes from altimetry but also of rivers and wetlands. Ref. [2] combined lake levels obtained from different missions with bathymetry and imagery to derive changes in lake water storage. Ref. [3] studied annual water level oscillations of the remote Lake Namco on the Tibetan Plateau, and Ref. [4] used conventional altimetry together with high resolution imagery to estimate lake water storage of small lakes. Ref. [5] used Geosat altimetry data to estimate river levels at different positions of the Amazon river. Ref. [6] validated water levels obtained from the different retrackers available from Envisat over the Amazon basin with in situ data, and Ref. [7] demonstrated that reliable water level estimates can be obtained from Envisat over narrow branches of the Mekong River by accounting for the hooking effect. Ref. [8] derived water level heights for both rivers and wetlands from TOPEX/Poseidon, and Ref. [9] used 10 Hz data from TOPEX/Poseidon to study water level changes over Louisiana vegetated wetlands between 1992 and 2002. Ref. [10] studied seasonal water level variability of boreal wetlands in Western Siberia from Envisat. Over time, the data quality and the methodology to process the data have greatly improved. Currently, root mean square error (RMSE) estimates of just a few cm are obtained for selected lakes when comparing with in situ data [11].

CryoSat-2 and the recently launched Sentinel-3 represent a new generation of altimetry missions. These satellites apply Synthetic Aperture Radar (SAR) technology [12], which entails a reduction of the footprint in the along-track direction to approximately 300 m [13]. The smaller footprint size allows for monitoring much smaller lakes more accurately than previously. CryoSat-2 covers the Earth up to 88 degree latitude and has a repeat period of 369 days. The number of satellite crossings over a given lake therefore depends on the lake extent in the east–west direction and the latitude [14]. Hence, smaller lakes are not visited sufficiently to capture the seasonal signal. On the other hand, significantly more lakes are visited. Recently, some studies regarding lake level estimation including new processing strategies of CryoSat-2 data have been carried out. Ref. [15] presented a new waveform retracker based on cross-correlation of a modeled CryoSat-2 waveform with the observed waveforms. Ref. [16] demonstrated that the SAR mode provides an increased precision for small lakes compared to conventional altimetry. Ref. [11] presented a novel SAR mode retracker, which utilizes information from several waveforms simultaneously, and [17] demonstrated that waveform classification might be a powerful tool to handle erroneous data. Ref. [14,18] used CryoSat-2 data to investigate the trend and seasonal signal of lakes on the Tibetan Plateau.

Here, we intend to quantify the quality of CryoSat-2 data in the SAR and SARIn modes for lake level estimation and prove its better performance over smaller lakes compared to conventional altimetry from Envisat. This has previously only been done in studies where a few lakes were investigated [16,17].

To quantify the quality of the lake levels derived from CryoSat-2, we perform a thorough investigation of the performance of CryoSat-2 compared to conventional altimetry as observed by Envisat. The study is based on a set of 145 lakes which are covered by both CryoSat-2 (SAR or SARIn mode) and Envisat (LRM). The lakes are located in Canada, Finland, and Denmark and have areas ranging from a few to several thousand km^2. A way to evaluate the data is to consider the standard deviation of the predicted water level for each crossing over a given lake. For each lake, the standard deviations are summarized by the median, which hereafter is referred to as the median of standard deviation (MSD). The MSD gives a measure of how accurately the water level is estimated, which subsequently determines how small water level variations that can be observed. We estimate the MSD for each lake and test its dependence on lake area, in order to evaluate the improvement available with the new altimetry modes. In situ data is available for selected Canadian lakes, which enables the evaluation of the ability to capture annual and interannual signals. Finally, the mean water level of Danish lakes is evaluated against accurate laser scanner data.

2. Deriving Water Levels from Satellite Altimetry

In satellite altimetry [19], the distance to the surface, the range R, is measured. This is done by emission of an electromagnetic transmitted pulse traveling with the speed of light. The reflected signal is subsequently received by the antenna on-board the satellite. The range is derived from the two-way travel time of the pulse. Assuming the altitude h of the satellite is known with respect to a reference ellipsoid, the surface elevation H relative to this ellipsoid is given by the following simple relation (see Figure 1):

$$H = h - R. \tag{1}$$

The range provided by the satellite is often referenced to the center of the range window and is therefore only an approximate estimate (see Figures 1 and 2). The range window is the area in the direction of the pulse where the satellite can pick up the reflected signal. For CryoSat-2, the range window is 60 and 240 m for for the SAR and SARIn modes, respectively. To estimate the exact range, on-ground processing, referred to as retracking, must be performed. Retracking is the procedure of identifying the surface on the leading edge of the waveform (see Figure 2). The waveform is the received power as a function of the power bins in the range window. In empirical retracking, the surface or retracking point is typically defined as the decimal bin along the leading edge, which is associated with a certain power threshold. The distance between the center bin and the retracking point in the waveform defines the retracking correction R_{retrack} (see Figure 2).

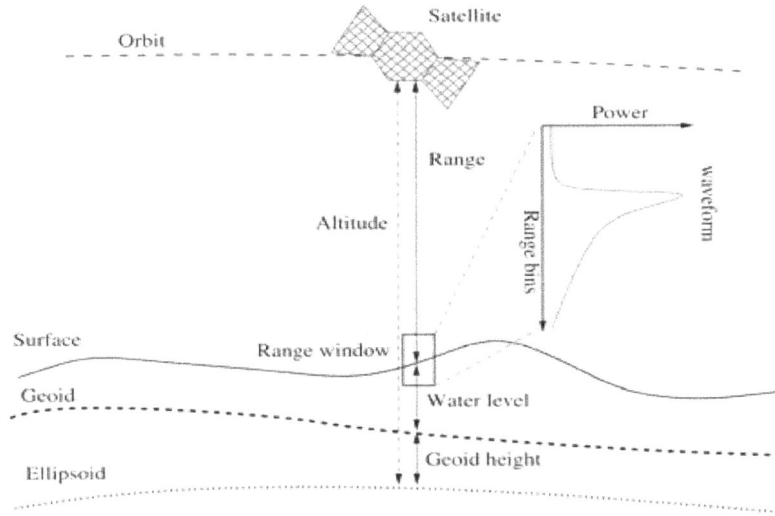

Figure 1. The principle of satellite altimetry.

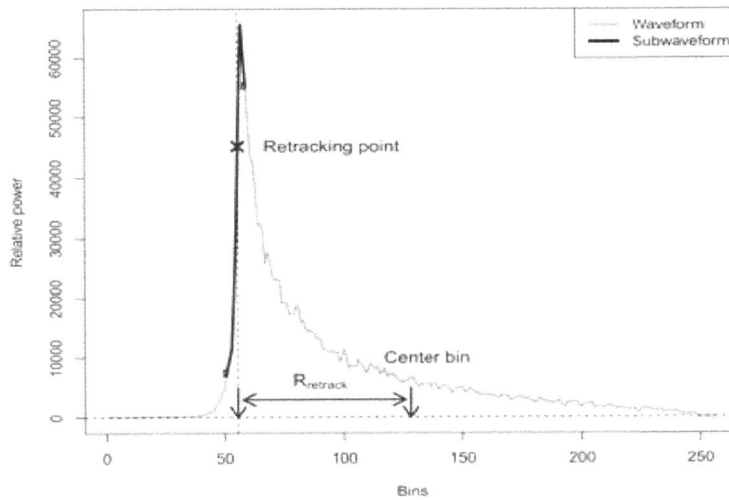

Figure 2. Explanation of the retracking correction.

The range must also be corrected for any path delay that occurs when the signal travels through the atmosphere and for geophysical signals that influence the elevation of the water surface. Hence, the range is corrected for the ionosphere, wet and dry troposphere, solid Earth tide, ocean loading tide, and geocentric polar tide, which are combined in the correction term R_{geo}. The water level above a reference geoid N is derived from the following expression:

$$H = h - (R + R_{\text{retrack}} + R_{\text{geo}}) - N. \tag{2}$$

3. Study Area

To evaluate the performance of both the SAR and SARIn modes, several relevant regions in Canada, Finland and Denmark are selected. These regions have a large concentration of lakes. There are 25 Danish lakes included in the study, which all are smaller than 40 km². These lakes are situated in a relatively flat terrain. In this area, CryoSat-2 is operating in the SAR mode. There is a total of 120 Finish and Canadian lakes, which are covered by CryoSat-2 in SARIn mode, and these lakes range in area from 51 to 27,816 km². A large fraction of the lakes has complex coastlines and several small islands. Figure 3 displays the study areas: A, Finland, B, Denmark, and C, Canada. The location of the lakes is marked with triangles.

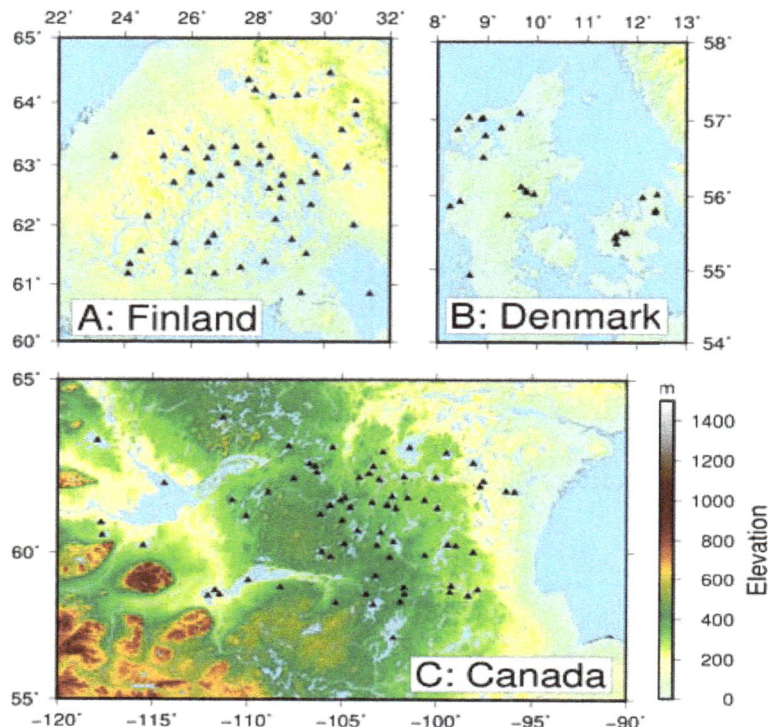

Figure 3. An overview of the lakes included in the study.

4. Data

We use the CryoSat-2 European Space Agency (ESA) L1b baseline C and the Envisat Radar Altimetry (RA) Geophysical Data Record (GDR) data products, which are thoroughly described in the following subsections. These products also include the geophysical corrections R_{geo} described above. The applied geoid model is the Earth Gravitational Model 2008 (EGM2008) [20]. To extract measurements from water returns, lake masks from the Global Lakes and Wetlands Database [21] and the Danish Geodata Agency [22] are applied.

4.1. Envisat

Envisat operated from 2002 to 2012 in a 35-day repeat cycle, with a distance between tracks of approximately 85 km at the Equator. The Radar Altimeter 2 (RA-2) onboard Envisat was a dual-frequency altimeter operating at Ku- and S-band, with the Ku-band channel being the primary altimetry radar and the additional S-band channel being used to correct for ionospheric effect. The Ku radar operated as a pulse-limited altimeter which emitted pulses at 1800 Hz, but with a subsequently averaging of 100 return pulses onboard the satellite, resulting in an 18 Hz product being transmitted to the ground stations. The pulse-limited altimeter gives circular footprints which are slightly elongated in the along-track direction due to the averaging of the return pulses. The size of the RA2 footprint was 10 km to 15 km depending on the height distribution within the illuminated surface area. In this study, we use the range measurements based on the Ice1 retracker [23], which is based on the Offset Center of Gravity (OCOG) retracker.

4.2 CryoSat-2

The CryoSat-2 satellite was launched in 2010. The SAR Interferometer Radar Altimeter (SIRAL) onboard CryoSat-2 is a single frequency Ku band altimeter capable of operating in three different modes: Low Resolution Mode (LRM), SAR mode, and SARIn mode. In LRM, the SIRAL operates like a conventional altimeter with properties comparable to RA2; however, to allow seamless switch between the different modes, it emits pulses at 1970 Hz. In SAR mode, the pulse repetition frequency (PRF) is increased to 17.8 kHz and pulses are emitted in bursts of 64 pulses. The high PRF ensures that the return pulses are correlated, and it is therefore possible to apply Doppler processing of the 64 pulses. In the Doppler processing, it is possible to divide the area illuminated by all 64 pulses into 64 areas in the along-track direction. The result is a footprint that is pulse limited in the across-track direction and Doppler limited in the along-track direction. The Doppler beams from different bursts that illuminate a selected area on the ground are then averaged to form the waveform. Since the along-track footprint is Doppler-limited, it is not dependent on the height distribution within the illuminated area. The SARIn mode is similar to the SAR mode but includes an additional receiving antenna that allows determination of the position of the reflecting surface in the across-track direction.

The CryoSat-2 data contains waveforms with 256 and 1024 bins for SAR and SARIn, respectively. The waveforms are retracked by an empirical sub-waveform retracker; the Narrow Primary Peak Threshold (NPPT) [24], which is part of the Lars Advanced Retracking System (LARS) [25]. In SARIn, it is possible to correct the range for off-nadir returns, and, in this study, this correction is performed according to [26].

4.3. In-Situ Data

Height measurements from a national survey were extracted for a subset of the Danish lakes. The survey was conducted in 2014 and 2015 with the aim to improve the Danish elevation model. The data set contains laser scanner data with a point density of four to five measurements per square meter. The heights are referenced to DVR90, but has been converted to heights above the WGS84 reference ellipsoid with the software "KMSTRANS" [27]. The error of the data is less than 5 cm in the vertical direction. The data is available from [22].

In situ data of the water level is freely available for several lakes in Canada from the Government of Canada [28]. Lakes in the study area, which are measured with both CryoSat-2 and Envisat and where in situ data is available, are Great Slave, Athabasca, Wollaston, Claire, Nonacho, and Reindeer. The water levels are referenced to different datums, e.g., the Geodetic Survey of Canada Datum.

5. Methods and Data Processing

Waveforms related to returns from inland water might be multi-peaked due to land contamination in the signal or from the presence of strong off-nadir signals. Such complex waveforms might result in noisy and potentially erroneous water levels, and it is essential to handle these in a robust manner.

To construct lake level time series, we follow the approach described in [16], in which a state-space model is used to reconstruct the time series. The model consists of a process part and an observation part. The process part intends to describe how the true water levels vary over time. It is implemented as a random walk, which implies that water levels measured within a short time span will tend to be more alike. The observation part describes how the measurements relate to the true water level. The measurement distribution is described by a mixture between a Gaussian and a Cauchy distribution. Compared to a pure Gaussian distribution, this describes the situation where a fraction of the measurements is wrong or extremely noisy. The heavier tails of the Cauchy distribution will have the effect of reducing the influence of such erroneous observations. The described state-space model represents a robust model in the sense that the estimated water levels are not substantially biased by erroneous observations. The process enables the model to exploit the temporal correlation in the true water levels. A detailed description of the model is found in [16].

The state-space model has been implemented in a software package "tsHydro" written in the open source language "R". The package is built via the R-Package Template Model Builder (TMB) [29], which is a tool to construct complex state-space models using Automatic Differentiation and the Laplace approximation to obtain accurate and stable optimization [30]. The package offers the user the possibility to easily estimate robust water levels. To construct time series, the user must provide an input file that contains the following columns, the time in decimal years, the track number and the raw water levels. The program returns the predicted water level at each time step together with its standard deviation. The package is freely available from Github [31].

Before applying the "tsHydro" package, a rough outlier criterion is applied. For each lake, the median of all water levels is estimated. Subsequently, water levels above and below the median ± 5 m are removed. A limit of 5 m is not recommended in general, since lake levels may vary several meters over time. However, for the lakes in this study, a limit of 5 m was found appropriate.

The MSD is used as a summary measure of the uncertainty for each data type (DT), CryoSat-2 or Envisat, at each lake. We wish to quantify and test if the different data types result in different levels of uncertainty. It is also expected that the lake area has an influences on the uncertainty, which must be taken into account. The lakes are divided into three groups (AG) defined by their area: small <100 km^2, medium 100–1000 km^2, or large >1000 km^2. Each uncertainty measurement, MSD, is described by the following standard two-way analysis of variance (ANOVA) model:

$$\log(\mathrm{MSD}_i) = \mu + \alpha(DT_i) + \beta(AG_i) + \gamma(DT_i, AG_i) + \epsilon_i. \tag{3}$$

Here, $i = 1, \ldots, N$, where N is the number of observations. μ is a common intercept. The model parameters α describe the main effect of the data types. The model parameters β describe the main effects of the lake area groups. The model parameters γ describe the interaction effect between the lake area group and data types. The interaction term describes how the effect of data types differs in the various lake area groups. If the hypothesis $H_0 : \gamma = 0$ is rejected (by a standard F-test), then the effect of data types is not the same in all lake area groups. The noise term for the logarithm of the MSDs is assumed to follow a normal distribution $\epsilon_i \sim N(0, \sigma^2)$.

6. Results

In this study, we have predicted CryoSat-2 and Envisat water levels for 145 lakes to evaluate the performance of the SAR and SARIn modes compared to conventional altimetry.

6.1. Evaluation of MSD, Uncertainty

The median of the standard deviations of the predicted water levels, MSD, which is a measure of the uncertainty, was evaluated for all lakes. Figure 4A displays the estimated MSD of CryoSat-2 and Envisat as a function of the lake area. The MSDs of the CryoSat-2 and Envisat results lie in the range of 1–8 cm and 1–28 cm, respectively. The MSD of the CryoSat-2 results is generally lower, where the most pronounced difference is seen for lakes with a small area. For large lakes, the MSD is similar for the two data sets. Figure 4B displays the MSD ratio, showing Envisat over CryoSat-2, as a function of the lake area. Values above and below 1 indicate lakes where the CryoSat-2 or Envisat results have the lowest MSD, respectively. For most lakes, this ratio demonstrates that the MSD of CryoSat-2 is less than half as the MSD of Envisat.

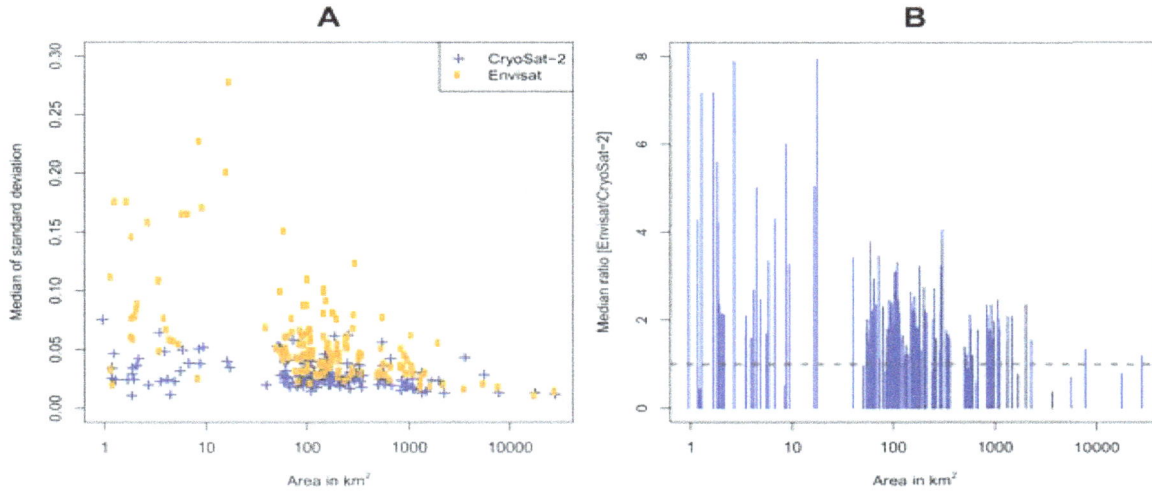

Figure 4. (**A**) the MSD of CryoSat-2 (blue) and Envisat (orange) as a function of Area; (**B**) the MSD ratio as a function of area. Values above and below 1 indicate ratios, where CryoSat-2 or Envisat have the lowest MSD, respectively.

6.2. The Significance of Lake Area with Respect to MSD

The model for the logarithm of the median of standard deviations (3) was validated by visual inspection of the residuals. The hypothesis that the difference between the two data types is the same for all three area groups was rejected by a standard F-test (p-value 0.006716). The difference between the two data types is different for the three area groups. For the smallest area group, the MSD was 2.2 times higher for Envisat than for CryoSat-2 with a 95% confidence interval of [1.9–2.7]. For the medium area group, the MSD was 1.7 times higher for Envisat with a confidence interval of [1.5–2.0]. Finally, for the largest area group, the MSD was 1.3 times higher for the Envisat, but the difference was not significant, as the confidence interval [0.9–1.8] included 1. A detailed description of the MSD distributions for the three area groups are shown in Figure 5.

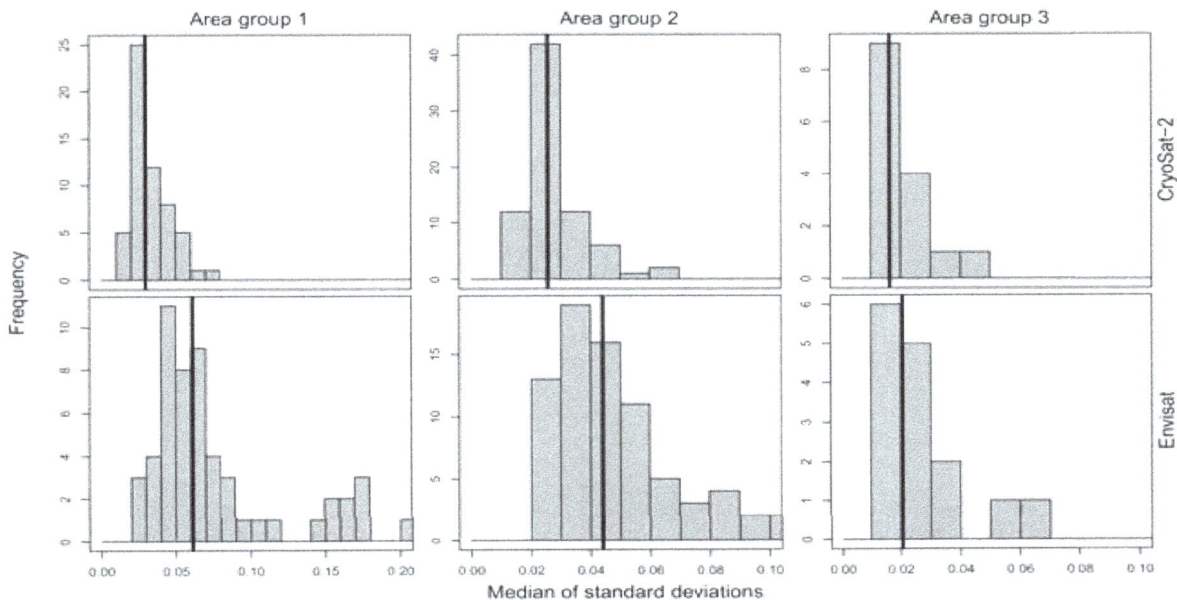

Figure 5. Histograms of MSD for CryoSat-2 and Envisat in the three area groups; group 1 (<100 km^2), group 2, (100–1000 km^2), and group 3 (>1000 km^2). The vertical black line indicates the median of the MSDs.

6.3. Comparison with In Situ Data

6.3.1. Canadian Lakes

The second measure of performance of CryoSat-2 and Envisat to measure water level variations is the agreement with the true water level. The true water level is represented by in situ measurements of the water level. In situ measurements are available for six Canadian lakes: Great Slave Lake, Lake Athabasca, Reindeer Lake, Lake Wollaston, Lake Claire, and Lake Nonacho. Since the satellite and the in situ data are referenced with respect to different datums, a bias in the water levels is estimated and subtracted from the satellite data. Figure 6 shows the estimated time series of the water level together with the in situ data for the six lakes. The circles represent the water level of the retracked data, while the crosses represent the model based predictions. In general, the predicted satellite-based time series follow the in situ data quite well. For the lakes Wollaston, Nonacho, and Reindeer, the CryoSat-2 based time series give a better representation of the water level variations than the Envisat based solution. This is quantified by RMSE estimates, which are listed in Table 1. For Great Slave Lake, both satellite based models reveal erroneous water level estimates, although the overall variation is well represented. These estimates result in an artificially increased RMSE value.

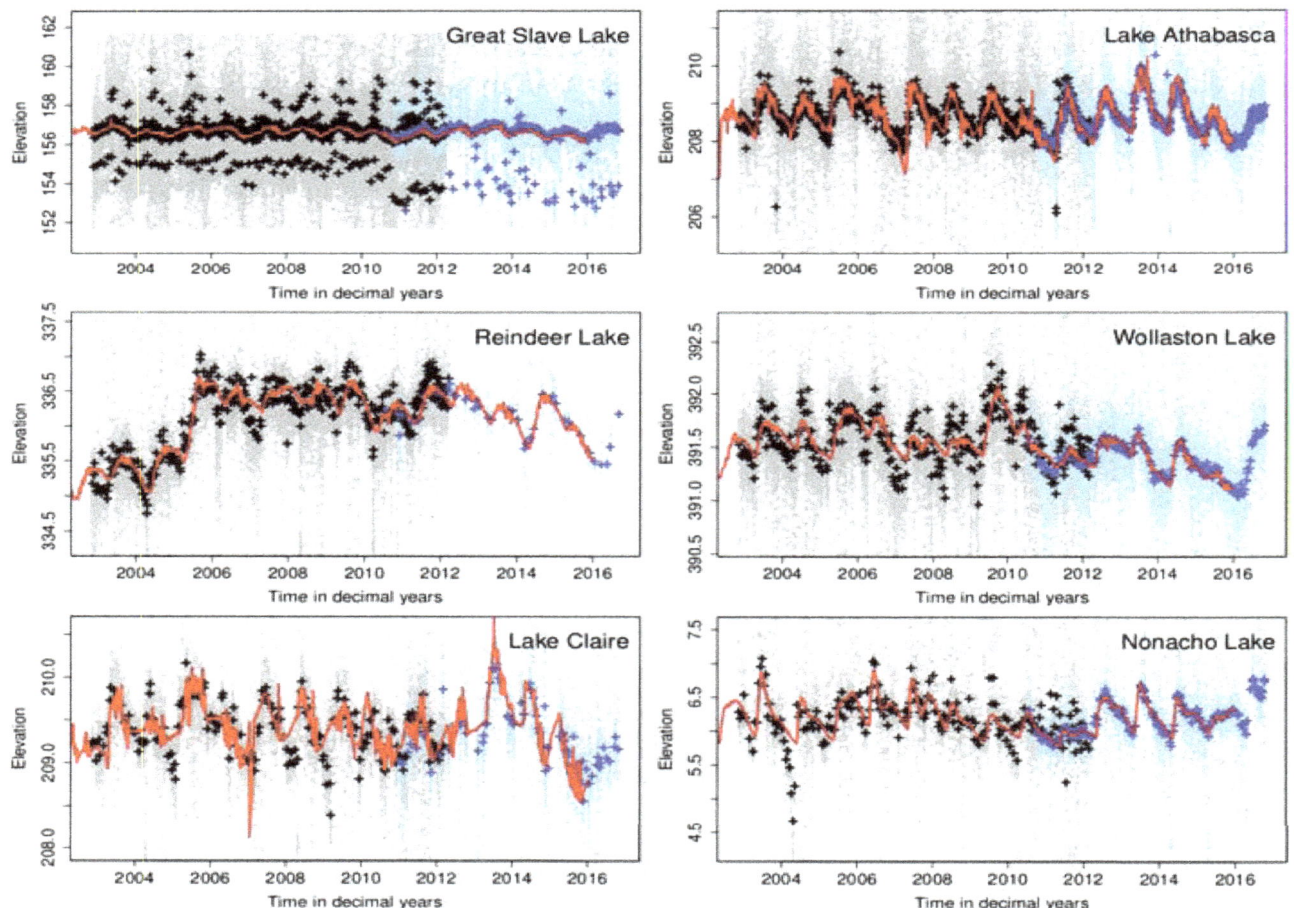

Figure 6. Water level time series for the six Canadian lakes: Great Slave, Athabasca, Wollaston, Claire, Nonacho, and Reindeer. The gray (Envisat) and blue (CryoSat-2) circles represent the raw retracked water levels. The black (Envisat) and blue (CryoSat-2) crosses represent the predicted water level, and the red line is the in situ water levels.

Table 1. RMSE values for CryoSat-2 and Envisat.

Lake	RMSE CryoSat [m]	RMSE Envisat [m]	Area [km^2]
Great Slave	0.68	0.54	27,816
Athabasca	0.19	0.25	7782
Reindeer	0.12	0.19	5597
Wollaston	0.05	0.17	2272
Claire	0.20	0.23	1326
Nonacho	0.06	0.24	847

6.3.2. Danish Lakes

Here, we compare the laser based heights with the mean water levels obtained from CryoSat-2 and Envisat. To account for the range bias between the two missions, the Envisat heights have been corrected with a bias of -0.69 cm [32] to be comparable with the CryoSat-2 heights. The mean water level for each lake is constructed as a weighted average of the predicted water levels for each crossing. Figure 7A displays the height with respect to the WGS84 reference ellipsoid for the laser, CryoSat-2, and Envisat data. The height estimates and their corresponding standard deviations are collected in Table 2. The agreement between the satellite based estimates and the laser scanner data is generally good, except for the lake Fårup Sø. For the lake Gudensø, there is a discrepancy between the CryoSat-2 and the Envisat estimates. Figure 7B displays the ratio of standard deviations. As indicated by Figure 7B, the CryoSat-2 based solutions generally have a smaller standard deviation.

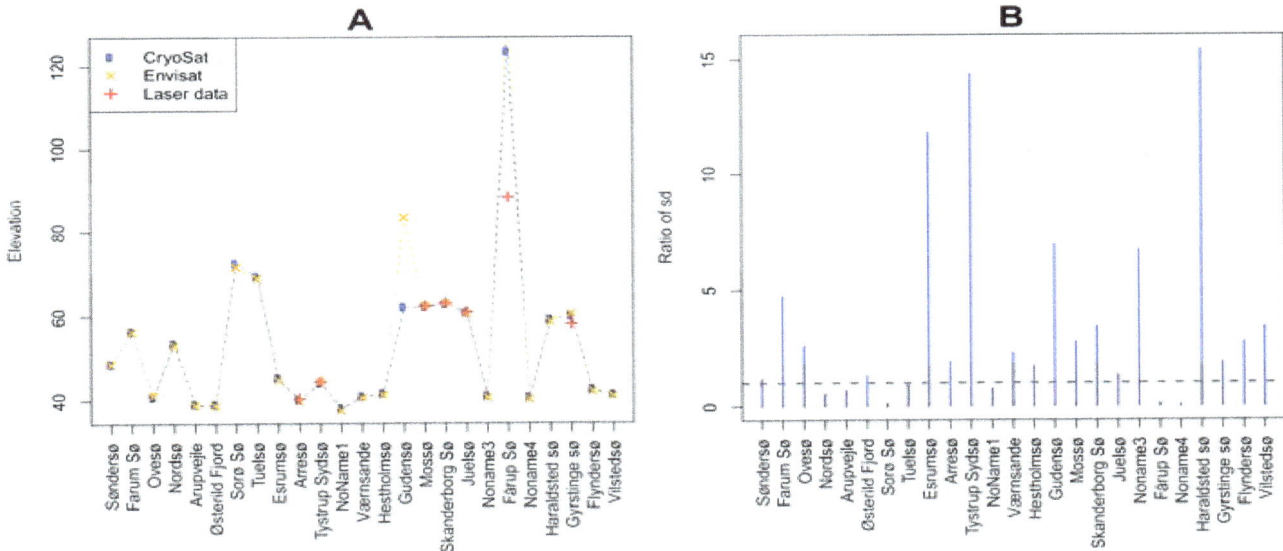

Figure 7. (A) the WGS84 elevations for the Danish lakes based on laser scanner data, CryoSat-2 and Envisat; (B) the ratio [Envisat/CryoSat-2] of the standard deviation of the estimated mean water level.

Table 2. Heights in meter above WGS84 for selected Danish lakes.

Lake	Laser	CryoSat-2 Height	CryoSat-2, sd	Envisat, Height	Envisat, sd	Area [km^2]
Arresø	40.41	40.10	0.004	39.67	0.009	39.67
Mossø	62.34	62.22	0.008	62.41	0.039	16.34
Skanderborgsø	63.01	62.82	0.008	63.12	0.030	8.67
Juelsø	60.91	60.75	0.011	60.38	0.025	8.43
Tystrup Sydsø	44.43	44.07	0.009	44.55	0.165	6.73
Gyrstinge Sø	58.20	59.94	0.011	60.53	0.021	2.04
Fårup Sø	88.17	126.18	0.053	133.44	0.714	0.96

7. Discussion

The water level for each crossing, based on CryoSat-2 and Envisat data, was estimated for 145 lakes with areas between 1 and 27,816 km². The MSDs of the CryoSat-2 and Envisat results were compared, and the predicted water levels were compared with in situ data. In the following, the applied methodology and the results are discussed in detail.

As expected, the new modes, SAR and SARIn, generally lead to an improved estimate of the water level compared to conventional altimetry. The analysis performed here has quantified that the effect is most pronounced for smaller lakes. For larger lakes, the lower uncertainty is insignificant due to the larger number of measurements. Here, it should be mentioned that, despite the high quality of the CryoSat-2 data as demonstrated in Figure 5, the uncertainty is also affected by the lake setting because topography and off-nadir signals may considerably increase the noise in the data. An example of this is seen for the Danish lake Fårup Sø in Figure 7. This lake has an area of just 0.96 km². The terrain surrounding this lake is relatively steep and in the vicinity smaller lakes located at a higher elevation are present. This configuration of terrain and surrounding lakes causes the water levels to be incorrectly estimated in the retracking process. However, by inspecting the retracked water levels, CryoSat-2 is actually able to capture the "correct" water level at some crossings (see Appendix A).

Estimating the water level for inland water bodies is challenging, since the raw retracked measurement can be noisy and erroneous (Figure 6), which easily influences the estimate. However, a robust method here that is able to account for erroneous observations in an objective manner was used. The estimates are, therefore, less sensitive to outlying observations (see Figure 6). For Great Slave Lake, a large fraction of erroneous water level estimates is present for both data sets. The applied method is clearly unable to detect the "correct" water level in this case. However, a closer inspection of the data reveals groups of erroneous data at these times (see Figure 8). In fact, at most of these times, no data at the "correct" level is present. The large fraction of erroneous data causes the state-space model to give the data a too high weight compared to the underlying process. This results in a wrong estimate of the water level. Situations like these are a weakness of the applied model. It is possible that a future extension of the model to account for the correlation between observations on the same track could reduce the weight of such sets of incorrect observations, which could give a more correct reconstruction.

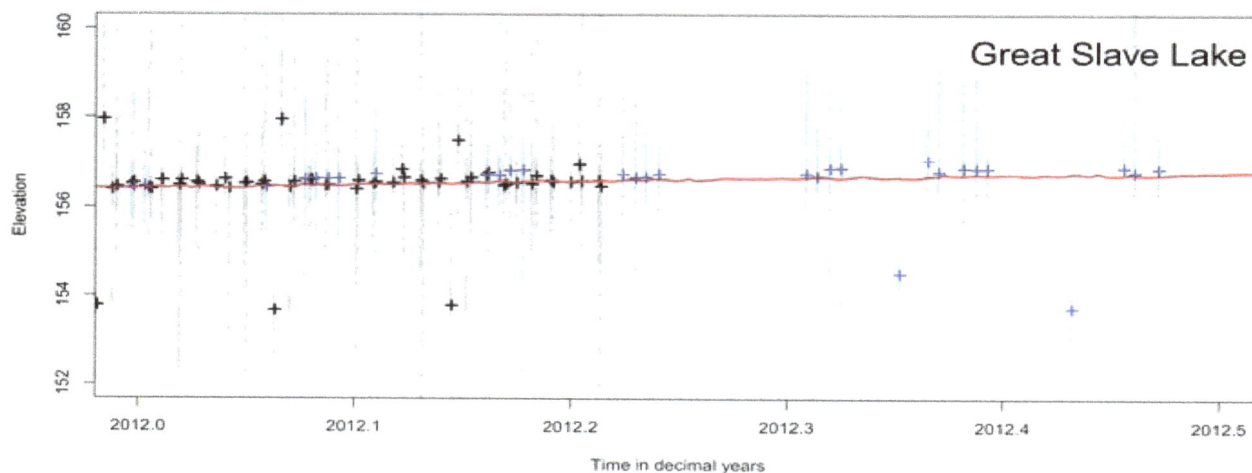

Figure 8. The water level time series of Great Slave Lake between January 2012 and May 2012. The gray (Envisat) and blue (CryoSat-2) circles represent the raw retracked water levels. The black (Envisat) and blue (CryoSat-2) crosses represent the predicted water level, and the red line is the in situ water levels.

Here, we have chosen to use the MSD as a measure of quality, since it represents the uncertainty of the estimated water level for a given crossing. The individual observations are often very noisy and mixed with outliers, hence the MSD of the estimated lake level is a more accurate measure with respect to the usefulness of the data. It is a measure of how detailed the water level can be described over time. The temporal variations of the lake level can be tracked in greater details when the MSD is at a low level.

For a subset of the Danish lakes, the satellite-based mean water levels above WGS84 were compared to laser scanner data collected between 2014 and 2015 (Figure 7). Both data sets showed a good agreement. The minor height difference might partly be explained by the retracking bias which can be of several cm or small variations in the inter-mission bias. Furthermore, the laser data were collected after the time period of the Envisat data. However, the water level variation of Danish lakes is small. Based on Google Earth, the lake Gudensø has an elevation similar to the lakes Mossø, Juelsø, and Skanderborg Sø. This indicates that the CryoSat-2 based height is closer to the "correct" height.

8. Conclusions

Based on the results found in this study, it can be concluded that the CryoSat-2 derived lake levels have a significant lower MSD compared to Envisat for lakes with an area smaller than 1000 km^2. Furthermore, the CryoSat-2 results show an overall better agreement with in situ data for the six Canadian lakes. The RMSE values are in the range of 5–68 cm and 17–54 cm for CryoSat-2 and Envisat, respectively. Both CryoSat-2 and Envisat based mean water levels agreed well with the laser scanner data. These results reveal a promising potential of Sentinel-3, which is operating in the SAR mode globally with a repeat period of 27 days. Hence, assuming that the data quality of Sentinel-3 resembles that of CryoSat-2, water level variations below 10 cm can potentially be captured for relatively small lakes.

Acknowledgments: We acknowledge the FP7 project Land and Ocean take up from Sentinel-3 (LOTUS) under Grant No. 313238 that has partly funded this work. Anders Nielsen is acknowledged for statistical guidance. Louise Sandberg Sørensen and Stine Kildegaard Rose are acknowledged for language editing. Finally, we acknowledge the anonymous reviewers for improving the manuscript.

Author Contributions: Karina Nielsen processed CryoSat-2 data, conducted the analysis, and wrote the majority of the manuscript. Lars Stenseng processed CryoSat-2 and contributed to manuscript writing. Ole B. Andersen and Per Knudsen contributed with discussions and manuscript writing.

Appendix A

Table A1 shows all the observed water levels of Fårup Sø from CryoSat-2 and Envisat.

Table A1. Water levels of Fårup Sø.

Time CryoSat-2	Height CryoSat-2	Time Envisat	Height Envisat
2011.259	126.881	2003.367	133.756
2011.259	161.100	2003.750	110.016
2011.616	112.106	2004.612	141.591
2011.616	105.081	2004.994	132.531
2013.631	147.994	2005.186	134.269
2013.631	147.963	2005.953	139.167
2014.640	112.315	2007.008	138.599
2014.640	88.106	2008.636	110.615
2015.292	88.038	2009.690	110.621
2015.292	88.019	2010.265	136.480
2015.648	125.739	2010.745	110.055
2015.648	125.606		
2016.656	148.264		
2016.656	148.020		

References

1. Birkett, C.M.B. The contribution of TOPEX/POSEIDON to the global monitoring of climatically sensitive lakes. *J. Geophys. Res.* **1995**, *100204*, 25179–25204.

2. Crétaux, J.F.; Birkett, C. Lake studies from satellite radar altimetry. *C. R. Geosci.* **2006**, *338*, 1098–1112.

3. Kropáček, J.; Braun, A.; Kang, S.; Feng, C.; Ye, Q.; Hochschild, V. Analysis of lake level changes in Nam Co in central Tibet utilizing synergistic satellite altimetry and optical imagery. *Int. J. Appl. Earth Obs. Geoinf.* **2012**, *17*, 3–11.

4. Baup, F.; Frappart, F.; Maubant, J. Combining high-resolution satellite images and altimetry to estimate the volume of small lakes. *Hydrol. Earth Syst. Sci.* **2014**, *18*, 2007–2020.

5. Koblinsky, C.J.; Clarke, R.T.; Brenner, A.C.; Frey, H. Measurement of river level variations with satellite altimetry. *Water Resour. Res.* **1993**, *29*, 1839–1848.

6. Frappart, F.; Calmant, S.; Cauhopé, M.; Seyler, F.; Cazenave, A. Preliminary results of ENVISAT RA-2-derived water levels validation over the Amazon basin. *Remote Sens. Environ.* **2006**, *100*, 252–264.

7. Boergens, E.; Dettmering, D.; Schwatke, C.; Seitz, F. Treating the Hooking Effect in Satellite Altimetry Data: A Case Study along the Mekong River and Its Tributaries. *Remote Sens.* **2016**, *8*, 91.

8. Birkett, C.M. Contribution of the TOPEX NASA Radar Altimeter to the global monitoring of large rivers and wetlands. *Water Resour. Res.* **1998**, *34*, 1223–1239.

9. Lee, H.; Shum, C.K.; Yi, Y.; Ibaraki, M.; Kim, J.W.; Braun, A.; Kuo, C.Y.; Lu, Z. Louisiana wetland water level monitoring using retracked TOPEX/POSEIDON altimetry. *Mar. Geod.* **2009**, *32*, 284–302.

10. Zakharova, E.A.; Kouraev, A.V.; Rémy, F.; Zemtsov, V.A.; Kirpotin, S.N. Seasonal variability of the Western Siberia wetlands from satellite radar altimetry. *J. Hydrol.* **2014**, *512*, 366–378.

11. Villadsen, H.; Deng, X.; Andersen, O.B.; Stenseng, L.; Nielsen, K.; Knudsen, P. Improved inland water levels from SAR altimetry using novel empirical and physical retrackers. *J. Hydrol.* **2016**, *537*, 234–247.

12. Raney, R.K. The delay/Doppler radar altimeter. *IEEE Trans. Geosci. Remote Sens.* **1998**, *36*, 1578–1588.

13. Wingham, D.J.; Francis, C.R.; Baker, S.; Bouzinac, C.; Brockley, D.; Cullen, R.; de Chateau-Thierry, P.; Laxon, S.W.; Mallow, U.; Mavrocordatos, C.; et al. CryoSat: A mission to determine the fluctuations in Earth's land and marine ice fields. *Adv. Space Res.* **2006**, *37*, 841–871.

14. Kleinherenbrink, M.; Lindenbergh, R.; Ditmar, P. Monitoring of lake level changes on the Tibetan Plateau and Tian Shan by retracking Cryosat SARIn waveforms. *J. Hydrol.* **2015**, *521*, 119–131.

15. Kleinherenbrink, M.; Ditmar, P.; Lindenbergh, R. Retracking Cryosat data in the SARIn mode and robust lake level extraction. *Remote Sens. Environ.* **2014**, *152*, 38–50.

16. Nielsen, K.; Stenseng, L.; Andersen, O.B.; Villadsen, H.; Knudsen, P. Validation of CryoSat-2 SAR mode based lake levels. *Remote Sens. Environ.* **2015**, *171*, 162–170.

17. Göttl, F.; Dettmering, D.; Müller, F.; Schwatke, C. Lake Level Estimation Based on CryoSat-2 SAR Altimetry and Multi-Looked Waveform Classification. *Remote Sens.* **2016**, *8*, 885.

18. Jiang, L.; Nielsen, K.; Andersen, O.B.; Bauer-Gottwein, P. Monitoring recent lake level variations on the Tibetan Plateau using CryoSat-2 SARIn mode data. *J. Hydrol.* **2017**, *544*, 109–124.

19. Fu, L.L.; Cazenave, A. *Satellite Altimetry and Earth Sciences: A Handbook of Techniques and Applications*; Academic: San Diego, CA, USA, 2001; p. 463.

20. Pavlis, N.K.; Holmes, S.A.; Kenyon, S.C.; Factor, J.K. The development and evaluation of the Earth Gravitational Model 2008 (EGM2008). *J. Geophys. Res. Solid Earth* **2012**, *117*, doi:10.1029/2011JB008916.

21. Lehner, B.; Döll, P. Development and validation of a global database of lakes, reservoirs and wetlands. *J. Hydrol.* **2004**, *296*, 1–22.

22. Kortforsyningen. Available online: http://kortforsyningen.dk/ (accessed on 23 May 2017).

23. Bamber, J.L. Ice sheet altimeter processing scheme. *Int. J. Remote Sens.* **1994**, *15*, 925–938.

24. Villadsen, H.; Andersen, O.B.; Stenseng, L.; Nielsen, K.; Knudsen, P. CryoSat-2 altimetry for river level monitoring—Evaluation in the Ganges–Brahmaputra River basin. *Remote Sens. Environ.* **2015**, *168*, 80–89.

25. Stenseng, L.; Andersen, O.B. Preliminary gravity recovery from CryoSat-2 data in the Baffin Bay. *Adv. Space Res.* **2012**, *50*, 1158–1163.

26. Armitage, T.W.K.; Davidson, M.W.J. Using the interferometric capabilities of the ESA Cryosat-2 mission to improve the accuracy of sea ice freeboard retrievals. *IEEE Trans. Geosci. Remote Sens.* **2014**, *52*, 529–536.

27. Kmstrans. Available online: https://bitbucket.org/KMS/kmstrans (accessed on 23 May 2017).

28. Government of Canada. Water Level and Flow. Available online: https://wateroffice.ec.gc.ca/ (accessed on 23 May 2017).
29. Kristensen, K.; Nielsen, A.; Berg, C.W.; Skaug, H.; Bell, B.M. TMB: Automatic differentiation and laplace approximation Authors. *J. Stat. Softw.* **2016**, *70*, 1–21.
30. Skaug, H.J.; Fournier, D.A. Automatic approximation of the marginal likelihood in non-Gaussian hierarchical models. *Comput. Stat. Data Anal.* **2006**, *51*, 699–709.
31. Github. Available online: https://github.com/cavios/tshydro (accessed 23 May 2017).
32. Bosch, W.; Dettmering, D.; Schwatke, C. Multi-Mission Cross-Calibration of Satellite Altimeters: Constructing a Long-Term Data Record for Global and Regional Sea Level Change Studies. *Remote Sens.* **2014**, *6*, 2255–2281.

9

A Comparative Study of GRACE with Continental Evapotranspiration Estimates in Australian Semi-Arid and Arid Basins: Sensitivity to Climate Variability and Extremes

Hong Shen [1,2,*], Marc Leblanc [1,3,4], Frédéric Frappart [5,6], Lucia Seoane [5], Damien O'Grady [1] (iD), Albert Olioso [7] and Sarah Tweed [3]

[1] Centre for Tropical Water Research & Aquatic Ecosystem Research (TropWATER), James Cook University, 4870 Cairns, Australia; marc.leblanc@univ-avignon.fr (M.L.); damien.ogrady@my.jcu.edu.au (D.O.)

[2] State Key Laboratory of Hydro-Science and Engineering, Department of Hydraulic Engineering, Tsinghua University, Beijing 100084, China

[3] Research Institute for the Development, UMR G-EAU, 34000 Montpellier, France; sarah.tweed@ird.geau.fr

[4] Hydrogeology Laboratory, UMR EMMAH, University of Avignon, 84000 Avignon, France

[5] Géosciences Environnement Toulouse (GET)—UMR5563, CNRS, IRD, Université de Toulouse UPS, OMP-GRGS, 14 Avenue E. Belin, 31400 Toulouse, France; frederic.frappart@legos.obs-mip.fr (F.F.); lucia.seoane@get.obs-mip.fr (L.S.)

[6] Laboratoire d'Etudes en Géophysique et Océanographie Spatiales (LEGOS)—UMR5566, CNES, CNRS, IRD, Université de Toulouse UPS, OMP-GRGS, 14 Avenue E. Belin, 31400 Toulouse, France

[7] UMR EMMAH, INRA—University of Avignon, 84000 Avignon, France; albert.olioso@inra.fr

* Correspondence: hongshen@mail.tsinghua.edu.cn

Abstract: This study examines the dynamics and robustness of large-scale evapotranspiration products in water-limited environments. Four types of ET products are tested against rainfall in two large semi-arid to arid Australian basins from 2003 to 2010: two energy balance ET methods which are forced by optical satellite retrievals from MODIS; a newly developed land surface model (AWRA); and one approach based on observations from the Gravity Recovery and Climate Experiment (GRACE) and rainfall data. The two basins are quasi (Murray-Darling Basin: 1.06 million km^2) and completely (Lake Eyre Basin: 1.14 million km^2) endorheic. During the study period, two extreme climatic events—the Millennium drought and the strongest La Niña event—were recorded in the basins and are used in our assessment. The two remotely-sensed ET products constrained by the energy balance tended to overestimate ET flux over water-stressed regions. They had low sensitivity to climatic extremes and poor capability to close the water balance. However, these two remotely-sensed and energy balance products demonstrated their superiority in capturing spatial features including over small-scale and complicated landscapes. AWRA and GRACE formulated in the water balance framework were more sensitive to rainfall variability and yielded more realistic ET estimates during climate extremes. GRACE demonstrated its ability to account for seasonal and inter-annual change in water storage for ET evaluation.

Keywords: evapotranspiration; GRACE; land surface model; water balance; energy balance; Australia

---

1. Introduction

Evapotranspiration (ET) governs water and energy exchange between the atmosphere and the Earth's surface [1]. Globally, more than half of the solar energy absorbed by the land surface is used for evaporation and transpiration [2]. Annually, approximately 60% of the water precipitated over land returns to the atmosphere via the ET process [3]. This percentage could even reach more than 90% in some semi-arid and arid regions such as inland Australia [4,5]. Accurate ET quantification, especially at a catchment or basin scale, is necessary for water resources allocation and irrigation schedule design [6,7]. It is also beneficial for understanding regional climate change and hydrological interactions [8,9].

Numerous continental or global ET datasets (or products) have been produced (e.g., [1,10–17]); they represent the prevailing methods relying on satellite retrievals [18–20], upscaling ground observations [1,9], or simulations from land surface models (LSMs) [16,21], global general circulation models [15,21], and atmospheric reanalysis [1,15]. The mechanisms lying behind these ET estimates/models are usually either based on the energy balance (e.g., Penman–Monteith equation, residual method of surface energy balance, and flux tower measurements) or water budget (e.g., catchment water balance method) [22]. Due to the discrepancy in model structure, forcing datasets, parameterization, upscaling schemes, calibration strategies, as well as validation sources, contemporary ET products/models in and/or between the same categories may have a poor agreement [16,23,24]. Virukollu et al. [1] reported that the largest uncertainties were found in transition zones between humid and dry regions. Hu et al. [25] also found large spatial inconsistency between two ET remote sensing (RS) datasets (MODIS and LSA-SAFMSG) in some semi-arid regions in Europe. Long et al. [16] showed that two RS ET estimations (from MODIS and AVHRR) behaved abnormally higher than LSM during extremely dry conditions.

In water-limited regions, ET estimates based on the energy balance method could be questionable, as they highlight the energy rather than the water to be the dominant factor in controlling the ET process. Some studies have commented that RS ET products/models may not be sensitive to soil moisture and water deficit [7,26,27]. The use of thermal infrared for forcing surface energy balance models with surface temperature was usually providing coherent estimation of ET and water stress [18,23,28–30]. However, up to now, no dataset of ET derived from thermal infrared data are available at the continental or global scale. In addition, eddy covariance tower observations (constrained by energy balance at field/paddock scales and assuming homogeneous landscapes within an image pixel) are commonly employed as an independent source in large-scale ET validation, which may not be able to fully uncover the deficiency in energy-balance-constrained (RS-based) ET products.

From the points above, semi-arid and arid basins are interesting regions to spot the potential weaknesses of current ET products; the catchment water balance method would be an ideal approach for ET estimation or validation under dry conditions. GRACE satellites make the change in terrestrial water storage (ΔS) detectable at inter- or intra-annual scales [31,32], which further assist the basin-scale ET inter-comparison [33,34]. In this study, we introduced ET estimates derived from regional GRACE solutions (with a higher spatial resolution and less "striping" noises) as an independent source, to compare with three other (two RS-based and one LSM-based) sets of ET products over two large Australian semi-arid and arid basins. Our objective is to assess their overall performance and sensitivity to rainfall variability. To our knowledge, few studies have specifically targeted the water-limited basins to examine the reliability of large-scale ET products.

2. Materials and Methods

2.1. Study Areas

The Murray-Darling Basin (MDB) and Lake Eyre Basin (LEB) were chosen as study areas (Figure 1). Both basins are sufficiently large to meet the minimum spatial resolution of GRACE. The Lake Eyre basin is endorheic, while the Murray-Basin can be considered as sub-closed since there has been no or very limited outflow to the ocean in the past two decades [35,36]. In both basins, ET is the dominating process redistributing the precipitation.

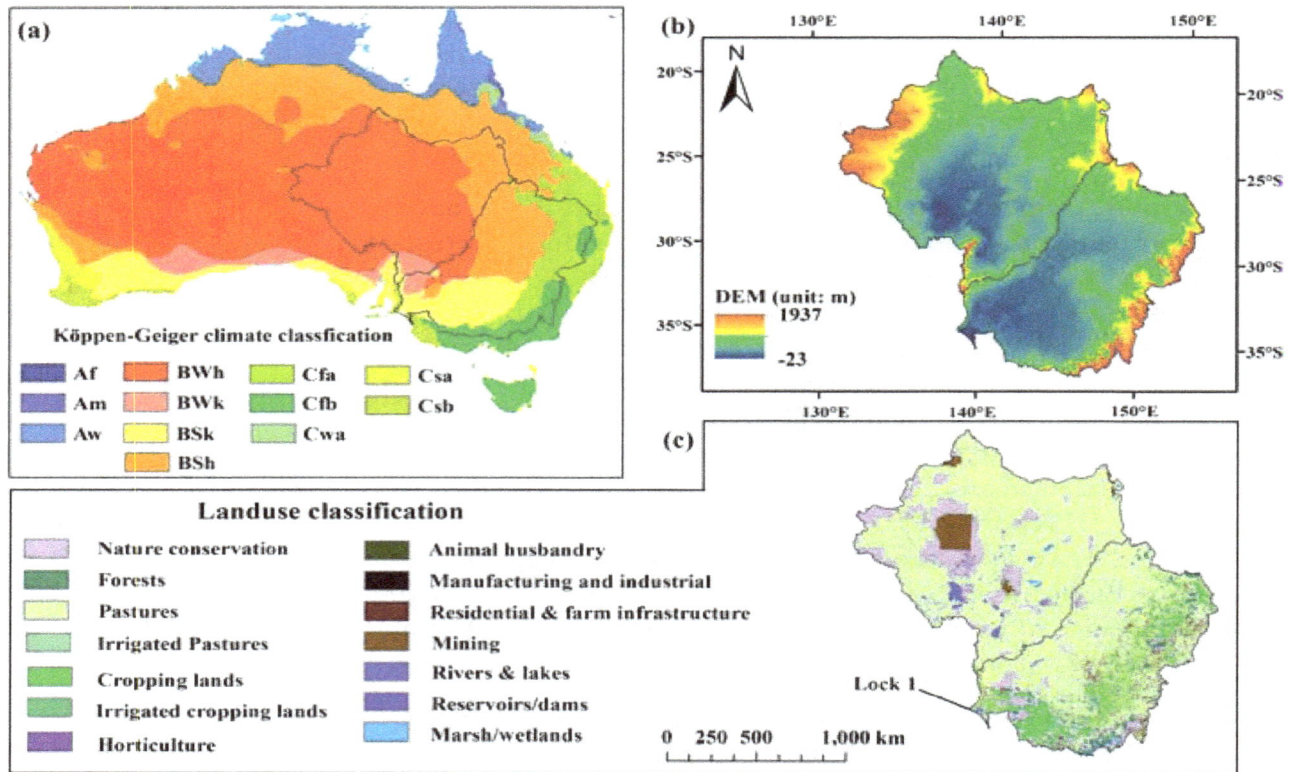

Figure 1. (**a**) Location map of the Murray-Darling and Lake Eyre Basin with climate zone according to the Koppen–Geiger classifications (adapted from [37]): A, equatorial; B, arid; C, warm temperate; W, desert; S, steppe; f, fully humid; s, summer dry; m, monsoonal; w, winter dry; h, hot arid; k, cold arid; a, hot summer; b, warm summer. (**b,c**) Digital elevation model (DEM) and land use maps were accessed from [38,39], respectively.

The Lake Eyre Basin (1.14 million km^2) encompasses 81.5% hot dessert (arid climate) and ~15% hot steppe (semi-arid climate). Rainfall within the basin varies widely and is usually unable to meet the atmospheric demand. The annual rainfall over the arid region of the LEB barely reaches 200 mm/year while the north-east edge receives rainfall of up to 700 mm/year under the influence of infrequent tropical storms [40]. The annual potential evaporation (PET) averaged from 1961 to 1990 is of 1453 mm/year (sourced from Australian Bureau of Meteorology data). Floods in the basin are ephemeral and extremely variable [41]. Flows and floodwaters that form during heavy rainfall events carry only 1% of total rainfall; a large fraction (~99%) of rainfall is lost through evaporation and transpiration [35].The Murray-Darling Basin (1.06 million km^2) contains a transition from subtropical to dry arid climate. Rainfall distributions within the basin vary greatly, decreasing from the south-eastern and eastern boundaries (between 600 and 800 mm) towards its western and north-western boundaries (between 100 and 300 mm). The average annual PET was 1236 mm/year (1961–1990). The basin consists of three large river systems and ~30,000 wetlands [42]. As the country's food bowl, ~80% of its area is used for agriculture [43]. Nearly half the surface water is redistributed by irrigation [35]. Many areas in the basin have little or no regular runoff [36,44]; the water levels were reduced to historically low levels during the Millennium Drought from mid-1990s to 2009 [45,46].

2.2. Datasets and Methods

2.2.1. Rainfall, Potential Evaporation, and Discharge Data

Rainfall data, provided by the Australian Bureau of Meteorology (BoM, Melbourne, Australia, data access: [47]), were used for two purposes: (1) to be used with GRACE's measurements to obtain basin-scale ET estimates through the water balance equation; (2) to evaluate ET datasets and identify their potential deficiencies. The rainfall data is a gridded daily product interpolated from weather station observations [48].

In addition, potential evaporation (hereafter, denoted as E_p) data was obtained from Australian Water Availability Project (AWAP, Canberra, Australia, data access: [49]). Within the Priestley–Taylor framework, E_p was forced by the gridded meteorological data obtained from BoM [50]. A comparison of rainfall with potential evaporation gives a first indication of areas that are mainly controlled by the availability of water versus energy. Both the rainfall and E_p datasets have a spatial resolution of $0.05° \times 0.05°$, and are available from January 2003 to December 2010.

We took the Lock 1 discharge measurements (see the location in Figure 1c) as the river outflow from the MDB. The data was originally archived as daily records (accessed from: [51]), and were further aggregated to monthly equivalent water height (mm/month) divided by total basin area. The LEB is a completely closed basin and no discharge data were required for this basin.

2.2.2. Model-Based ET Estimates

Three continental modelled ET products were used in this study: PT-CMRS, PM-Mu (a.k.a. MOD16), and AWRA-L. They vary in approach, data inputs, and ground calibration (see Table 1). To make these datasets consistent in time and space, ET results sourced from the three models were converted into monthly values from 2003 to 2010 at $0.05°$ (~5 km) spatial resolution. A brief summary of the forcing datasets, mechanism, and the major features for each ET model is given in Table 1.

- PT-CMRS model

The model. forced by MODIS retrievals and gridded meteorological datasets, was modified based on the Priestley–Taylor (PT) formula ([52]; see Equation (1)) to dynamically represent the actual ET variations over the Australian continent (hereafter, denoted as "PT-CMRS").

$$\mathrm{ET_{PT\text{-}CMRS}} = \alpha \frac{\Delta}{\Delta + \gamma}(R_n - G), \tag{1}$$

where R_n is the surface net radiation(MJ m^{-2} d^{-1}); G is the soil heat flux (MJ m^{-2} d^{-1}); Δ is the slope of the curve relating saturation water vapor pressure to temperature (kPa K^{-1}); γ is the psychometric constant (kPa K^{-1}). The prominent advantage associated with the PT method is that it does have limited input data requirement; wind speed is not compulsory in the model.

The developers replaced the original empirical constant $\alpha = 1.26$ with a flexible scaling factor formulated by the Enhanced Vegetation Index (EVI) and the Global Vegetation Moisture Index (GVMI) [53]. This scheme allows GVMI to separate surface water bodies against bare soil when EVI is low and to detect vegetation water content when EVI is high. Also, the PT-CMRS model accounts for precipitation interception. This model was calibrated at seven flux sites in Australia and validated in 227 catchments. Readers can refer the details about the PT-CMRS ET model from [45]. Data can be downloaded from: [54].

Table 1. Summary of the ET datasets used in this study except for GRACE.

Name	Algorithm	Resolution		Forcing Data		Calibration in Australia
		Temporal	Spatial (°)	Ground Meteorological Inputs	Remote Sensing Inputs	
PT-CMRS	MODIS[1]-based retrievals to scale the PT method	8-day	0.002	SILO[2] (temperature + radiation) datasets	MOD13Q1 + MOD09A1	7 flux sites (two forests, two savannah, a grassland, a floodplain, and a lake)
PM-Mu	MODIS-based retrievals to force the PM method	monthly	0.05	GMAO[3] datasets	MOD12Q1 + MOD13A2 + MOD15A2 + MOD43C1	None
AWRA	Water and energy constrained model	daily	0.05	SILO and BAWAP[4] datasets	AVHRR[5] NDVI[6]	4 flux sites and up to 326 catchments

Notes: [1] Moderate Resolution Imaging Spectroradiometer; [2] SILO datasets were produced by the Department of Environment and Resource Management, Australia; [3] GMAO dataset produced by the NASA Global Modeling and Assimilation Office; [4] BAWAP datasets were produced by the Bureau of Meteorology, Australia; [5] Advanced Very High Resolution Radiometer; [6] Normalized Difference Vegetation Index.

• PM-Mu model ([10,13])

Based on Cleugh's method [10], the MOD 16 ET data set developed by Mu ([13]; denoted here as 'PM-Mu') -integrated MODIS-retrieved leaf area index (LAI) into the Penman-Monteith (PM) equation ([55]; see Equation (2)), and improved the estimation of surface resistance and soil evaporation [13].

$$\text{ET}_{PM-Mu} = \frac{(R_n - G)\Delta + \rho_a c_p (e_s - e_a)/r_a}{\lambda[\Delta + \gamma(1 + r_s/r_a)]}, \tag{2}$$

where ρ_a is the mean air density at constant pressure (kg m^{-3}); c_p is the specific heat of air at constant pressure (MJ kg^{-1} K^{-1}); $(e_s - e_a)$ is the water vapor pressure deficit between the saturated air pressure and the actual air pressure (kPa); r_s and r_a represent the surface/aerodynamic resistance (m s^{-1}); λ is latent heat of evaporation (MJ m^{-2} d^{-1}).

Further modifications were described in [19]. The dataset is available at global scale. Results have been tested at global and flux-site scales. Data can be downloaded from: [56]. The PM model is process-based and constrained by energy balance. It requires considerably more input data than the PT model; some of this data (especially the wind speed) are barely available over large basins or regions. However, meteorological data have been obtained from atmospheric model reanalysis or gridded meteorological datasets.

• ET estimates from AWRA-L land surface model

The AWRA (Australian Water Resources Assessment) ET output (denoted as "AWRA") was produced by a multi-model system simulating hydrological processes and dynamics in landscape, river and groundwater systems, and water use all over Australia [57]. This system is jointly developed by the BoM and the Commonwealth Scientific and Industrial Research Organization (CSIRO). Either the PM or the PT functions were used, depending on the availability of wind speed data [58]. Within each cell, ET is summarized as:

$$E = E_i + E_t + E_g + E_s + E_r, \tag{3}$$

where E_t is vegetation transpiration; E_i, E_r, E_g, and E_s are evaporation from rainfall interception, open water bodies, groundwater, and soil profiles, respectively. AWRA ET estimates balance the requirement between water/energy conservation and data unavailability. However, the model neglects lateral water flows, which leads to an underestimation of ET over areas receiving inflows, such as wetlands, floodplains, and irrigated farmland. Gauged runoff and eddy covariance flux tower observations were used to visually fit some components and parameters of the model.

2.2.3. ET Estimates from GRACE Rainfall and Discharge Observations

A number of studies have derived ET estimates from GRACE observations in large basins, e.g., [11,33,34].

Here we computed ET estimates using regional GRACE solutions which are characterized by reduced north-to-south striping (using constrained regularization) and no contamination from other parts of the world [59,60]. Several studies over different continents—South America [61], Australia [62], and Africa [63]—demonstrated that this regional approach offers a reduction of both north-south striping due to the distribution of GRACE satellite tracks and temporal aliasing of correcting models that are present in the global GRACE solutions. GRACE regional solutions were available at a spatial resolution of 2° × 2° from 4 July 2003 to 3 December 2010 at intervals of 10 days [62]. Change in the basin water storage (ΔS) is calculated as the difference between two successive GRACE terrestrial water storage anomalies (TWSA) against the average (ΔTWS):

$$\Delta TWS = \Delta S = TWS_{t2} - TWS_{t1}. \tag{4}$$

Using the water balance equation at basin-scale, ET was obtained as the difference between the total amount of rainfall over the basin (P), the river discharge at the outlet (Q) and the change in ΔS over a specific time period Δt:

$$\text{ET} = P - Q - \Delta S, \tag{5}$$

To be consistent with the above modelled ET datasets, calculations were performed over 30-day time period. Missing data from 20 January 2004 to 29 January 2004 were linearly interpolated. Combining GRACE ΔTWS observations and BoM rainfall datasets, monthly basin-average ET estimates over the MDB were computed using Equation (5) with all variables set. However, due to the fact that Q is unavailable at gridded-cell scale, GRACE ET maps were approximated as ($P - \Delta S$) with Q assumed to be negligible. This operation was also applied to the LEB.

2.2.4. Evaluation of ET Estimates Using the Budyko Diagram Scheme

Water balance at basin-scale is governed by Equation (5). In the Budyko framework, the nature of the annual water balance is determined by the ratio E/P (evaporation efficiency) as a function of E_p/P (drought index or climatic aridity) accounting for the partition of rainfall between evaporation and runoff [64]:

$$\frac{E}{P} = \left\{ \left[1 - \exp\left(-\frac{E_p}{P}\right) \right] \frac{E_p}{P} \tan h\left(\frac{P}{E_p}\right) \right\}^{0.5} \tag{6}$$

In extremely dry cases, if a basin is provided with sufficient evaporative energy but limited precipitation, then E will approximate to P; conversely, E will be mainly determined by E_p in a wet basin. These form two asymptotes to constrain the Budyko curve in a boundary. A Budyko diagram is usually used at long-term average scale. Since GRACE could provide the annual basin water storage change information, we assumed that the Budyko diagram could be valid at inter-annual time step. Annual E and P were presented as Ea and Pa, respectively.

Noticing that the indices of Ea/Pa and E_p/Pa could be sensitive to the different sources of Pa and E_p, we assumed that E_p/Pa would not vary widely as E_p in MDB and LEB is much larger than Pa; Pa is based on measurements from rainfall stations.

3. Results

3.1. Spatial Evaluation

Figure 2 shows the spatial distribution across the two basins of average annual ET for each dataset and average annual rainfall over the 2003–2010 period. The estimated annual ET maps are spatially consistent with corresponding rainfall maps. The highest ET rates were found in south-eastern MDB (>800 mm/year) where the climate is more humid and intense canopy transpiration occurs. The northern parts of MDB and LEB also present a high ET rate of 400–600 mm/year, which is attributed to the strong ET fluxes caused by tropical rainstorms during the summer. Controlled by an arid climate, 100–200 mm/year of rainfall, the central part of the LEB and the western MDB show the lowest ET rates of all four datasets.

However, we identified some differences between ET estimates. Over the extremely dry regions of the central LEB (rainfall: 100–200 mm/year), ET estimates from AWRA and GRACE were more reasonable than the values provided by two energy balance constrained methods (PM-Mu and PT-CMRS) which predicted 200–300 mm/year of water lost via ET processes over the central LEB. Over humid regions, PM-Mu estimated an annual ET flux lower than the other three methods.

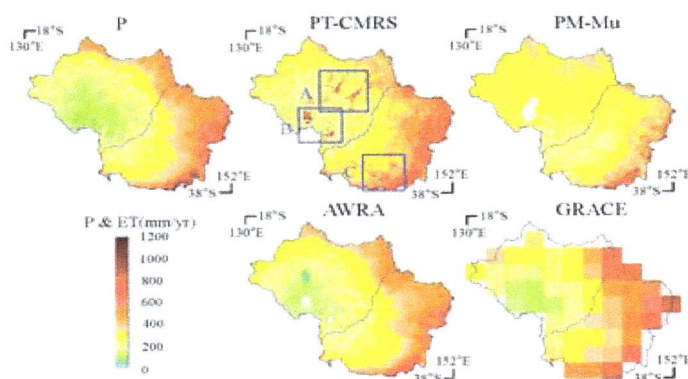

Figure 2. Spatial distribution of mean annual rainfall (P) and mean annual ET derived from PT-CMRS, PM-Mu, AWRA, and GRACE across the Murray-Darling and the Lake Eyre Basins. Boxes A, B, and C show areas that are zoomed in Figure 3.

Figure 3 zooms in three smaller regions (boxes A, B, and C shown in Figure 2) that are prone to inundation (A and B) or irrigation (C). The red patches in contrast to the dry background in PT-CMRS represent the high-ET patterns over irrigated areas as well as flood plains composed of open water bodies and dense forests, demonstrating the dataset's ability to capture relatively small ET features and heterogeneities over those landscapes. This ability is mostly attributed to the integration of MODIS-based EVI and GVMI indices in PT-CMRS [53]. Although it was also driven by MODIS optical products, PM-Mu did not clearly present those small-scale ET features in space. AWRA also considered ET flux from canopy cover and open waters as well as saturated soil. Nevertheless, its coarse gridding system and fractional index parameterization for different land covers make the model homogenize those landscape features.

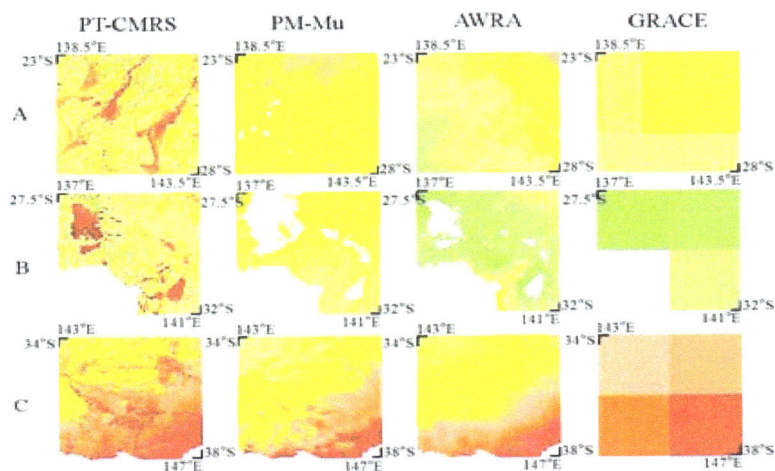

Figure 3. Zoom in mean annual ET values from the four models over three smaller regions (boxes A, B, and C shown in Figure 2) that are prone to inundation or irrigation.

3.2. Temporal Evaluation

3.2.1. Seasonal Variations

Table 2 shows the mean seasonal ET estimates and corresponding rainfall averaged over the period 2003–2010 and the fractions of ET to rainfall. The four methods used to estimate ET in this study were able to capture the seasonal ET patterns in each basin. LEB is much drier (has little rainfall against high potential evaporation requirement) than the MDB. Therefore, it is likely that the ET fluxes in the LEB is lower than in the MDB through the four seasons. However, in a comparison of the magnitude of seasonal ET to rainfall, the four methods behaved inconsistently. Again, seasonal PM-Mu estimates were constantly lower than other ET datasets in MDB, with the lowest ratio of 0.4 found in winter. The ratios of ET/P are frequently above 1.0 in PT-CMRS and PM-Mu during the dry seasons in LEB; the largest ratios (1.5 for PT-CMRS and 1.6 for PM-Mu) are observed in autumn, indicating that estimated ET is greater than the rainfall by 20–30 mm on average. This is also clearly shown by the large gap in Figure 4b. Both methods PM-Mu and PT-CMRS actually did not use the input data directly related to water balance (as rainfall or groundwater storage variations) or to water stress (as thermal infrared data) but mainly relied on the evolution of vegetation as monitor from remote sensing data. Inclusion of thermal data could have picked some low and high moisture conditions. LAI alone may not capture the variation on soil moisture, typically when the bare soil evaporation becomes significant after rainfall or under extreme dry conditions. By comparison, constrained by rainfall in its algorithm, mean seasonal ETs estimated by AWRA and GRACE were closer to the rainfall levels, with their ratios falling into a range of 0.7–1.2.

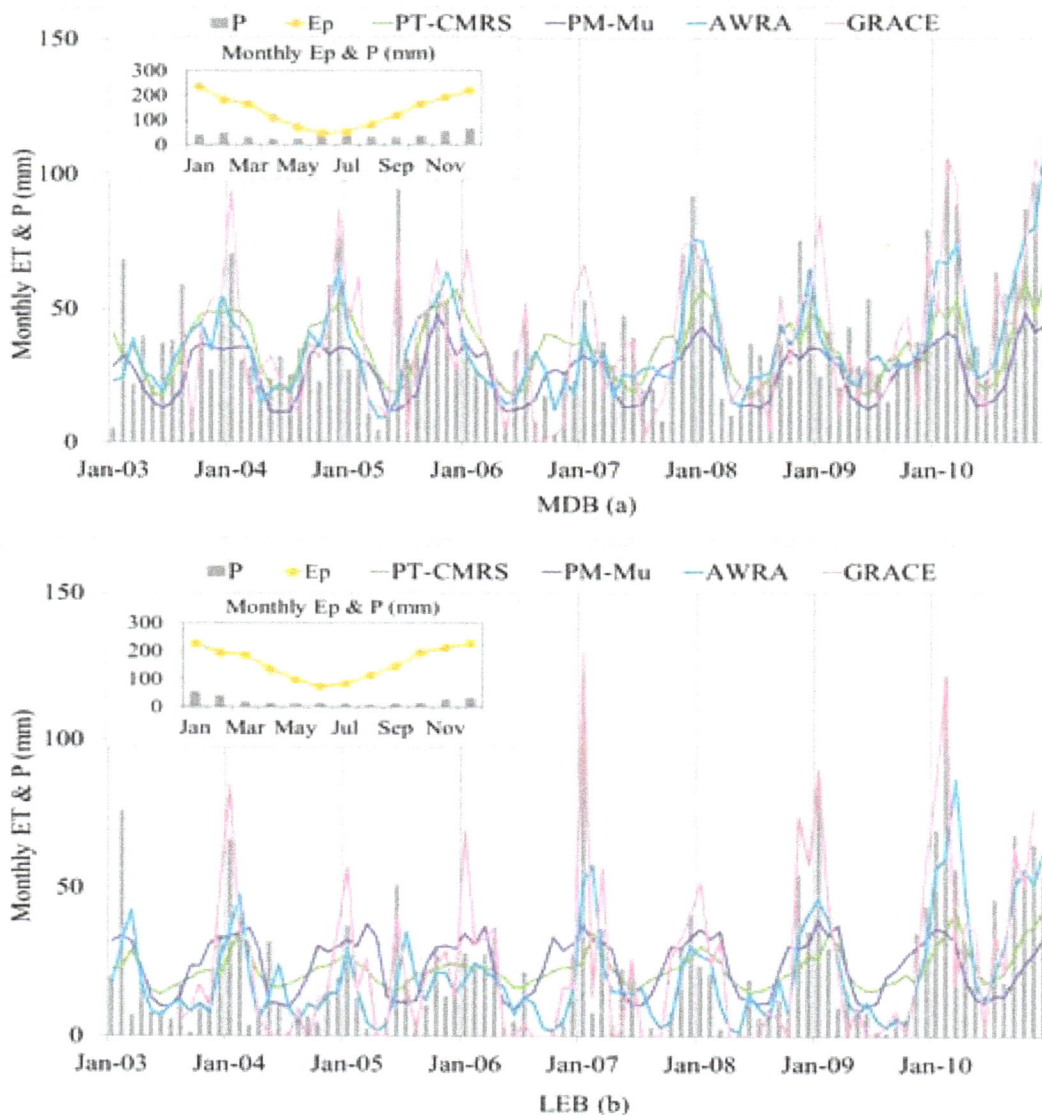

Figure 4. Monthly ET time series derived from the four ET datasets with rainfall and potential evaporation (E_p) over the MDB (**a**) and LEB (**b**) from 2003 to 2010.

Table 2. Mean seasonal ET from the four ET models and their ratio to rainfall (denoted as P) for the period 2003–2010.

Basin	ET Model	Spring (September–November)			Summer (December–February)			Autumn (March–May)			Winter (June–August)		
		ET (mm)	P (mm)	Ratio	ET (mm)	P (mm)	Ratio	ET (mm)	P (mm)	Ratio	ET (mm)	P (mm)	Ratio
LEB	PT-CMRS	68.7	57.9	1.2	83.1	132.4	0.6	68.9	45.2	1.5	51.4	36.9	1.4
	PM-Mu	80.1		1.4	99.9		0.8	72.5		1.6	36.3		1.0
	AWRA	50.1		0.9	102.5		0.8	55.5		1.2	40.3		1.1
	GRACE	63.5		1.1	165.0		1.2	49.9		1.1	24.2		0.7
MDB	PT-CMRS	129.6	125.6	1.0	137.4	156.3	0.9	87.3	81.7	1.1	72.7	112.0	0.7
	PM-Mu	102.2		0.8	105.8		0.7	65.9		0.8	47.2		0.4
	AWRA	126.5		1.0	145.5		0.9	77.3		1.0	85.4		0.8
	GRACE	120.1		1.0	192.8		1.2	77.1		0.9	79.8		0.7

In Figure 5, ET estimates during the peak Millennium Drought and the La Niña spell are compared with mean monthly values averaged over August 2003 to July 2010. The peak of the Millennium Drought occurred in August to December 2006 for the MDB and January to May 2008 for the LEB; monthly rainfall reduced by 20–40 mm on average. The La Niña caused abnormally high rainfall over the MDB (December 2009–May 2010) and the LEB (November 2009–April 2010). Four ET methods responded to the climate extremes dynamically. The smaller the anomalies were, the more similar the modelled ET estimates were to their multi-year average, and vice versa. Small anomalies were found in PT-CMRS and PM-Mu, which implies that ET estimates reproduced by these two energy balance constrained models do not apparently vary during the extremely wet or dry spells. AWRA and GRACE performed sensitively to climatic extremes. In spite of some spikes, the GRACE-based ET product tended to provide a relatively high/low ET with a magnitude consistent with rainfall.

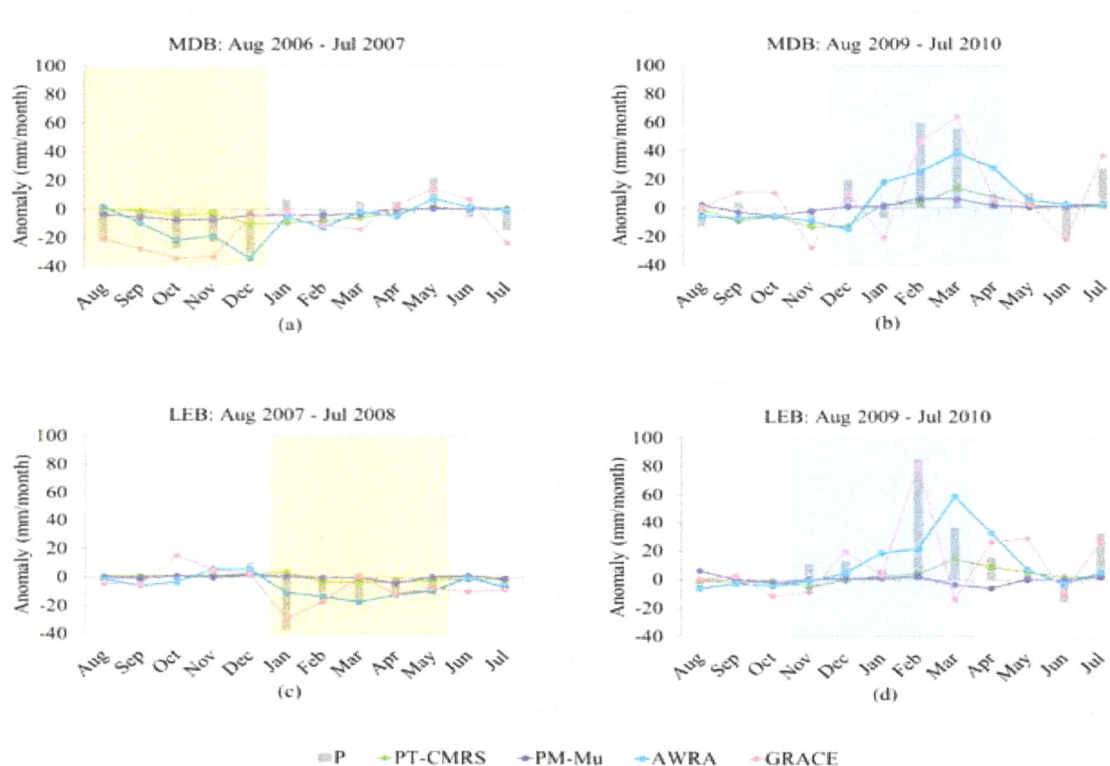

Figure 5. Monthly ET anomalies during extreme climatic events: peak of Millennium drought and La Niña computed as deviation to monthly averages for the whole study period (August 2003 to July 2010). MDB (**a**) and LEB (**c**) during the peak of the Millennium drought; MDB (**b**) and LEB (**d**) during the La Niña period.

3.2.2. Inter-Annual Variations

Inter-annual comparisons were performed at basin-scale to test the response of the four ET datasets. Figure 4 shows monthly ET estimated by the four ET methods over the MDB and LEB from 2003 to 2010 (GRACE ET starts from August 2003), against the rainfall and potential evapotranspiration (E_p) for the same period. At monthly time steps, the correlation coefficients (R) were computed to measure the agreement between each ET pair as well as their sensitivity to rainfall.

Monthly E_p rates (ranging from 50 to 250 mm/month in MDB and from 80 to 250 mm/month in LEB) were always higher than the rainfall levels throughout the study period, indicating water-limited conditions. In that case, ET fluxes in our study areas is predominantly controlled by water availability rather than by energy. Two energy-based models show a quite low linear correlation with rainfall; $R_{PT-CMRS} = 0.35$ and $R_{PM-Mu} = 0.41$ for the MDB, and $R_{PT-CMRS} = 0.51$ and $R_{PM-Mu} = 0.37$ for the LEB.

Remarkable differences can be found during wet ($ET_{PT-CMRS}$ and ET_{PM-Mu} < P) or dry spells ($ET_{PT-CMRS}$ and ET_{PM-Mu} > P; particularly in LEB). By comparison, AWRA has high correlation with rainfall in both basins, 0.7 < R < 0.8. AWRA and GRACE ET estimates were closer to rainfall levels (especially GRACE) due to the fact that they were forced by rainfall data. They also had a good consistency with each other: R_{MDB} = 0.67 and R_{LEB} = 0.69. Although PT-CMRS had a higher linear relationship with AWRA ($0.77 \leq R_{MDB/LEB} \leq 0.79$), the amplitude of the PT-CMRS was significantly lower than AWRA during the extreme seasons. In general, the two energy-balanced products exhibited similar patterns at monthly time steps, and displayed large differences with AWRA and GRACE-based ET in dry and wet periods. PM-Mu was systematically lower than other ET sources in MDB.

Figure 6 presents the capability of each ET dataset in capturing ET variations in different hydrological years. PM-Mu constantly provided the lowest ET estimates over the MDB; a common short of ~100 mm/year was found when compared to PT-CMRS. Compared to AWRA and GRACE estimates, PT-CMRS provided larger values over the MDB in dry hydrological years but lowest values during wet years (e.g., August 2009–July 2010; impacted by La Niña event). ET overestimation by the two energy balance constrained models was commonly observed in the LEB, with annual ET flux twice or larger than the rainfall during the extremely dry hydrological year of August 2004–July 2005 and August 2007–July 2008. Moreover, these two estimates did not exhibit any response to rain variations from year to year and provided almost constant annual values. Constrained by the water balance equation, AWRA and GRACE showed a good sensitivity to annual rainfall variations both in timing and magnitude. GRACE-derived ET estimates at annual scales showed larger inter-annual variations than AWRA over the MDB (with ET larger than rain for some years), while very similar results were obtained over the LEB.

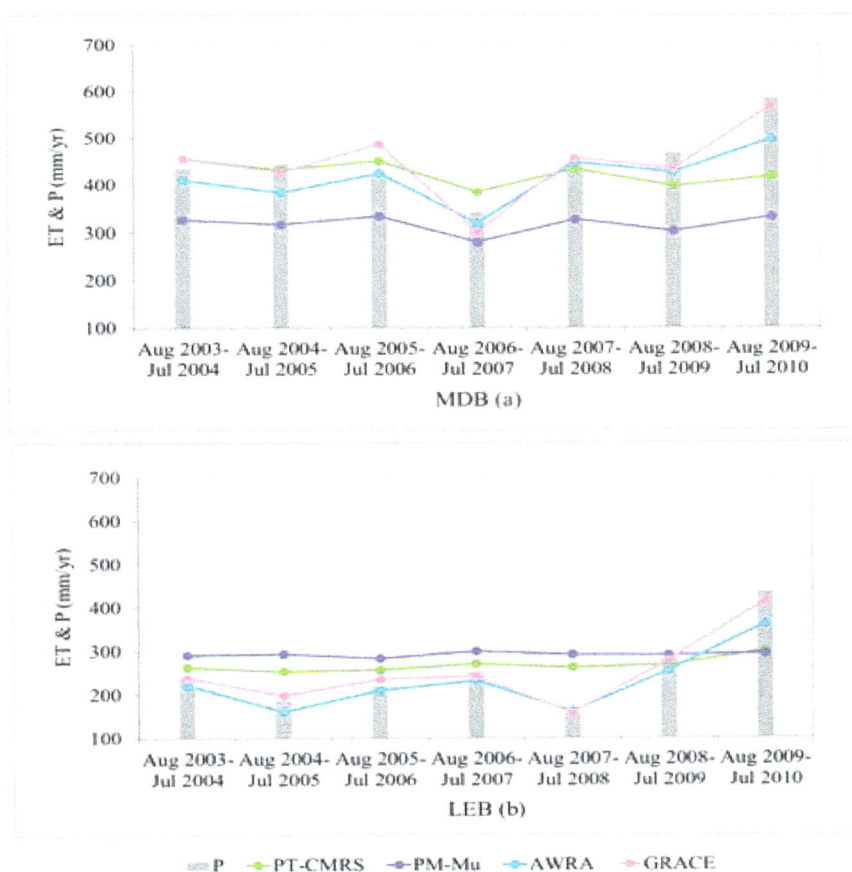

Figure 6. Annual rainfall and ET computed in each hydrological year from August 2003 to July 2010 for the MDB (**a**) and the LEB (**b**).

A Budyko diagram was used to confirm that ET fluxes in our two basins were determined by water availability rather than the energy factor. Estimates were plotted in the Budyko diagrams to examine the variation of the water closure property across hydrological years (Section 2.2.4).

The aridity indices in the Budyko diagrams (Figure 7) confirm that LEB is more arid than MDB (LEB: $4.0 < E_p/Pa < 11.5$; MDB: $2.7 < E_p/Pa < 5.0$). In MDB, Ea/Pa ratios estimated by PM-Mu were sitting in a range 0.5–0.8, demonstrating a systematically low percentage of rainfall transforming into ET. For the same basin, PT-CMRS had some Ea/Pa values lying beyond y = 1 in some dry years (larger E_p/Pa), manifesting a slight outperformance of PT-CMRS ET during drought conditions. Such a phenomenon was more apparent in the LEB. The dryer the year in the LEB ($E_p/Pa > 10$), the more unrealistically the PM-Mu and PT-CMRS estimates; Ea/Pa ratios reached 1.8 (PM-Mu) and 1.7 (PT-CMRS) when LEB received the lowest annual rainfall (158.6 mm/year) from August 2007 to July 2008. By comparison, Ea/Pa predicted by the two water-balance-constrained methods (GRACE and AWRA) were much closer to the asymptotic curve. AWRA and GRACE predicted 85.1% and 97.5% of rainfall lost via evaporation in both basins during the wettest year August 2009–July 2010, respectively; that is to say, 15% (from AWRA) and 2.5% (from GRACE) of rainfall would be converted into runoff and exit the basin or be stored in the aquifer.

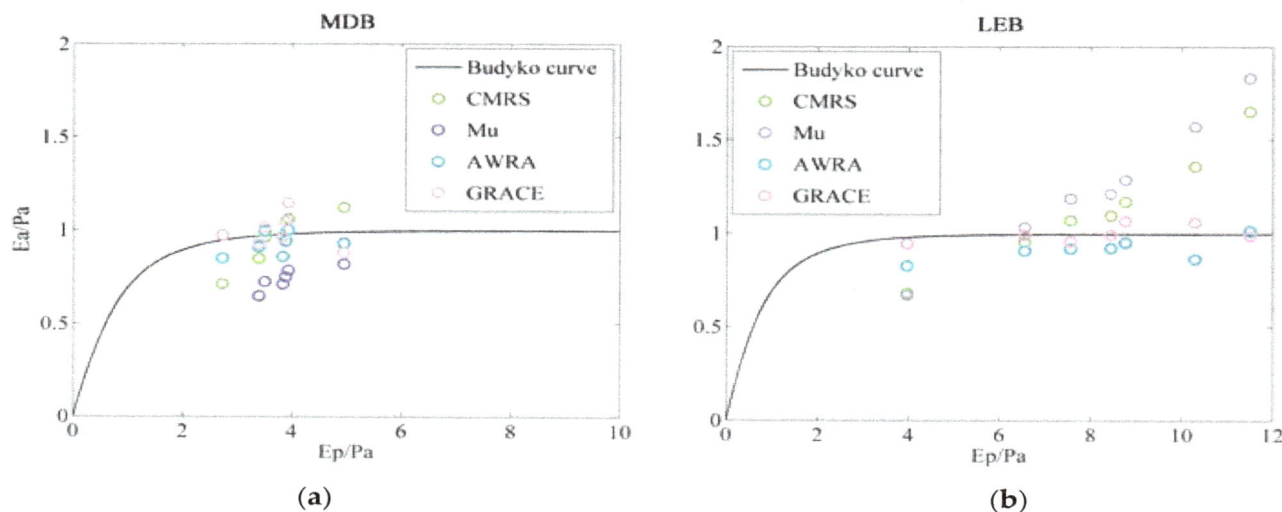

Figure 7. Budyko diagrams for the MDB (**a**) and the LEB (**b**). Ea represents the annual ET estimated by PT-CMRS, PM-MU, AWRA and GRACE. Pa is annual rainfall. E_p is potential evapotranspiration based on the Priestley–Taylor method. All the terms were computed for each hydrological annual year from 2003 to 2010.

4. Discussion

Catchment/basin water balance method provides the simplest way to solve ET estimation by meeting only three water budget components as its input variables. However, the component of ΔS is usually the hardest part to access or measure. In most cases, the common way to address the absence of ΔS is to regard it as negligible over a long-term period [31,32,65]. But when it comes to the annual or inter-annual scales, neglecting of ΔS would lead to an imbalance of water budget equation [66–68]. In fact, short-term ΔS can be a crucial component in water budget, particularly when a basin meets climate extremes. Time series of ΔS anomaly over the MDB and the LEB exhibits large annual (several tenths of mm of equivalent water height) and inter-annual variations (with a decrease from 2003 to 2009 and an increase since) between 2003 and 2010 (Figure 8). Table 3 presents GRACE-derived annual ΔS in different water years over the MDB and LEB: remarkable changes in water storage occurred during August 2005–July 2006 (a water deficit of −20.3 and −70.4 mm/year occurred in LEB and MDB, respectively) and August 2009–July 2010 for LEB only (a water gain by 31.8 mm/year). Such a water loss or gain in ΔS would take up 10–20% of total annual rainfall for both basins. By including GRACE-estimated ΔS, the closure of the basin water balance equation is improved; what's more, the phase and amplitude of the annual and seasonal cycle of ET can be ascertained [69].

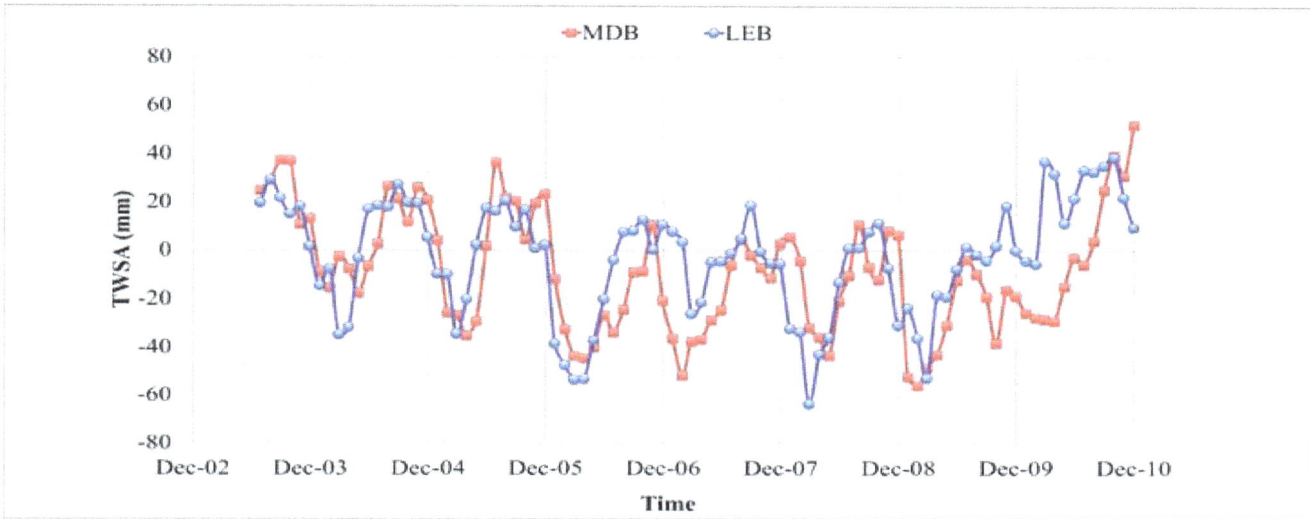

Figure 8. Time series of terrestrial water storage anomaly (TWSA) between 2003 and 2010 from GRACE regional solutions over the MDB (red) and the LEB (blue).

Table 3. Annual changes in terrestrial water storage (ΔS) over the MDB and the LEB between 2003 and 2010. Annual ΔS was computed as the TWS difference between the end month (August) and the start month (July) within a hydrological year (unit: mm/year).

Basin	August 2003 –July 2004	August 2004 –July 2005	August 2005 –July 2006	August 2006 –July 2007	August 2007 –July 2008	August 2008 –July 2009	August 2009 –July 2010	Average
LEB	−0.9	−2.3	−20.3	2.7	2.2	0.6	31.8	2.0
MDB	−21.9	33.6	−70.4	27.7	−4.2	−6.7	−2.2	−4.4

In terms of the uncertainty of GRACE-based ET estimates, they are largely dependent on the quality of GRACE TWSA and thus computed ΔS. A previous study using regional GRACE solutions over Australia found an uncertainty of 19.1 mm when computing the standard deviation of the TWSA over a xeric region [52]. Another major source of uncertainty associated with GRACE ET estimates is attributed to the quality of rainfall data. Usually, arid and semi-arid basins have poor rainfall monitoring networks. In our case, MDB has much denser monitoring networks than LEB does; the latter has a large unpopulated region without rainfall stations [63]. At basin scales, the RMSE calculated between BoM and TRMM 3B43 rainfall datasets are 3.0 and 27.4 mm/month for MDB and LEB, respectively. By blending gauge records with ancillary rainfall data, such as radar and satellite measurements or an ensemble method, may improve the accuracy of rainfall input data [63,64].

Optical satellite imagery allows landscape conditions such as vegetated surface, biome types, and surface standing waters to be distinguishable at unprecedentedly high spatiotemporal resolution (e.g., 0.05° and 8/16 days for MODIS), which facilitates an enhancement in simulating large-scale ET processes. For example, an integration of RS vegetation index (VIs, including LAI, NDVI, and EVI) allows for a separation between plant transpiration and soil evaporation, as well as for adding the canopy rainfall interception component into ET modelling. In our study, PM-Mu employs a complex and process-based scheme to address vegetation transpiration and soil evaporation separately; the new version sets up numerous thresholds to define canopy stomatal conditions over biomes and wet canopy [19]. PT-CMRS combines the EVI and GVMI indexes into a simple and dynamic scaling strategy of the Priestley–Taylor framework to deal with transpiration or evaporation from vegetation, open water bodies, and bare soil; it does consider canopy rainfall interception but merely as a scaled precipitation component. In addition, PT-CMRS ET estimates heavily rely on ground calibration (none of the flux tower sites are distributed in inland dry areas) [53]. Obviously, there is a tradeoff between accuracy and parsimoniousness in ET modelling, as well as a consideration of data availability from ground observations.

In arid and semiarid environments, soil evaporation often makes up the majority of the total ET due to low vegetation coverage. This means that ET models should be able to reflect the relation with soil moisture conditions. Land surface models like AWRA partition precipitation to soil moisture and groundwater systems (other LSMs may not), and thus put a soil moisture constraint onto ET. However, the soil water constraint of energy-based ET products, such as PM-Mu and PT-CMRS, tends to be weak; most models tend to determine ET fluxes by energy factor, in particular based on net radiation estimation rather than by water constraint. Long et al. [16] mentioned that such RS-based ET models only make soil moisture implicitly linked to VIs and atmospheric variables. PM-Mu has soil moisture information indirectly linked to the LAI/NDVI and vapor pressure deficit (VDP); so does PT-CMRS, replaced by EVI and GVMI. These models implicitly assume that vegetation develops in agreement with water availability, which may be true over a long term or "standard" climatic period, but can fail in conditions when drought or rainfall present situations far from average. These models also rely on parameters (e.g., describing stomatal response) that are difficult to assess at large scale, as well as on the accuracy of land use maps. In semi-arid and arid regions, such as MDB and LEB, rain may fall and evaporate directly from bare soil, hence without major influence on vegetation indices and derived ET estimates. That is a potential explanation as to why satellite ET datasets like PT-CMRS and PM-Mu are less sensitive to rainfall/soil moisture and cannot balance well the water budget in these arid and semi-arid environments. Other RS-based ET models rely on the use of thermal infrared data that provide a strong link to water stress and moisture availability and which may improve significantly energy balance estimates. However, today there are no available operational ET products based on thermal infrared, in particular because such models are difficult to implement over large areas such as a continent or even a country. They either rely on a highly accurate characterization of spatial variation of climatic variables (in particular air temperature) or require a very homogeneous climatic zone for implementation.

5. Conclusions

This study, examines the dynamics of continental ET products in water-limited environments. Four ET datasets derived from PT-CMRS, PM-Mu, AWRA, and GRACE were compared in the Murray-Daring and Lake Eyre Basins, against rainfall variations and during climate extremes. Two energy balance constrained ET methods, PT-CMRS and PM-Mu, which are forced by optical satellite retrievals, have poor response to high water variability; they provided unrealistically high ET values (beyond the rainfall) during the drought and low values when rainfall became exceptionally high. This problem can be attributed to the lack of water constraint following water balance. In contrast, AWRA and GRACE are forced by rainfall data, demonstrating their dynamics in addressing the spatial/temporal rainfall variability over water-limited environments. Absolute error associated with each ET product is not directly measurable but a measure of error may be established via a basin-scale comparison with the ensemble mean.

Our results imply that: (1) current ET models based on the energy balance and derived using optical data are not accurate over water-limited areas and that they may need to include further water constraint(s) for ET estimation over arid and semi-arid regions; (2) GRACE observations provide a valuable tool for quantifying basin-scale water storage change at annual or sub-annual scales, as well as an independent source for large-scale ET mapping and validation; (3) a promising way to enhance ET estimation over large water-limited basins/regions may rely on merging high-resolution (but poor in basin-water-balance-constrained) optical RS-based ET products with well water-balance-constrained (but coarse in spatial resolution) GRACE ET estimates [20,70].

Acknowledgments: The authors would like to thank the Murray-Darling Basin Authority and the BoM for providing some of the datasets used in this study. We also thank four anonymous reviewers and the journal's academic editor for their valuable comments that helped us in improving the quality of our manuscript.

Author Contributions: Each co-author contributed to the research published in this paper.

References

1. Vinukollu, R.K.; Meynadier, R.; Sheffield, J.; Wood, E.F. Multi-model, multi-sensor estimates of global evapotranspiration: Climatology, uncertainties and trends. *Hydrol. Process.* **2011**, *25*, 3993–4010. [CrossRef]

2. Trenberth, K.E.; Smith, L.; Qian, T.; Dai, A.; Fasullo, J. Estimates of the Global Water Budget and Its Annual Cycle Using Observational and Model Data. *J. Hydrometeorol.* **2007**, *8*, 758–769. [CrossRef]

3. L'vovich, M.I.; White, G.F. Use and transformation of terrestrial water systems. In *The Earth as Transformed by Human Action*; Turner, B.L., Ed.; Cambridge University Press: New York, NY, USA, 1990; pp. 235–252.

4. Chirouze, J.; Boulet, G.; Jarlan, L.; Fieuzal, R.; Rodriguez, J.C.; Ezzahar, J.; Er-Raki, S.; Bigeard, G.; Merlin, O.; Garatuza-Payan, J.; et al. Inter-comparison of four remote sensing based surface energy balance methods to retrieve surface evapotranspiration and water stress of irrigated fields in semi-arid climate. *Hydrol. Earth Syst. Sci. Discuss.* **2013**, *10*, 895–963. [CrossRef]

5. Australia. National Water Commission. *Australian Water Resources 2005: A Baseline Assessment of Water Resources for the National Water Initiative*; National Water Commision: Canberra, Australia, 2006.

6. Gowda, P.H.; Senay, G.B.; Howell, T.A.; Marek, T.H. Lysimetric Evaluation of Simplified Surface Energy Balance Approach in the Texas High Plains. *Appl. Eng. Agric.* **2009**, *25*, 665–669. [CrossRef]

7. Senay, G.B.; Leake, S.; Nagler, P.L.; Artan, G.; Dickinson, J.; Cordova, J.T.; Glenn, E.P. Estimating basin scale evapotranspiration (ET) by water balance and remote sensing methods. *Hydrol. Process.* **2011**, *25*, 4037–4049. [CrossRef]

8. Huntington, T.G. Evidence for intensification of the global water cycle: Review and synthesis. *J. Hydrol.* **2006**, *319*, 83–95. [CrossRef]

9. Jung, M.; Reichstein, M.; Ciais, P.; Seneviratne, S.I.; Sheffield, J.; Goulden, M.L.; Bonan, G.; Cescatti, A.; Chen, J.; de Jeu, R.; et al. Recent decline in the global land evapotranspiration trend due to limited moisture supply. *Nature* **2010**, *467*, 951–954. [CrossRef] [PubMed]

10. Cleugh, H.A.; Leuning, R.; Mu, Q.; Running, S.W. Regional evaporation estimates from flux tower and MODIS satellite data. *Remote Sens. Environ.* **2007**, *106*, 285–304. [CrossRef]

11. Ferguson, C.R.; Sheffield, J.; Wood, E.F.; Gao, H. Quantifying uncertainty in a remote sensing-based estimate of evapotranspiration over continental USA. *Int. J. Remote Sens.* **2010**, *31*, 3821–3865. [CrossRef]

12. Leuning, R.; Zhang, Y.Q.; Rajaud, A.; Cleugh, H.; Tu, K. A simple surface conductance model to estimate regional evaporation using MODIS leaf area index and the Penman-Monteith equation. *Water Resour. Res.* **2008**, *44*, W10419. [CrossRef]

13. Mu, Q.; Zhao, M.; Heinsch, F.A.; Liu, M.; Tian, H.; Running, S.W. Evaluating water stress controls on primary production in biogeochemical and remote sensing based models. *J. Geophys. Res.* **2007**, *112*, G01012. [CrossRef]

14. Tang, Q.; Peterson, S.; Cuenca, R.H.; Hagimoto, Y.; Lettenmaier, D.P. Satellite-based near-real-time estimation of irrigated crop water consumption. *J. Geophys. Res. Atmos.* **2009**, *114*. [CrossRef]

15. Mueller, B.; Seneviratne, S.I.; Jimenez, C.; Corti, T.; Hirschi, M.; Balsamo, G.; Ciais, P.; Dirmeyer, P.; Fisher, J.B.; Guo, Z.; et al. Evaluation of global observations-based evapotranspiration datasets and IPCC AR4 simulations. *Geophys. Res. Lett.* **2011**, *38*. [CrossRef]

16. Long, D.; Longuevergne, L.; Scanlon, B.R. Uncertainty in evapotranspiration from land surface modeling, remote sensing, and GRACE satellites. *Water Resour. Res.* **2014**, *50*, 1131–1151. [CrossRef]

17. Vinukollu, R.K.; Wood, E.F.; Ferguson, C.R.; Fisher, J.B. Global estimates of evapotranspiration for climate studies using multi-sensor remote sensing data: Evaluation of three process-based approaches. *Remote Sens. Environ.* **2011**, *115*, 801–823. [CrossRef]

18. Allen, R.G.; Tasumi, M.; Morse, A.; Trezza, R.; Wright, J.L.; Bastiaanssen, W.; Kramber, W.; Lorite, I.; Robison, C.W. Satellite-Based Energy Balance for Mapping Evapotranspiration With Internalized Calibration (METRIC)—Applications. *J. Irrig. Drain. Eng.* **2007**, *133*, 395–406. [CrossRef]

19. Mu, Q.; Zhao, M.; Running, S.W. Improvements to a MODIS global terrestrial evapotranspiration algorithm. *Remote Sens. Environ.* **2011**, *115*, 1781–1800. [CrossRef]

20. Tang, R.; Li, Z.-L.; Jia, Y.; Li, C.; Sun, X.; Kustas, W.P.; Anderson, M.C. An intercomparison of three remote sensing-based energy balance models using Large Aperture Scintillometer measurements over a wheat–corn production region. *Remote Sens. Environ.* **2011**, *115*, 3187–3202. [CrossRef]

21. Syed, T.H.; Webster, P.J.; Famiglietti, J.S. Assessing variability of evapotranspiration over the Ganga river basin using water balance computations. *Water Resour. Res.* **2014**, *50*, 2551–2565. [CrossRef]

22. Li, Z.; Tang, R.; Wan, Z.; Bi, Y.; Zhou, C.; Tang, B.; Yan, G.; Zhang, X. A Review of Current Methodologies for Regional Evapotranspiration Estimation from Remotely Sensed Data. *Sensors* **2009**, *9*, 3801–3853. [CrossRef] [PubMed]

23. Tang, R.; Li, Z.L.; Sun, X. Temporal upscaling of instantaneous evapotranspiration: An intercomparison of four methods using eddy covariance measurements and MODIS data. *Remote Sens. Environ.* **2013**, *138*, 102–118. [CrossRef]

24. Badgley, G.; Fisher, J.B.; Jiménez, C.; Tu, K.P.; Vinukollu, R. On Uncertainty in Global Terrestrial Evapotranspiration Estimates from Choice of Input Forcing Datasets. *J. Hydrometeorol.* **2015**, *16*, 1449–1455. [CrossRef]

25. Hu, G.; Jia, L.; Menenti, M. Comparison of MOD16 and LSA-SAF MSG evapotranspiration products over Europe for 2011. *Remote Sens. Environ.* **2015**, *156*, 510–526. [CrossRef]

26. Gokmen, M.; Vekerdy, Z.; Verhoef, A.; Verhoef, W.; Batelaan, O.; Tol, C.V.D. Integration of soil moisture in SEBS for improving evapotranspiration estimation under water stress conditions. *Remote Sens. Environ.* **2012**, *121*, 261–274. [CrossRef]

27. Ruhoff, A.L.; Paz, A.R.; Collischonn, W.; Aragao, L.E.O.C.; Rocha, H.R.; Malhi, Y.S. A MODIS-Based Energy Balance to Estimate Evapotranspiration for Clear-Sky Days in Brazilian Tropical Savannas. *Remote Sens.* **2012**, *4*, 703–725. [CrossRef]

28. Olioso, A.; Chauki, H.; Courault, D.; Wigneron, J.P. Estimation of evapotranspiration and photosynthesis by assimilation of remote sensing data into SVAT models. *Remote Sens. Environ.* **1999**, *68*, 341–356. [CrossRef]

29. Carlson, T.N.; Taconet, O.; Vidal, A.; Gillies, R.R.; Olioso, A.; Humes, K. An overview of the workshop on thermal remote sensing held at La Londe les Maures, France, September 20–24, 1993. *Agric. For. Meteorol.* **1995**, *77*, 141–151. [CrossRef]

30. Boulet, G.; Mougenot, B.; Lhomme, J.P.; Fanise, P.; Lilichabaane, Z.; Olioso, A.; Bahir, M.; Rivalland, V.; Jarlan, L.; Merlin, O. The SPARSE model for the prediction of water stress and evapotranspiration components from thermal infra-red data and its evaluation over irrigated and rainfed wheat. *Hydrol. Earth Syst. Sci.* **2015**, *12*, 7127–7178. [CrossRef]

31. Zhang, L.; Dawes, W.R.; Walker, G.R. Response of mean annual evapotranspiration to vegetation changes at catchment scale. *Water Resour. Res.* **2001**, *37*, 701–708. [CrossRef]

32. Yang, D.; Shao, W.; Yeh, J.F.; Yang, H.; Kanae, S.; Oki, T. Impact of vegetation coverage on regional water balance in the nonhumid regions of China. *Water Resour. Res.* **2009**, *45*, 450–455. [CrossRef]

33. Rodell, M.; Famiglietti, J.S.; Chen, J.; Seneviratne, S.I.; Viterbo, P.; Holl, S.; Wilson, C.R. Basin scale estimates of evapotranspiration using GRACE and other observations. *Geophys. Res. Lett.* **2004**, *31*. [CrossRef]

34. Ramillien, G.; Frappart, F.; Guntner, A.; Ngo-Duc, T.; Cazenave, A.; Laval, K. Time variations of the regional evapotranspiration rate from Gravity Recovery and Climate Experiment (GRACE) satellite gravimetry. *Water Resour. Res.* **2006**, *42*. [CrossRef]

35. Ladson, A. *Hydrology: An Australian Introduction*; Oxford University Press: Melbourne, Australia, 2008.

36. Leblanc, M.; Tweed, S.; Van Dijk, A.; Timbal, B. A review of historic and future hydrological changes in the Murray-Darling Basin. *Glob. Planet. Chang.* **2012**, *80–81*, 226–246. [CrossRef]

37. Peel, M.C.; Finlayson, B.L.; Mcmahon, T.A. Updated world map of the Köppen-Geiger climate classification. *Hydrol. Earth Syst. Sci.* **2007**, *4*, 439–473. [CrossRef]

38. Digital Elevation Model (DEM) of MDB and LEB. Available online: http://www.ga.gov.au/metadata-gateway/metadata/record/72760/ (accessed on 8 March 2017).

39. Landuse Map of MDB and LEB. Available online: http://data.daff.gov.au/anrdl/metadata_files/pb_luausg9abll20160616_11a.xml (accessed on 8 March 2017).

40. McMahon, T.A.; Murphy, R.E.; Peel, M.C.; Costelloe, J.F.; Chiew, F.H.S. Understanding the surface hydrology of the Lake Eyre Basin: Part 1—Rainfall. *J. Arid Environ.* **2008**, *72*, 1853–1868. [CrossRef]

41. McMahon, T.A.; Murphy, R.E.; Peel, M.C.; Costelloe, J.F.; Chiew, F.H.S. Understanding the surface hydrology of the Lake Eyre Basin: Part 2—Streamflow. *J. Arid Environ.* **2008**, *72*, 1869–1886. [CrossRef]

42. Van Dijk, A.I.J.M.; Renzullo, L.J.; Wada, Y.; Tregoning, P. A global water cycle reanalysis (2003–2012) reconciling satellite gravimetry and altimetry observations with a hydrological model ensemble. *Hydrol. Earth Syst. Sci.* **2014**, *18*, 2955–2973. [CrossRef]

43. Long, D.; Pan, Y.; Zhou, J.; Chen, Y.; Hou, X.; Hong, Y.; Scanlon, B.R.; Longuevergne, L. Global analysis of spatiotemporal variability in merged total water storage changes using multiple GRACE products and global hydrological models. *Remote Sens. Environ.* **2017**, *192*, 198–216. [CrossRef]

44. Zhang, Y.Q.; Chiew, F.H.S.; Zhang, L.; Leuning, R.; Cleugh, H.A. Estimating catchment evaporation and runoff using MODIS leaf area index and the Penman-Monteith equation. *Water Resour. Res.* **2008**, *44*, W10420. [CrossRef]

45. Leblanc, M.J.; Tregoning, P.; Ramillien, G.; Tweed, S.O.; Fakes, A. Basin-scale, integrated observations of the early 21st century multiyear drought in southeast Australia. *Water Resour. Res.* **2009**, *45*. [CrossRef]

46. Van Dijk, A.I.J.M.; Beck, H.E.; Crosbie, R.S.; de Jeu, R.A.M.; Liu, Y.Y.; Podger, G.M.; Timbal, B.; Viney, N.R. The Millennium Drought in southeast Australia (2001–2009): Natural and human causes and implications for water resources, ecosystems, economy, and society. *Water Resour. Res.* **2013**, *49*, 1040–1057. [CrossRef]

47. BoM. Available online: http://www.bom.gov.au/ (accessed on 15 May 2014).

48. Jeffrey, S.J.; Carter, J.O.; Moodie, K.B.; Beswick, A.R. Using spatial interpolation to construct a comprehensive archive of Australian climate data. *Environ. Model. Softw.* **2001**, *16*, 309–330. [CrossRef]

49. Australian Water Availability Project (AWAP). Available online: http://www.csiro.au/awap/ (accessed on 16 March 2014).

50. Raupach, M.R.; Briggs, P.R.; Haverd, V.; King, E.A.; Paget, M.J.; Trudinger, C.M. *Australian Water Availability Project (AWAP), CSIRO Marine and Atmospheric Research Component: Final Report for Phase 3*; Technical Report No. 013; Centre for Australian Weather and Climate Research: Canberra, Australia, 2009.

51. WaterConnect. Available online: https://www.waterconnect.sa.gov.au/ (accessed on 2 April 2014).

52. Priestley, C.H.B.; Taylor, R.J. On the Assessment of Surface Heat Flux and Evaporation Using Large-Scale Parameters. *Mon. Weather Rev.* **1972**, *100*, 81–92. [CrossRef]

53. Guerschman, J.P.; Van Dijk, A.I.J.M.; Mattersdorf, G.; Beringer, J.; Hutley, L.B.; Leuning, R.; Pipunic, R.C.; Sherman, B.S. Scaling of potential evapotranspiration with MODIS data reproduces flux observations and catchment water balance observations across Australia. *J. Hydrol.* **2009**, *369*, 107–119. [CrossRef]

54. WIRADA. Available online: http://remote-sensing.nci.org.au/u39/public/html/wirada/index.shtml (accessed on 17 March 2014).

55. Monteith, J.L. Evaporation and environment. *Symp. Soc. Exp. Biol.* **1965**, *19*, 205–234. [PubMed]

56. MOD16. Available online: http://www.ntsg.umt.edu/project/mod16 (accessed on 27 May 2014).

57. Van Dijk, A.I.J.M.; Renzullo, L.J. Water resource monitoring systems and the role of satellite observations. *Hydrol. Earth Syst. Sci.* **2011**, *15*, 39–55. [CrossRef]

58. Van Dijk, A.I.J.M. *Landscape Model (Version 0.5) Technical Description*; AWRA Technical Report 3; Water Information Research and Development Alliance/CSIRO Water for a Healthy Country Flagship: Canberra, Australia, 2010.

59. Ramillien, G.; Biancale, R.; Gratton, S.; Vasseur, X.; Bourgogne, S. GRACE-derived surface water mass anomalies by energy integral approach: Application to continental hydrology. *J. Geod.* **2011**, *85*, 313–328. [CrossRef]

60. Ramillien, G.; Seoane, L.; Frappart, F.; Biancale, R.; Gratton, S.; Vasseur, X.; Bourgogne, S. Constrained Regional Recovery of Continental Water Mass Time-variations from GRACE-based Geopotential Anomalies over South America. *Surv. Geophys.* **2012**, *33*, 887–905. [CrossRef]

61. Frappart, F.; Seoane, L.; Ramillien, G. Validation of GRACE-derived terrestrial water storage from a regional approach over South America. *Remote Sens. Environ.* **2013**, *137*, 69–83. [CrossRef]

62. Seoane, L.; Ramillien, G.; Frappart, F.; Leblanc, M. Regional GRACE-based estimates of water mass variations over Australia: Validation and interpretation. *Hydrol. Earth Syst. Sci. Discuss.* **2013**, *10*, 5355–5395. [CrossRef]

63. Ramillien, G.; Frappart, F.; Seoane, L. Application of the Regional Water Mass Variations from GRACE Satellite Gravimetry to Large-Scale Water Management in Africa. *Remote Sens.* **2014**, *6*. [CrossRef]

64. Budyko, M.I. *Climate and Life*; Academic Press: New York, NY, USA, 1974.

65. Xu, X.; Liu, W.; Scanlon, B.R.; Zhang, L.; Pan, M. Local and global factors controlling water-energy balances within the Budyko framework. *Geophys. Res. Lett.* **2013**, *40*, 2013GL058324. [CrossRef]

66. Potter, N.J.; Lu, Z. Interannual variability of catchment water balance in Australia. *J. Hydrol.* **2009**, *369*, 120–129. [CrossRef]

67. Gulden, L.E.; Rosero, E.; Yang, Z.L.; Rodell, M.; Jackson, C.S.; Niu, G.Y.; Yeh, P.J.F.; Famiglietti, J. Improving land-surface model hydrology: Is an explicit aquifer model better than a deeper soil profile? *Geophys. Res. Lett.* **2007**, *34*, L09402. [CrossRef]

68. Zhang, L.; Potter, N.; Hickel, K.; Zhang, Y.; Shao, Q. Water balance modeling over variable time scales based on the Budyko framework—Model development and testing. *J. Hydrol.* **2008**, *360*, 117–131. [CrossRef]

69. Andam-Akorful, S.A.; Ferreira, V.G.; Awange, J.L.; Forootan, E.; He, X.F. Multi-model and multi-sensor estimations of evapotranspiration over the Volta Basin, West Africa. *Int. J. Climatol.* **2015**, *35*, 3132–3145. [CrossRef]

70. Shen, H. Satellite Gravimetry in Water-Limited Environments: Applications and Spatial Enhancement. Ph.D. Thesis, James Cook University, Townsville, Australia, 28 November 2014.

10

Size Distribution, Surface Coverage, Water, Carbon and Metal Storage of Thermokarst Lakes in the Permafrost Zone of the Western Siberia Lowland

Yury M. Polishchuk [1,2], Alexander N. Bogdanov [1], Vladimir Yu. Polishchuk [3,4], Rinat M. Manasypov [5,6], Liudmila S. Shirokova [6,7], Sergey N. Kirpotin [5] and Oleg S. Pokrovsky [7,*]

[1]   Ugra Research Institute of Information Technologies, Mira str., 151, Khanty-Mansiysk 628011, Russia; yupolishchuk@gmail.com (Y.M.P.); albo06@yandex.ru (A.N.B.)
[2]   Institute of Petroleum Chemistry SB RAS, 4 Akademichesky av., Tomsk 634021, Russia
[3]   Institute of Monitoring of Climate and Ecological Systems SB RAS, 10/3 Akademichesky av., Tomsk 634021, Russia; gloeocapsa@mail.ru
[4]   Tomsk Polytechnic University, Lenina av., 30, Tomsk 634004, Russia
[5]   BIO-GEO-CLIM Laboratory, Tomsk State University, Lenina av., 36, Tomsk 634004, Russia; rmmanassypov@gmail.com (R.M.M.); kirp@mail.tsu.ru (S.N.K.)
[6]   Federal Center for Integrated Arctic research, Institute of Ecological Problem of the North, 23 Nab Severnoi Dviny, Arkhangelsk 163000, Russia; lshirocova@ya.ru
[7]   GET UMR 5563 CNRS University of Toulouse (France), 14 Avenue Edouard Belin, 31400 Toulouse, France
*   Correspondence: oleg.pokrovsky@get.omp.eu

Academic Editors: Frédéric Frappart and Luc Bourrel

Abstract: Despite the importance of thermokarst (thaw) lakes of the subarctic zone in regulating greenhouse gas exchange with the atmosphere and the flux of metal pollutants and micro-nutrients to the ocean, the inventory of lake distribution and stock of solutes for the permafrost-affected zone are not available. We quantified the abundance of thermokarst lakes in the continuous, discontinuous, and sporadic permafrost zones of the western Siberian Lowland (WSL) using Landsat-8 scenes collected over the summers of 2013 and 2014. In a territory of 105 million ha, the total number of lakes >0.5 ha is 727,700, with a total surface area of 5.97 million ha, yielding an average lake coverage of 5.69% of the territory. Small lakes (0.5–1.0 ha) constitute about one third of the total number of lakes in the permafrost-bearing zone of WSL, yet their surface area does not exceed 2.9% of the total area of lakes in WSL. The latitudinal pattern of lake number and surface coverage follows the local topography and dominant landscape zones. The role of thermokarst lakes in dissolved organic carbon (DOC) and most trace element storage in the territory of WSL is non-negligible compared to that of rivers. The annual lake storage across the WSL of DOC, Cd, Pb, Cr, and Al constitutes 16%, 34%, 37%, 57%, and 73%, respectively, of their annual delivery by WSL rivers to the Arctic Ocean from the same territory. However, given that the concentrations of DOC and metals in the smallest lakes (<0.5 ha) are much higher than those in the medium and large lakes, the contribution of small lakes to the overall carbon and metal budget may be comparable to, or greater than, their contribution to the water storage. As such, observations at high spatial resolution (<0.5 ha) are needed to constrain the reservoirs and the mobility of carbon and metals in aquatic systems. To upscale the DOC and metal storage in lakes of the whole subarctic, the remote sensing should be coupled with hydrochemical measurements in aquatic systems of boreal plains.

Keywords: remote sensing; size; surface; volume; thermokarst; carbon; metal

1. Introduction

The quantification of the abundance, size distribution, and water storage of lakes and reservoirs has critical importance for the evaluation of carbon and nutrient storage, and the potential of greenhouse gas (GHG) exchange between the Earth's surface and the atmosphere. For these reasons, several detailed studies have documented the lake number and size distribution on the scale of our planet [1–3]. Thus, the use of medium resolution Landsat-7 images has allowed the creation of a global database of lakes and water reservoirs, including all lakes larger than 0.2 ha or 45 m × 45 m [3,4]. The total number of lakes was estimated as being 117 million, with an overall surface area of 5×10^6 km^2 (500×10^6 ha), which corresponds to 3.7% of the non-glaciated land area. The number of small lakes (0.2–1 ha) is around 90 million, whereas their overall area is equal to 25×10^6 ha, which is only 5% of the overall lake surface coverage. The small lakes exhibited a deviation from the general power dependence between the lake size and the lake number [4,5]; as a result, the extrapolation of power law to smaller lakes may overestimate the lake number [3]. At the same time, the small lakes subjected to full freezing in winter or evaporation in summer, with a short residence time of water, play a crucial role in the integration of the carbon and other elements transported from the watershed [6–8], which is particularly important in high latitude regions, which are the most vulnerable to climate change [9,10].

The Arctic and subarctic permafrost-bearing regions exhibit the maximal changes in the terrestrial freshwater budget, although the hydrological responses to environmental changes strongly differ across the boreal and subarctic regions of the subarctic [11]. In particular, in the tundra, continuous permafrost development strongly influences water fluxes and storage, whereas in boreal plains, slow surface and subsurface water movement produces extensive wetlands [11]. Once the permafrost becomes discontinuous to sporadic in the south, this allows significant groundwater feeding of rivers [12] and, presumably, lakes [13].

In this regard, the boreal and tundra plains are extremely important for a lake inventory study because of the high coverage of the watershed area by these lakes (up to 70% in some western Siberian river watersheds [14,15]), and fast temporal dynamics of thermokarst lake landscapes, reflecting on-going climate change in their watersheds [16–20]. The latter brings about a shorter residence time of lakes, whose size changes, especially at southern latitudes, due to the disappearance of sporadic and isolated permafrost. It is also worth noting the primary role of lakes in controlling greenhouse gas exchange with the atmosphere, both in permafrost-free [21,22] and permafrost-bearing regions [9,23–26].

Over the past decades, the formation of thermokarst lakes and thaw ponds due to permafrost degradation was documented in Alaska, Canada, Europe, and Siberia [27–31]. The majority of available studies had a rather limited geographic coverage (<10,000 km^2, [16,32,33]), or described relatively small regions within larger territories [29]. The high resolution studies, down to 0.1 ha lake size, dealt with even smaller territories (700 km^2 in ref. [34]; 4 km^2 in ref. [35]; 1.4 km^2 in ref. [36]), whereas the large geographic coverages, on the scale of one hundred thousand to million km^2, were limited to large lakes (>5 ha in North American Arctic, ref. [37,38] and 10–40 ha in western Siberia, ref. [39,40]).

Several studies of thermokarst lakes in Alaska, Yukon, Scandinavia, and Siberia, were focused on monitoring the change in the lake area over the past 30–40 years, within relatively small regions [18,29,41–45]. Remote sensing studies of the permafrost zone of western Siberia demonstrated that the number of newly formed small thermokarst lakes (0.5–5 ha) over the past three decades exceeds, by a factor of 20, the number of large lakes which tend to disappear during the same period [46]. Recently, the dynamics of the number and surface area of thermokarst lakes in the discontinuous permafrost zone of western Siberia, over the period of 1973 to 2009, has been studied within the watersheds of the Nadym and Pur rivers [47]. According to these authors, the temporal evolution of large size (>10 ha) lakes, whose number constituted 78%–85% of all lakes, exhibited a variation within 10%. The size distribution of thermokarst lakes followed the power law, both in eastern [30] and western [47,48] Siberia. In particular, Polishchuk et al. [48] presented the results of the number of appearing and disappearing lakes in western Siberia between the 1970s and the present time, and the laws of statistical distribution of very small lakes (<0.5 ha) on several test sites of the western Siberian Lowland (WSL). In contrast, the present study encompasses a much larger territory of western Siberia and provides, for the first time, a full inventory of large to small lakes (>0.5 ha).

The main goal of the present study was to establish the law of the lake number and area distribution for the whole WSL territory and to bring together the hydrology and hydrochemistry using available data on lake depth and carbon and metal concentration in the lake water. Towards this goal, we classified and analyzed medium resolution Landsat-8 scenes, which provided complete coverage of the WSL (105×10^6 ha or 1.05 million km^2). The first specific objective of this study was to increase the resolution of the lake size to 0.5 ha for the whole territory of permafrost-affected WSL, in order to compare the results with the global database [3]. Indeed, in view of the disproportionally high importance of small thermokarst water bodies relative to medium and large lakes in GHG emission and C storage [49–51], a rigorous quantification of the number and area of thermokarst lakes is very timely. The second objective of this work was to assess the water, carbon, and metal storage in thermokarst lakes. Recent progress in the quantification of depth, and area and lake size-averaged concentrations of major and trace elements in western Siberian thermokarst lakes [49,50,52–54], allows a first-order evaluation of the water and element stocks in lakes and a characterization of the role of lakes in element storage, relative to rivers draining the same territory.

2. Materials and Methods

2.1. Study Area

The studied region is located within a tundra and forest-tundra zone of the northern part of the western Siberia lowland (1.05×10^6 km^2). In the northern part of the WSL, the sporadic, discontinuous, and continuous permafrost zones share 31.7%, 29.1%, and 39.2% of the overall territory, respectively (Figure 1). The mean annual temperature (MAT) ranges from -0.5 °C in the permafrost-free region (Tomskaya region) to -9.5 °C in the north (Yamburg), and the annual precipitation ranges from 400 to 460 mm. For the period of the end of July–August in the central part of the studied zone (Novuy Urengoy), the average low daily temperature was 17.4 °C and 10.9 °C in 2013 and 2014, respectively. The average high temperature was 23.4 and 14.9, respectively. A detailed physico-geographical description, hydrology, lithology, and list of the soils can be found in earlier works [55,56] and in our recent limnological and pedological studies [50,53,54,57]. The WSL has rather homogenous landscape conditions (palsa peat bogs, forest-tundra, and polygonal tundra), lithology (Pliocene sands and silts), soil cover (1–1.5 m thick peat, half of soil profile is frozen), and runoff (200–250 mm·year^{-1}), across a large gradient of permafrost coverage [58–60].

The bioclimatic sub-zones of permafrost-affected WSL regions gradually change northward, from the northern taiga zone (38×10^6 ha) to forest tundra (13×10^6 ha), southern tundra (30×10^6 ha), and northern tundra (24×10^6 ha). A detailed GIS survey of the WSL allowed the quantification of the regional distribution of major wetland types and complexes [56]. According to those authors, the sporadic permafrost zone, north of the Ob River (61°–63° N), is dominated by sphagnum bogs with pools and an open stand of trees with abundant forested (treed) shrubs- and moss-dominated mires. The discontinuous permafrost zone (63° to 67° N) of the forested tundra and southern tundra essentially comprises high palsa and flat palsa mires, mixed with palsa-hollow and pool-hollow patterned mires. Flat-palsa and hollow-pool flat-palsa bogs are also abundant in this region. Finally, further to the north, within the continuous permafrost zone of the southern and northern tundra (67°–73° N), the landscape is dominated by patterned (hollow and hollow-pool) flat-palsa bogs, polygonal mires combined with grass and moss-dominated mires [56]. Thermokarst lakes are highly specific water bodies of the permafrost zone of the WSL: they have a shallow depth and are rarely connected to the hydrological network [61]. Most lakes of the northern part of the WSL freeze solid in winter and have frozen sediments at their bottom throughout the year [54]. The region is dominated by the presence of thermokarst lakes having a <100 ha surface area [42,62,63].

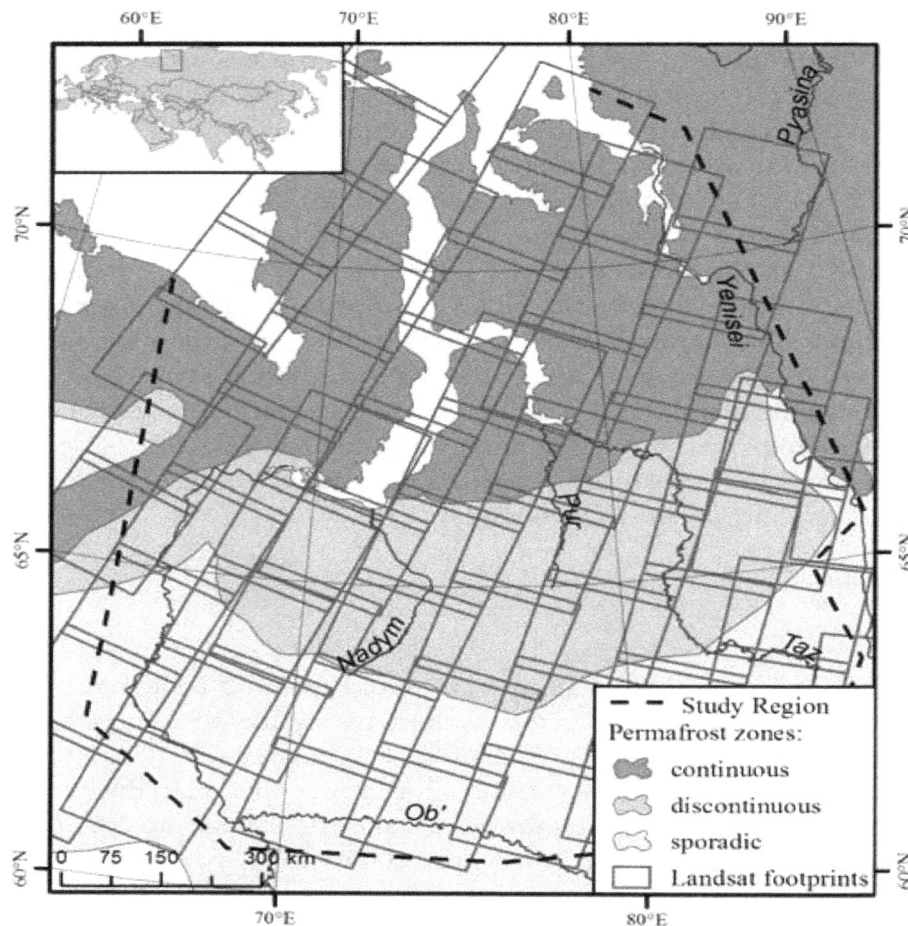

Figure 1. Location of three permafrost zones in the study territory of western Siberia. The position of the permafrost provenances in the western Siberian Lowland (WSL) is based on extensive geocryological work in this region [59].

The lakes are located within peat flat-mound bogs (ridge-hollow complex, palsas, and polygonal tundra); the bottom sediments of the lakes are dominated by peat detritus. An active thermokarst occurs due to the thawing of syngenetic and epigenetic segregation ice, ice wedges, and ice layers in the deep (>2 m) horizons, and is primarily due to the ice thawing of the active layer (<2 m). The thermokarst activity produces depressions, subsidences, and ponds, which are usually separated by flat mound peat bogs up to 2 m in height [64]. The largest thermokarst lakes that are located within the peat bog are km-size, with a depth of 0.5–1.5 m [64,65]. The overwhelming majority of lakes in all three permafrost-bearing zones of the WSL (sporadic, discontinuous, and continuous) have a thermokarst origin, i.e., thawing of frozen peat and clay surface horizons [42,53,61,66]. As a result, most thermokarst lakes of the WSL exhibit quite a shallow depth, ≤1 m, in contrast to the deeper lakes of other Arctic regions, originating from surface disturbance, the melting of ground ice, and ice wedges (e.g., ref. [67]).

2.2. Remote Sensing Analysis

Satellite imagery from a Landsat-8 Operational Land Imager (OLI, with 30 m resolution), available at USGS Global Visualization Viewer [68], were used to map the lake distribution. We used medium-resolution Landsat-8 images collected at the end of July–August 2013–2014. This period corresponds to the minimal coverage of the territory by lakes and minimal seasonal variation of the lake water level. Besides, this is the period of total disappearance of ice coverage throughout all of the studied area. The sampling of the lake water and lake depth measurements was also performed during this period of the year. Note that, from the view point of optical remote sensing, there is no difference between thermokarst and non-thermokarst lakes, because their reflection yield for Landsat-8 OLI is similar. However, according to published expert estimations of several zones of the WSL [42], the majority of lakes on the WSL are of thermokarst origin.

For the total territory of 1.05 million km², we superposed 134 images (Figure 1). We used not only the images with <10% of cloud, but all of the images that were useful to fill the gaps in the coverage. Nevertheless, for the mosaic, only the cloudless parts of the images were used. In the case where several images were available for the same territory, the one which had the lowest cloud coverage in the middle of the summer was used. The mosaic consisted of images taken in 2013 and 2014, which were combined because the single year data could not provide full coverage of the territory.

The river waters were excluded from the analyses via the creation of the river mask. The data of the river location were taken from national river water cadasters and open street maps. The open ocean and marine coastal zones were also excluded from the analyses. The treatment of satellite images was performed using standard tools of ArcGIS 10.3 software [69], which included classification, vectorization, and surface area quantification. The automatic identification of lakes employed the Fmask algorithm developed for Landsat images, which allows resolving the lakes under some cloud coverage [70]. First, for the mosaic of Landsat 8 imagery, the cloud masks were defined for individual images. Then, the cloud masks were removed from the images and replaced by cloud-free fragments taken from other adjacent images. The minimal lake size was chosen as 0.5 ha, based on following. The space resolution of Landsat-8 images is 30 m. Because the pixel size of the image is equal to 30 m × 30 m (900 m²), in the area of 0.5 ha, one can distinguish 5.55 pixels in size. This number of pixels is sufficient for the reliable identification of lakes from the background digital noise of the image. According to the works of our group and Bryksina [18,31,41,46,48], the uncertainty of the lake area measurement using remote sensing is a few percent. Note that the thermokarst lakes and thaw ponds of western Siberia are different from the glacial lakes of other boreal and subarctic regions. The latter are often developed on the moraine till and crystalline rocks, and present highly irregular shapes (skinny or elongated along the glacier direction). In contrast, due to the homogeneity of the soil substrate in western Siberia (1–3 m thick frozen peat), the thermokarst processes in the peat bog of this region always produce the isomeric (round, and much less common, oval) isolated water bodies [14,31,50,61]. According to our field and topographical map-based measurements across a sizeable latitudinal gradient of western Siberia, the share of lakes having an irregular shape is less than 10% [49,53,54].

To assess the latitudinal dependence of thermokarst lake properties, the studied territory was divided into latitudinal zones of 0.5° wide. Such a division of the territory was consistent with the latitudinal gradient of the permafrost and landscape features of the WSL [56,71]. Using ArcGIS 10.3 software [69], we first measured the area and number of lakes on each 0.5°-zone of mosaic of the Landsat 8 imagery. First, we conducted the vectorization and then we determined the area of lakes using the standard procedure of all GIS software. This allowed the quantification of the number, surface area, and volume of lakes, the density of lakes, and the degree of land surface coverage by lakes, as described below. The total lake area (S_{tot}) in each 0.5-degree zone was computed from Equation (1):

$$S_{tot} = \sum_{i}^{n} S_i \tag{1}$$

where S_i is the surface of the i-th lake and n is the number of lakes. The lake fraction is calculated as the ratio of S_{tot}/S_o, where S_o is the area of each 0.5-degree zone. The lake density was computed as the number of lakes (n) per unit of area, n/S_o. In order to estimate the stock of carbon and related elements in lakes, the lake volume (V) was computed following Equation (2):

$$V = \sum_i^n h_i \times S_i \qquad (2)$$

where h_i is the averaged depth of the i-th lake, which primarily depends on the lake size (see Section 2.3).

We used a power dependence between the number and the surface area of lakes in the WSL (correlation coefficient $r = 0.99$, $p < 0.001$), in accordance with global distribution law [2]:

$$k = A \times s^B \qquad (3)$$

where k is the relative number of lakes in the histogram intervals, s is the lake surface area, and A and B are the empirical constants that depend on permafrost and landscape context, respectively.

2.3. Lake Depth and Hydrochemistry

The depth of thermokarst lakes in the WSL depends on their size. The detailed depth mapping of ~50 lakes larger than 2 ha having a solid bottom (frozen sand or silt) was performed via a Humminbird GPS-echosounder from a PVC boat, along several transects of the lake. The minimal depth of probing was 20 cm. The depth of the PVC boat submersion was between 5 and 10 cm, and necessary corrections for the sensor position were made. In lakes shallower than 50 cm, a manual depth measurement with a calibrated stick was performed. The small lakes (0.5–2 ha), having essentially frozen peat at the bottom with a high amount of porous organic detritus, were monitored via the manual probing of the water depth, across the lake transect or in the middle of the lake. Based on available field measurements of the depth and surface areas of ~150 thermokarst lakes from the sporadic to continuous permafrost zone [14,31,49,50,52–54,61,72–77], the depth was approximated to be equal to 0.54 ± 0.25 m (2 s.d.) for lakes smaller than 2 ha, and 0.85 ± 0.25 m (2 s.d.) for lakes ≥ 2 ha. The average uncertainty of these values of h_i is 20% for $n = 150$. One has to note that the two discrete numbers of lake depth used in this study for a survey of 727,700 lakes is a first-order approximation. However, all ~150 thermokarst lakes studied by our group over the last nine years in the WSL, across three permafrost zones, were extremely similar and exhibited an average depth of 1.0 ± 0.5 m. This is a particular feature of the WSL thermokarst lakes located within the polygonal tundra, the peat palsa bog, and the ridge-lake-bog complexes.

The total stock of dissolved organic carbon, and major and trace elements in the lakes of the permafrost zone of the WSL, was evaluated based on the available dependencies between the lake surface area and the dissolved components of the lake water, obtained during extensive sampling campaigns in July 2010, 2012, and July–August 2013–2014 [50,52–54,75,76]. For this, water samples were collected from the lake surface (0.3 to 0.5 m) in pre-cleaned polypropylene containers and filtered on-site or within 4 h after sampling through disposable acetate cellulose filter units (0.45 μm poresize, 33 mm diameter), using sterile plastic syringes and vinyl gloves. An ultraclean sampling procedure was used [78]. The filtered samples were stored at 4–5 °C in the dark, before analysis. The concentrations of dissolved organic and inorganic carbon (DOC and DIC, respectively), cations, and trace elements (TEs), were measured using routine methods for analyzing boreal water samples in the GET laboratory (Toulouse) [79,80]. The DOC and DIC were measured using a Shimadzu Total Organic Carbon Analyzer TOC 6000 with an uncertainty of 5%. The trace metals were measured using inductively coupled plasma-mass spectrometry (ICP-MS, Agilent 7500 CE and Element XR), with indium and rhenium as the internal standards and a precision better than $\pm 5\%$.

For an estimation of the stock of DOC and metals in lakes, the area-averaged values of element concentration in the lake water of discontinuous to continuous permafrost zones [53], complemented with data from discontinuous [49,54,74] and discontinuous to sporadic permafrost regions [50,52], were used. The available databases included a sufficient number of lakes of different sizes, so that the lake size-element concentration dependencies could be obtained for each latitudinal range. Specifically, we sampled ~100 lakes in the discontinuous permafrost zone, ~50 lakes in the continuous permafrost zone, and 30 lakes in the sporadic permafrost zone. For most elements, the concentrations were weakly sensitive ($p > 0.05$) to the lake size for lakes >0.5 ha. The exception was dissolved organic carbon (DOC), whose concentration was approximated by the power dependence [DOC, mg/L] = $190 \times S(m^2)^{-0.26}$ for lakes of 0.5 to 50 ha and is assumed equal to 10 mg/L for lakes >50 ha. Some elements exhibited a clear latitudinal trend in the thermokarst lakes of the WSL, from south to north (e.g., Ca, ref. [53]). This trend has been taken into account via an equation of polynomial dependence between the element concentration (C_i) and the latitude (° N), applied to each latitudinal range in which the stock of water was evaluated:

$$C_i = a + b \times N + c \times N^2 \qquad (4)$$

where a, b, and c are the empirical coefficients, specific for each element.

3. Results

The number and surface area of lakes as a function of latitude are shown in Figure 2A,B, and the lake density and relative coverage of the surface area are illustrated in Figure 3A,B respectively. Presented in these plots are the average values of the territory of each 0.5°-wide latitudinal zones. In the region of 61°–65° N (sporadic to discontinuous permafrost), there are large (a factor of two to three) non-systematic variations of all physical parameters of the lakes. In the zone of 65° to 69° N (discontinuous to continuous permafrost), the lake number and the relative coverage decrease with the latitude, whereas north of 69°–70° N, the lake number and surface area decrease with the latitude increase.

Note that the irregular oscillations of lake density and area coverage, visible in the 0.5°-wide latitudinal zones, are linked to the spatial non-homogeneity of the thermokarst lake distribution. They disappear after a smoothing procedure in wider (2°) latitudinal zones (red dashed line in Figure 3). The results of measured lake parameters within the full WSL permafrost-affected territory (1.05 million km²) are listed in Table 1 and the map of the WSL coverage by lakes is given in Figure 4. There are 0.73 million lakes larger than 0.5 ha, with a total lake surface area of 5.97 million ha. It can be seen from Table 1 that the lake density and lake relative coverage increase by 19.3% and 13.8% from sporadic to continuous permafrost, respectively. The increase in the total lake number and their overall surface area from sporadic to continuous permafrost is equal to 42% and 48%, respectively.

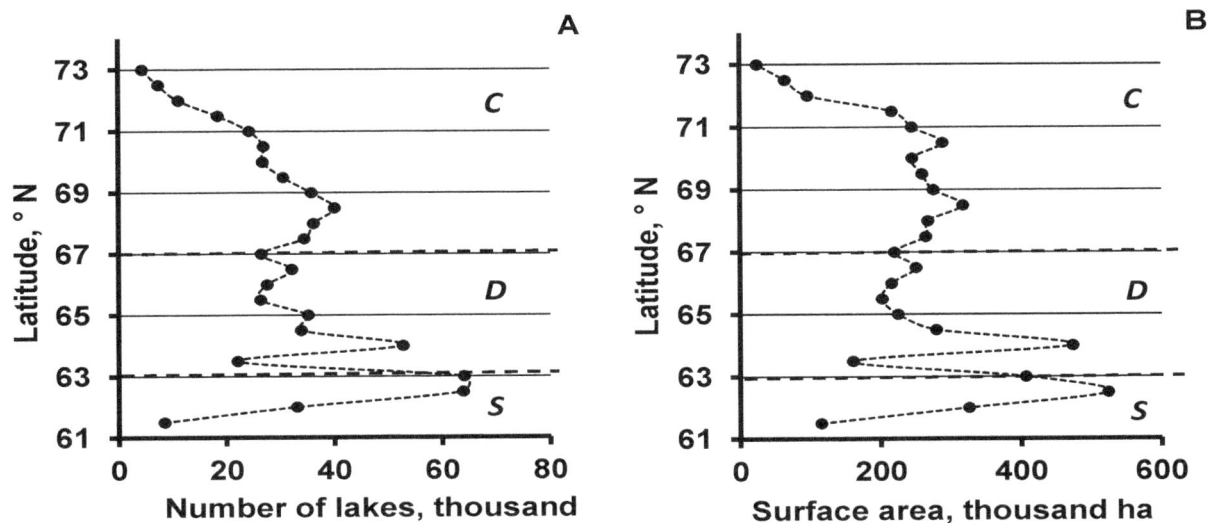

Figure 2. The total number of lakes (**A**) and their surface area (**B**) along the latitude. The dashed lines mark sporadic (*S*), discontinuous (*D*), and continuous (*C*) permafrost zones.

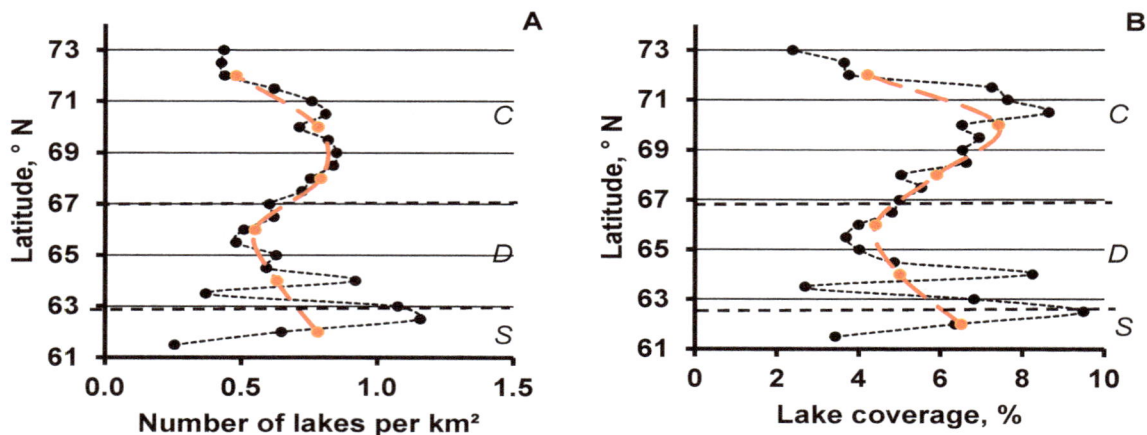

Figure 3. Dependence of the lake density (**A**) and relative coverage of the surface (**B**) on the latitude. The dashed black lines mark sporadic (*S*), discontinuous (*D*), and continuous (*C*) permafrost zones. The dashed red lines represent the smoothing of lake number and coverage using a 2-degree grid.

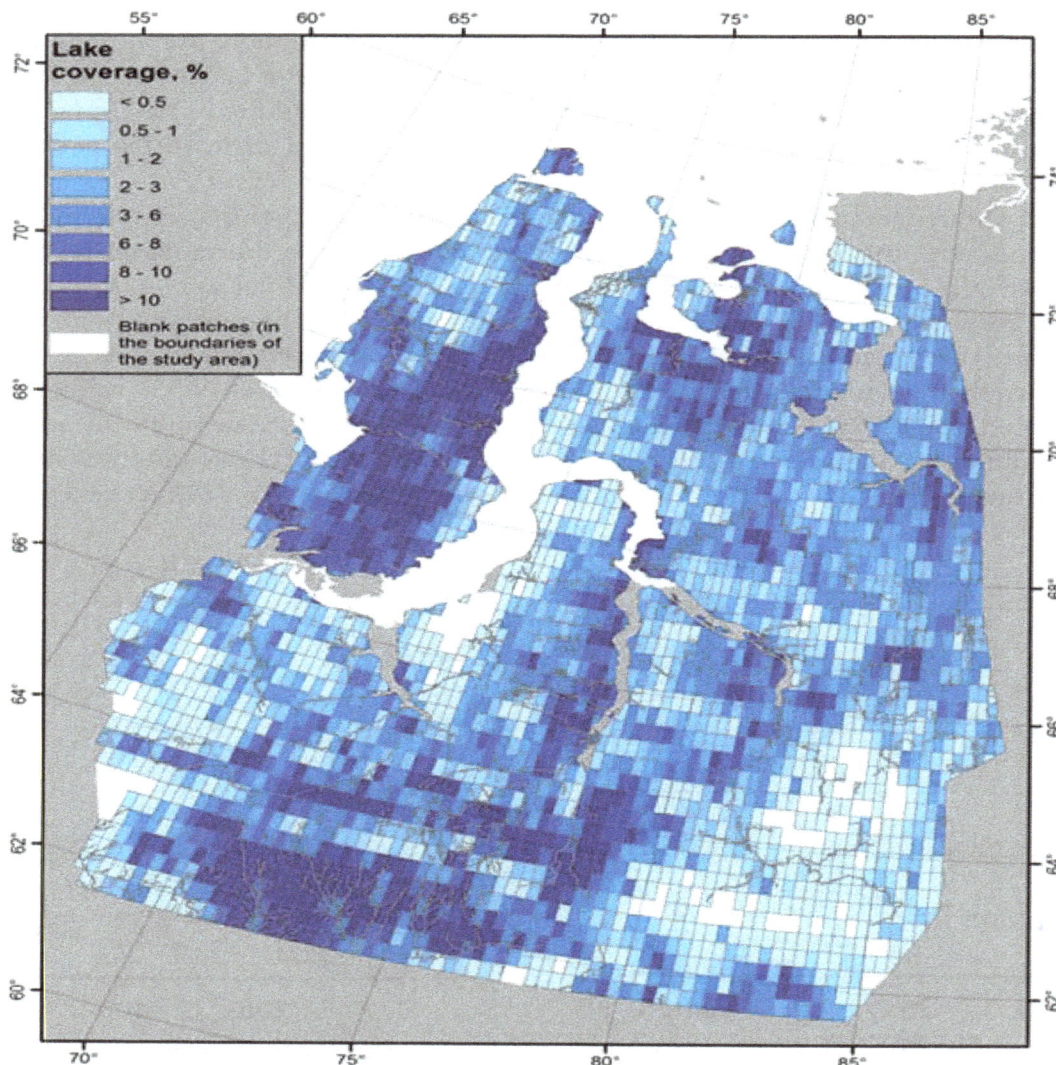

Figure 4. Synthetic map of the WSL coverage by lakes based on Landsat-8 scenes. The cell size is 0.25 degree in latitude and longitude.

Table 1. Thermokarst field parameters in different permafrost zones.

Permafrost	Total Number of Lakes, Thousand	Total Area of Lakes, Million ha	Lake Coverage of the Area, %	Lake Density, Number of Lake per km²
All territory	727.7	5.97	5.69	0.69
Continuous	305.0	2.59	6.28	0.74
Discontinuous	207.7	1.65	5.41	0.68
Sporadic	215.0	1.75	5.26	0.65

The partitioning of the lake number and surface area among different size ranges is listed in Table 2. The main contribution to the overall area and volume (about 85% and 87%, respectively) is provided by medium and large lakes (>5 ha), with the largest share of the total area and volume (15.5% and 16%, respectively) being kept by lakes whose size is between 20 and 50 ha. The number of lakes increases with a decrease in the size, but the overall surface area and water stock decrease for lakes <20 ha. Thus, the small lakes (0.5–1.0 ha) provide only 3% of the overall area, with less than 2% of the total water volume. It is therefore expected that the overall area of numerous lakes smaller than 0.5 ha will be lower than 3%, although high-resolution images are necessary to confirm this trend. Empirical dependencies of lake number as a function of lake size for the three permafrost zones of western Siberia are illustrated in Figure 5. The empirical coefficients of Equation (3) (A and B) for each permafrost zone of the WSL territory are listed in Table 3.

Taking into account the volume of the lake water in each lake size range and across the WSL territory (Table 3), the total amount of each dissolved (<0.45 μm fraction) element in all of the thermokarst lakes (>0.5 ha) of the WSL were estimated (Table 4). The typical uncertainty of these values ranges from ±20% to ±30%, with the exception of some elements (Zn, Cr, Ni, and Ba) exhibiting ±50% of the average value, due to a significant latitudinal trend and lake size dependence of element concentration in the lake water.

Table 2. Lake number, lake area, and volume for different size ranges.

Size Range, ha	Number of Lakes		Their Surface Area		Water Stock	
	Lakes	%	ha	%	km³	%
0.5–1	240,582	33.056	173,768	2.9	0.938	1.9
1–2	171,309	23.540	247,303	4.1	1.335	2.7
2–5	152,240	20.920	481,651	8.1	4.094	8.3
5–10	72,091	9.910	507,488	8.6	4.314	8.7
10–20	43,443	5.970	609,713	10.2	5.183	10.5
20–50	30,081	4.130	926,364	15.5	7.874	15.9
50–100	10,354	1.420	717,571	12.0	6.099	12.3
100–200	4636	0.640	638,175	10.7	5.425	11.0
200–500	2227	0.310	666,245	11.2	5.663	11.5
500–1000	511	0.070	352,499	5.9	2.996	6.1
1000–2000	169	0.020	233,803	3.9	1.987	4.0
2000–5000	57	0.010	162,680	2.7	1.384	2.8
5000–10,000	19	0.003	125,671	2.1	1.068	2.2
10,000–20,000	9	0.001	123,066	2.1	1.046	2.1
Total:	**727,728**	**100.000**	**5,965,997**	**100.0**	**49.40**	**100.0**

Table 3. Parameters of Equation (3) for three permafrost zones of the WSL.

Zone	A	B
continuous	15.96×10^9	−2.224
discontinuous	6.82×10^9	−2.154
sporadic	1.68×10^9	−2.065

Table 4. Dissolved organic and inorganic carbon (DOC and DIC, respectively), and major and trace element stocks in the thermokarst lakes of the Western Siberia Lowland (105 million ha). The major and trace elements are listed in the order of increasing atomic number (periodic table).

Element	C, µg/L	Stock, Ton	Element	C, µg/L	Stock, Ton
DOC	20,000 ± 10,000	500,000 ± 150,000	Zn	10 ± 5	500 ± 250
DIC	430 ± 100	22,200 ± 5000	As	0.63 ± 13	31 ± 6
B	3 ± 1	150 ± 50	Rb	0.3 ± 0.1	15 ± 5
Mg	190 ± 40	9400 ± 2000	Sr	6 ± 2	300 ± 100
Al	120 ± 20	6000 ± 1000	Zr	0.10 ± 0.03	4.9 ± 0.5
Si	300 ± 100	15,000 ± 5000	Mo	0.05 ± 0.02	2.5 ± 0.5
K	235 ± 60	12,000 ± 3000	Ba	3.0 ± 1.5	150 ± 75
Ca	700 ± 500	30,000 ± 20,000	Cd	0.02 ± 0.005	0.99 ± 0.25
V	0.6 ± 0.2	30 ± 10	La	0.20 ± 0.06	9.9 ± 3.0
Cr	1.0 ± 0.5	50 ± 25	Ce	0.10 ± 0.03	4.9 ± 0.5
Mn	20 ± 3	900 ± 150	Nd	0.10 ± 0.03	4.9 ± 0.5
Fe	200 ± 50	10,000 ± 4000	Pb	0.26 ± 0.05	12.8 ± 2.5
Co	0.10 ± 0.025	4.9 ± 1.2	Th	0.015 ± 0.005	0.74 ± 0.24
Ni	0.4 ± 0.2	20 ± 10			
Cu	0.55 ± 0.15	27 ± 7			

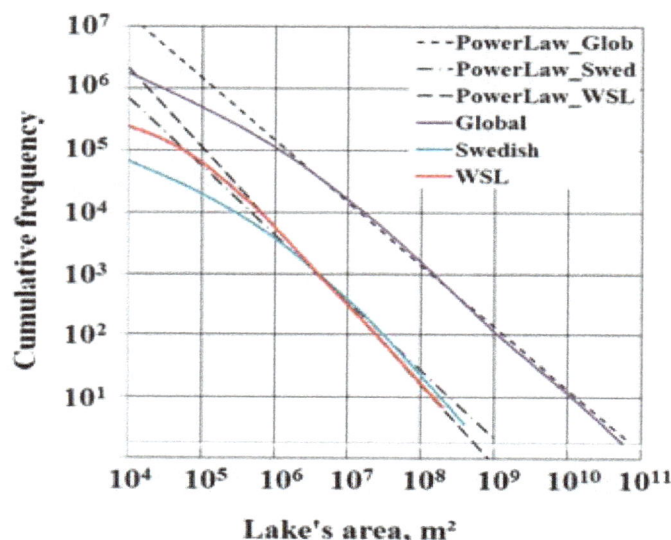

Figure 5. Relationship between the cumulative frequency (the number of lakes versus lake area) of lakes and the lake surface area for the whole territory of WSL (this study, red line), in comparison with lake distribution in the world (Global, dark blue line, [81]) and in Sweden (light blue line, [81]).

4. Discussion

4.1. Thermokarst Lake Area and Land Surface Coverage

Overall, the inventory of medium and large thermokarst lakes of the WSL demonstrates an agreement of size distribution and surface coverage of the lakes in this region, compared to the rest of the world. The latitudinal pattern of the number of lakes and their surface area is tightly linked to the topography and landscape conditions of the northern part of the WSL, located within the sporadic to continuous permafrost zone. Between 61° and 64° N, the northern taiga is represented by sphagnum-dominated bogs, with pools and an open stand of trees [56]. The maximal latitudinal variability of lake coverage is observed within the watershed divide Sibirskie Uvaly (around 63° N), where the number and proportion of lakes are strongly controlled by minor variations of local topography, such as the alternation of ridge-mire-lake complexes and taiga zones. Further to the north, the lake coverage remains fairly constant between 64.5° and 71° N, corresponding to the development of the peat palsa plateau with palsa-hollow patterned mires. The landscape here is highly homogeneous with a dominance of watershed divides of small and medium rivers, offering large flat surfaces suitable for the development of a thermokarst. Finally, a strong decrease in the lake number and area northward of 71° N may be linked to the dominance of polygonal-roller and polygonal-fissure mires, combined with grass and moss-dominated mires [56]. Presumably, the thermokarst processes in the polygonal mires of continuous permafrost zone are less developed than those in the peat palsas plateau, dominating the discontinuous permafrost zone.

Unlike the database that comprises all lakes of the Earth's surface [3], the present study addresses the distribution of thermokarst lakes (>0.5 ha) of the full WSL permafrost-affected territory. A consideration of very small thaw ponds (0.005 to 0.02 ha) in thermokarst-affected regions of the WSL increases the relative surface coverage by lakes to 10%–40%, with an average value of 20%, as shown using Canopus-V data on 18 test sites from 400 to 4000 ha each [82]. However, the decreasing of the minimum lake size to less than 0.1 ha over the whole area of the WSL goes beyond the goals of the present study. It is important to note that the distribution of these very small thaw ponds may deviate from the power dependence (Equation (3)), as reported in global databases [3,5]. The similarity of the B value (Equation (3)) among all three permafrost zones suggests a relatively weak variation of thermokarst lake size distribution patterns across the permafrost gradient in the WSL.

Noteworthy is the dramatic difference between the lake coverage of the WSL permafrost-affected territory estimated in this study (5% to 6% of the area) and the proportion of wet zones in the WSL river watersheds, assessed by ENVISAT radar altimetry (40% to 60% of the watershed area during open water period of the year [15]). These authors defined wet zones as various objects that are either constant in time (rivers, lakes, wetlands) or have seasonal variability (floodplains). It follows that the actual coverage of the WSL river watersheds by shallow (<0.1–0.5 m depth) surface water may be significantly higher than the "net" lake area. However, the estimations of the effect of flooding on land coverage by water and the lake abundance (i.e., see ref. [83] for review), or the water level fluctuation in lakes induced by evapotranspiration variation [84], were beyond the scope of this study.

4.2. Stock of DOC and Metals in Thermokarst Lakes of the WSL

The specificity of thermokarst lakes of the WSL is their low depth ($\leq$1 m), which allowed, for the first time, a reasonable inventory of the water volume and thus an evaluation of the stocks of dissolved components (Table 4). The typical range of water residence time in the thermokarst lakes of western Siberia is between 0.5 and 1.5 years [54]. The overwhelming majority of these lakes are not connected to the rivers, being isolated water bodies, protected by an impermeable permafrost layer both from the bottom (frozen sand and silt), and from the border (frozen peat). Probably for these reasons, the on-going dynamics of thermokarst lake abundances and surface areas are not yet reflected in the hydrological balance of large rivers in Western Siberia [40,47]. The stock of dissolved components in lakes on the permafrost-affected WSL territory can be compared to that delivered by all rivers of the WSL from the same territory to the Arctic Ocean. For this, watershed size-averaged, year-round fluxes of carbon, and major and trace elements assessed in previous works [79,80], can be used. A diagram of element stock in thermokarst lakes, relative to that in rivers of the WSL, is presented in Figure 6. Three groups of elements can be distinguished: (i) major and trace elements, whose storage in lakes is less than 10% of that in rivers (DIC, Mg, Ca, Sr, Ba, K, Si, B, Fe, Co, Ni, Mn, and Ce); (ii) elements presenting non-negligible storage in lakes (20% $\pm$ 10% of that in rivers): DOC, Rb, Zn, Cu, V, Mo, Zr, As, Nd, and Th; and finally (iii) elements having significant, 30 to 70% storage in lakes, relative to rivers: Cd, Pb, La, Cr, and Al. It can be seen from this classification that major cations, DIC, B, Si, and metals subjected to significant redox transformations (Mn, Fe, and Ce), are essentially present in the rivers because they are delivered by groundwater feeding or shallow subsurface flux [79,80]. The groundwater and subsoil feeding are very low in lakes which have frozen peat on the border and frozen sediments at the bottom, throughout the year [54]. The elements exhibiting strong affinity to organic matter (Al, Cr, and rare earth elements (REEs)) and metal toxicants (Cd and Pb) enriched in moss, exhibit sizeable storage in lakes of the WSL territory. This is consistent with the surface and suprapermafrost flow that deliver the solutes to the lakes.

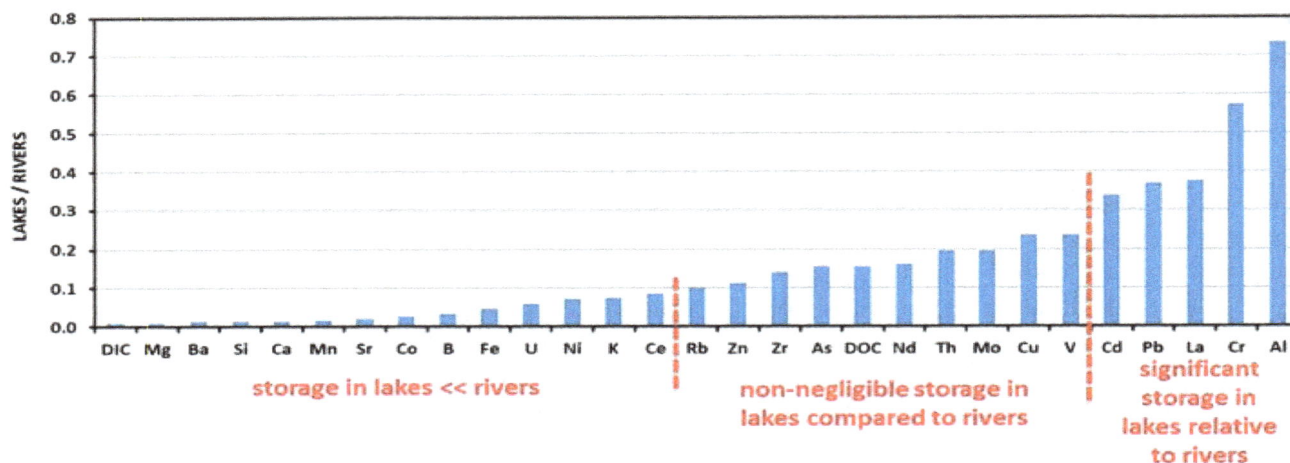

Figure 6. The mass ratio of element stocks in all thermokarst lakes covering the territory of 105 million ha to annual element delivery to the Arctic ocean by WSL rivers from the same territory.

The thermokarst lakes are typically 1–2 pH units more acidic than the surrounding rivers [50,80], and this may enhance the solubility and mobility of many low-soluble trivalent hydrolysates (Al, Fe^{3+}, rare earth elements), from the lake sediment to the water column. Another important source of solutes to the lakes is surface flow from surrounding peat bogs covered by mosses and lichens, consistent with the essentially allochthonous source of DOM in thermokarst lakes [10,50]. A high concentration of DOC, combined with an enrichment in Pb and Cd of the surrounding moss cover [57], may be responsible for the sizeable proportion of metal toxicants in lakes compared to rivers. Given that the DOC, Fe, and Al concentrations in the smallest (<100–1000 m^2) thermokarst depressions and permafrost subsidences are 3–10 times higher than that in lakes >0.5 ha inventoried in this work [50], and that most trace elements including metal micronutrients are present in the form of organic and organo-mineral colloids [61], the role of small thermokarst water bodies in element stock in surface waters and potential delivery to the hydrological network, may be particularly important and are currently strongly underestimated. For this, coupled land/water observations at a very high spatial resolution [85] may help to constrain the reservoirs and the mobility of carbon, metals, and greenhouse gases in adjacent aquatic and terrestrial biomes.

The role of small (<1000 m^2) thermokarst lakes is especially important for the regulation of DOC and greenhouse gas exchanges with surrounding reservoirs (hydrological network and atmosphere). According to available observations of the discontinuous to sporadic permafrost zone of western Siberia, the smallest thaw ponds (10 m × 10 m to 33 m × 33 m) and depressions (1–100 m^2) exhibit an order of magnitude higher concentration of CO_2, up to two orders of magnitude higher methane concentration, and a factor of three to ten higher concentrations of DOC and related metals [49,50,52]. As such, even with their contribution to the total lake surface area of 1%, these small bodies of water may display carbon storage and GHG flux to the atmosphere, which will be comparable to those of large and medium lakes. This hypothesis is verified in the non-permafrost European wetlands: the peatland open water pools are known to act as a net source of CO_2 to the atmosphere [86,87]. The importance of small (100–200 m^2) water bodies for CO_2 emission has recently been reported in the polygonal tundra of the Lena Delta observatory [51].

The upscaling of our estimation of the DOC and metal storage in lakes, relative to the river input, requires detailed knowledge of other lakes of the subarctic, since the riverine fluxes DOC, DIC, and most major elements of the circumpolar region, are fairly well defined [88–90]. The extrapolation of results from well-studied lakes of the Mackenzie Delta region of Canada [9,91–94], the Yakutia alasses and yedoma lakes [20,95,96], to much larger territories of boreal plains such as the WSL peatlands, and North-Siberia and Yana-Indigirka lowlands, remains unwarranted. The lakes of these lowlands may stand apart from other studied lakes of the subarctic, in view of their high peat context, low pH, shallow depth, and very low salt content. At present, a large-scale comparison of carbon and metal storage in thermokarst lakes and riverine fluxes of elements in the subarctic can only be provided for the western Siberia lowland.

5. Conclusions

A remote sensing analysis of thermokarst lakes (>0.5 ha) in the sporadic, discontinuous, and continuous permafrost zone of the western Siberia lowland demonstrated that the number of lakes smaller than 1 ha exceeds 33% of the total lake number, whereas their total surface area is only 2.9% of the total surface of WSL lakes. Within the full range of studied lake sizes and areas, a power dependence between the number of lakes and their surface area, consistent with the world-wide trend, is observed. The dependence of the lake number and surface coverage on the latitude exhibits: (i) a highly variable pattern (strong oscillations) between 61° and 63° N, within the watershed divide Sibirskie Uvaly, due to the variable topography of ridge-lake-bog complexes within the sporadic permafrost zone; (ii) stable values of lake fraction between 64° and 71° N of the peat palsa plateau and the discontinuous permafrost context; and finally (iii) abruptly decreasing the lake fraction northward, north of 71° N, within the continuous permafrost zone of the polygonal tundra. The obtained laws of lake number and surface area distribution allow the calculation of the total surface area and volume of water. This yielded the dissolved metal and carbon stocks in surface aquatic systems of the permafrost-affected zone of the WSL.

The stock of C and most metals in thermokarst lakes of the WSL does not exceed 10%–20% of the riverine flux of the territory. However, the role of lakes in the storage of Al, Cr, Cd, and Pb is comparable to, or even higher than, the transport of these elements by rivers. A low pH and high DOC in WSL thermokarst lakes compared to other regions of the subarctic may be responsible for such an important role of the WSL lakes in toxic metal storage. A high-resolution (0.01–0.1 ha) inventory of small thermokarst lakes, most susceptible to permafrost thaw in key representative zones of the WSL, will aid in accounting for short-term changes in water, carbon, and metal stocks, under climate warming scenarios. The extrapolation of obtained results to the whole circumpolar region is hampered by the lack of information on other thermokarst lakes from large (million km^2-scale) territories, notably the boreal plains.

Acknowledgments: Support from the RSF (RNF) grant No. 15-17-10009 "Evolution of thermokarst lake ecosystems in the context of climate change" (50%, hydrochemistry), the BIO-GEO-CLIM grant No. 14.B25.31.0001 of Russian Ministry of Science and Education and Tomsk State University, FCP "Kolmogorov" No. 14.587.21.0036, and RFBR grant No. 15-45-00075 and No. 16-35-60085 mol_a_dk, are acknowledged.

Author Contributions: Yury M. Polishchuk, Alexander N. Bogdanov and Vladimir Yu. Polishchuk conceived and designed the measurements of lake area; Alexander N. Bogdanov and Vladimir Yu. Polishchuk performed the lake area assessment; Yury M. Polishchuk, Alexander N. Bogdanov, Vladimir Yu. Polishchuk and Oleg S. Pokrovsky analyzed the data; Rinat M. Manasypov and Liudmila S. Shirokova contributed the data on lakes and river chemistry; Sergey N. Kirpotin provided the landscape-based assessment; Yury M. Polishchuk and Oleg S. Pokrovsky wrote the paper.

References

1. Lehner, B.; Doll, P. Development and validation of a global database of lakes, reservoirs and wetlands. *J. Hydrol.* **2004**, *296*, 1–22. [CrossRef]
2. Downing, J.A.; Prairie, Y.T. The global abundance and size distribution of lakes, ponds, and impoundments. *Limnol. Oceanogr.* **2006**, *51*, 2388–2397. [CrossRef]

3.   Verpoorter, C.; Kutser, T.; Seekel, D.A.; Tranvik, L.J. A global inventory of lakes based on high resolution satellite imagery. *Geophys. Res. Lett.* **2014**, *41*, 1–7. [CrossRef]

4.   Verpoorter, C.; Kutser, T.; Tranvik, L.J. Automated mapping of water bodies using Landsat multispectral data. *Limnol. Oceanogr. Meth.* **2012**, *10*, 1037–1050. [CrossRef]

5.   Seekell, D.A.; Carpenter, S.R.; Pace, M.L. Conditional heteroscedasticity as a leading indicator of ecological regime shifts. *Am. Nat.* **2011**, *178*, 442–451. [CrossRef] [PubMed]

6.   Schindler, D.W.; Bayley, S.E.; Parker, B.R.; Beaty, K.G.; Cruikshank, D.R.; Schindler, E.U.; Stainton, M.P. The effects of climatic warming on the properties of boreal lakes and streams at the Experimental Lakes Area, northwestern Ontario. *Limnol. Oceanogr.* **1996**, *41*, 1004–1017. [CrossRef]

7.   Jankowski, T.; Livingstone, D.M.; Buhler, H.; Forster, R.; Niederhauser, P. Consequences of the 2003 European heat wave for lake temperature profiles, thermal stability, and hypolimnetic oxygen depletion: Implications for a warmer world. *Limnol. Oceanogr.* **2006**, *51*, 815–819. [CrossRef]

8.   Graham, M.D.; Vinebrooke, R.D. Extreme weather events alter planktonic communities in boreal lakes. *Limnol. Oceanogr.* **2009**, *54*, 2481–2492. [CrossRef]

9.   Tank, S.E.; Lesack, L.F.W.; Hesslein, R.H. Northern delta lakes as summertime CO_2 absorbers within the Arctic landscape. *Ecosystems* **2009**, *12*, 144–157. [CrossRef]

10. Tank, S.E.; Lesack, L.F.W.; Gareis, J.A.L.; Osburn, C.L.; Hesslein, R.H. Multiple tracers demonstrate distinct sources of dissolved organic matter to lakes of the Mackenzie Delta, western Canadian Arctic. *Limnol. Oceanogr.* **2011**, *56*, 1297–1309. [CrossRef]

11. Bring, A.; Fedorova, I.; Dibike, Y.; Hinzman, L.; Mård, J.; Mernild, S.H.; Prowse, T.; Semenova, O.; Stuefer, S.L.; Woo, M.-K. Arctic terrestrial hydrology: A synthesis of processes, regional effects, and research challenges. *J. Geophys. Res. Biogeosci.* **2016**, *121*, 621–649. [CrossRef]

12. Frey, K.E.; Siegel, D.I.; Smith, L.C. Geochemistry of west Siberian streams and their potential response to permafrost degradation. *Water Resour. Res.* **2007**, *43*, W03406. [CrossRef]

13. Roach, J.; Griffith, B.; Verbyla, D.; Jones, J. Mechanisms influencing changes in lake area in Alaskan boreal forest. *Glob. Chang. Biol.* **2011**, *17*, 2567–2583. [CrossRef]

14. Kirpotin, S.; Polishchuk, Y.; Bryksina, N.; Sugaipova, A.; Kouraev, A.; Zakharova, E.; Pokrovsky, O.S.; Shirokova, L.; Kolmakova, M.; Manassypov, R.; et al. West Siberian palsa peatlands: Distribution, typology, cyclic development, present day climate-driven changes, seasonal hydrology and impact on CO_2 cycle. *Int. J. Environ. Stud.* **2011**, *68*, 603–623. [CrossRef]

15. Zakharova, E.A.; Kouraev, A.V.; Rémy, F.; Zemtsov, V.A.; Kirpotin, S.N. Seasonal variability of the Western Siberia wetlands from satellite radar altimetry. *J. Hydrol.* **2014**, *512*, 366–378. [CrossRef]

16. Marsh, P.; Russell, M.; Pohl, S.; Haywood, H.; Onclin, C. Changes in thaw lake drainage in the Western Canadian Arctic from 1950 to 2000. *Hydrol. Process.* **2009**, *23*, 145–158. [CrossRef]

17. MacDonald, L.A.; Turner, K.W.; Balasubramaniam, A.M.; Wolfe, B.B.; Hall, R.I.; Sweetman, J.N. Tracking hydrological responses of a thermokarst lake in the Old Crow Flats (Yukon Territory, Canada) to recent climate variability using aerial photographs and paleolimnological methods. *Hydrol. Process.* **2012**, *26*, 117–129. [CrossRef]

18. Polishchuk, Y.; Kirpotin, S.; Bryksina, N. Remote study of thermokarst lakes dynamics in West-Siberian permafrost. In *Permafrost: Distribution, Composition and Impacts on Infrastructure and Ecosystems*; Pokrovsky, O.S., Ed.; Nova Science Publishers: Hauppauge, NY, USA, 2014; pp. 173–204.

19. Coleman, K.A.; Palmer, M.J.; Korosi, J.B.; Kokelj, S.V.; Jackson, K.; Hargan, K.E.; Mustaphi, C.J.C.; Thienpont, J.R.; Kimpe, L.E.; Blais, J.M.; et al. Tracking the impacts of recent warming and thaw of permafrost peatlands on aquatic ecosystems: A multi-proxy approach using remote sensing and lake sediments. *Boreal Environ. Res.* **2015**, *20*, 363–377.

20. Boike, J.; Grau, T.; Heim, B.; Günther, F.; Langer, M.; Muster, S.; Gouttevin, I.; Lange, S. Satellite-derived changes in the permafrost landscapes of Central Yakutia, 2000–2011: Wetting, drying, and fires. *Glob. Planet. Chang.* **2016**, *139*, 116–127. [CrossRef]

21. Tranvik, L.J.; Downing, J.A.; Cotner, J.B.; Loiselle, S.A.; Striegl, R.G.; Ballatore, T.J.; Dillon, P.; Finlay, K.; Fortino, K.; Knoll, L.B.; et al. Lakes and reservoirs as regulators of carbon cycling and climate. *Limnol. Oceanogr.* **2009**, *54*, 2298–2314. [CrossRef]

22. Catalán, N.; Marcé, R.; Kothawala, D.N.; Tranvik, L.J. Organic carbon decomposition rates controlled by water retention time across inland waters. *Nat. Geosci.* **2016**, *9*, 501–504. [CrossRef]

23. Walter, K.M.; Smith, L.C.; Chapin, F.S. Methane bubbling from northern lakes: Present and future contributions to the global methane budget. *Philos. Trans. R. Soc. A* **2007**, *365*, 1657–1676. [CrossRef] [PubMed]

24. Laurion, I.; Vincent, W.F.; MacIntyre, S.; Retamal, L.; Dupont, C.; Francus, P.; Pienitz, R. Variability in greenhouse gas emissions from permafrost thaw ponds. *Limnol. Oceanogr.* **2010**, *55*, 115–133. [CrossRef]

25. Walter Anthony, K.M.; Anthony, P. Constraining spatial variability of methane ebullition in thermokarst lakes using point-process models. *J. Geophys. Res.* **2013**, *118*, 1015–1034. [CrossRef]

26. Walter Anthony, K.M.; Zimov, S.A.; Grosse, G.; Jones, M.C.; Anthony, P.M.; Chapin, F.S., III; Finlay, J.C.; Mack, M.C.; Davydov, S.; Frenzel, P.; et al. A shift of thermokarst lakes from carbon sources to sinks during the Holocene epoch. *Nature* **2014**, *511*, 452–456. [CrossRef] [PubMed]

27. Zuidhoff, F.S.; Kolstrup, E. Changes in palsa distribution in relation to climate change in Laivadalen, Northern Sweden, especially 1960–1997. *Permafr. Periglac.* **2000**, *11*, 55–69. [CrossRef]

28. Luoto, M.; Seppala, M. Thermokarst ponds as indicator of the former distribution of palsas in Finnish Lapland. *Permafr. Periglac.* **2003**, *14*, 19–27. [CrossRef]

29. Riordan, B.; Verbyla, D.; McGuire, A.D. Shrinking ponds in subarctic Alaska based on 1950–2002 remotely sensed images. *J. Geophys. Res. Biogeosci.* **2006**, *111*, G04002. [CrossRef]

30. Grosse, G.; Romanovsky, V.; Walter, K.; Morgenstern, A.; Lantuit, H.; Zimov, S. Distribution of Thermokarst Lakes and Ponds at Three Yedoma Sites in Siberia. In Proceedings of the 9th Intern Conference on Permafrost, Fairbanks, Alaska, 29 June–3 July 2008; pp. 551–556.

31. Kirpotin, S.; Polishchuk, Y.; Bryksina, N. Abrupt changes of thermokarst lakes in Western Siberia: Impacts of climatic warming on permafrost melting. *Int. J. Environ. Stud.* **2009**, *66*, 423–431. [CrossRef]

32. Plug, L.J.; Walls, C.; Scott, B.M. Tundra lake changes from 1978 to 2001 on the Tuktoyaktuk Peninsula, western Canadian Arctic. *Geophys. Res. Lett.* **2008**, *35*, L03502. [CrossRef]

33. Jepsen, S.M.; Voss, C.I.; Walvoord, M.A.; Minsley, B.J.; Rover, J. Linkages between lake shrinkage/expansion and sublacustrine permafrost distribution determined from remote sensing of interior Alaska, USA. *Geophys. Res. Lett.* **2013**, *40*, 882–887. [CrossRef]

34. Jones, B.M.; Grosse, G.; Arp, C.D.; Jones, M.C.; Walter Anthony, K.M.; Romanovsky, V.E. Modern thermokarst lake dynamics in the continuous permafrost zone, northern Seward Peninsula, Alaska. *J. Geophys. Res.* **2011**, *116*, G00M03. [CrossRef]

35. Sannel, A.B.K.; Kuhry, P. Warming-induced destabilization of peat plateau/thermokarst lake complexes. *J. Geophys. Res.* **2011**, *116*, G03035. [CrossRef]

36. Necsoiu, M.; Dinwiddie, C.L.; Walter, G.R.; Larsen, A.; Stothoff, S.A.; Necsoiu, M. Multi-temporal image analysis of historical aerial photographs and recent satellite imagery reveals evolution of water body surface area and polygonal terrain morphology in Kobuk Valley National Park, Alaska. *Environ. Res. Lett.* **2013**, *8*, 025007. [CrossRef]

37. Carroll, M.L.; Townshend, J.R.; DiMiceli, C.M.; Noojipady, P.; Sohlberg, R.A. A new global raster water mask at 250 m resolution. *Int. J. Digital Earth* **2009**, *2*, 291–308. [CrossRef]

38. Carroll, M.L.; Townshend, J.R.G.; DiMiceli, C.M.; Loboda, T.; Sohlberg, R.A. Shrinking lakes of the Arctic: Spatial relationships and trajectory of change. *Geophys. Res. Lett.* **2011**, *38*, L20406. [CrossRef]

39. Smith, L.; Sheng, Y.; Macdonald, G.; Hinzman, L. Disappearing Arctic Lakes. *Science* **2005**, *308*, 1429. [CrossRef] [PubMed]

40. Karlsson, J.M.; Lyon, S.W.; Destouni, G. Thermokarst lake, hydrological flow and water balance indicators of permafrost change in Western Siberia. *J. Hydrol.* **2012**, *464–465*, 459–466. [CrossRef]

41. Dneprovskaya, V.P.; Bryksina, N.A.; Polishchuk, Y.M. Study of thermokarst change in discontinuous permafrost zone of western Siberia using satellite images. *Study Earth Space* **2009**, *4*, 88–96. (In Russian)

42. Kravtzova, V.I.; Bystrova, A.G. Study of thermokarst lake size in different regions of Russia over last 30 years. *Earth Cryosphere* **2009**, *13*, 12–26. (In Russian)

43. Turner, K.W.; Turner, K.W.; Wolfe, B.B.; Edwards, T.W.D. Characterizing the role of hydrological processes on lake water balances in the Old Crow Flats, Yukon Territory, Canada, using water isotope tracers. *J. Hydrol.* **2010**, *386*, 103–107. [CrossRef]

44. Anderson, L.; Birks, J.; Rover, J.; Guldager, N. Controls on recent Alaskan lake changes identified from water isotopes and remote sensing. *Geophys. Res. Lett.* **2013**, *40*, 3413–3418. [CrossRef]

45. Lantz, T.C.; Turner, K.W. Changes in lake area in response to thermokarst processes and climate in Old Crow Flats, Yukon. *J. Geophys. Res. Biogeosci.* **2015**, *120*, 513–524. [CrossRef]

46. Bryksina, N.A.; Polishchuk, Y.M. Analysis of changes in the number of thermokarst lakes in permafrost of Western Siberia on the basis of satellite images. *Earth Cryosphere* **2015**, *19*, 100–105. (In Russian)

47. Karlsson, J.M.; Lyon, S.W.; Destouni, G. Temporal Behavior of lake size-distribution in a thawing permafrost landscape in Nothwestern Siberia. *Remote Sens.* **2014**, *6*, 621–636. [CrossRef]

48. Polishchuk, Y.M.; Bryksina, N.A.; Polishchuk, V.Y. Remote analysis of changes in the number and distribution of small thermokarst lakes by sizes in cryolithozone of Western Siberia. *Izv. Atmos. Ocean. Phys.* **2015**, *51*, 999–1006. [CrossRef]

49. Pokrovsky, O.S.; Shirokova, L.S.; Kirpotin, S.N.; Audry, S.; Viers, J.; Dupre, B. Effect of permafrost thawing on the organic carbon and metal speciation in thermokarst lakes of Western Siberia. *Biogeosciences* **2011**, *8*, 565–583. [CrossRef]

50. Shirokova, L.S.; Pokrovsky, O.S.; Kirpotin, S.N.; Desmukh, C.; Pokrovsky, B.G.; Audry, S.; Viers, J. Biogeochemistry of organic carbon, CO_2, CH_4, and trace elements in thermokarst water bodies in discontinuous permafrost zones of Western Siberia. *Biogeochemistry* **2013**, *113*, 573–593. [CrossRef]

51. Abnizova, A.; Siemens, J.; Langer, M.J.; Boike, J. Small ponds with major impact: The relevance of ponds and lakes in permafrost landscapes to carbon dioxide emissions. *Glob. Biogeochem. Cycles* **2012**, *26*, GB2041. [CrossRef]

52. Pokrovsky, O.S.; Shirokova, L.S.; Kirpotin, S.N.; Kulizhsky, S.P.; Vorobiev, S.N. Impact of western Siberia heat wave 2012 on greenhouse gases and trace metal concentration in thaw lakes of discontinuous permafrost zone. *Biogeosciences* **2013**, *10*, 5349–5365. [CrossRef]

53. Manasypov, R.M.; Pokrovsky, O.S.; Kirpotin, S.N.; Shirokova, L.S. Thermokarst lake waters across permafrost zones of Western Siberia. *Cryosphere* **2014**, *8*, 1177–1193. [CrossRef]

54. Manasypov, R.M.; Vorobyev, S.N.; Loiko, S.V.; Kritzkov, I.V.; Shirokova, L.S.; Shevchenko, V.P.; Kirpotin, S.N.; Kulizhsky, S.P.; Kolesnichenko, L.G.; Zemtzov, V.A.; et al. Seasonal dynamics of organic carbon and metals in thermokarst lakes from the discontinuous permafrost zone of western Siberia. *Biogeosciences* **2015**, *12*, 3009–3028. [CrossRef]

55. Kremenetski, K.V.; Velichko, A.A.; Borisova, O.K.; MacDonald, G.M.; Smith, L.C.; Frey, K.E.; Orlova, L.A. Peatlands of the West Siberian Lowlands: Current knowledge on zonation, carbon content, and Late Quaternary history. *Quat. Sci. Rev.* **2003**, *22*, 703–723. [CrossRef]

56. Peregon, A.; Maksyutov, S.; Yamagata, Y. An image-based inventory of the spatial structure of West Siberian wetlands. *Environ. Res. Lett.* **2009**, *4*, 045014. [CrossRef]

57. Stepanova, V.M.; Pokrovsky, O.S.; Viers, J.; Mironycheva-Tokareva, N.P.; Kosykh, N.P.; Vishnyakova, E.K. Major and trace elements in peat profiles in Western Siberia: Impact of the landscape context, latitude and permafrost coverage. *Appl. Geochem.* **2015**, *53*, 53–70. [CrossRef]

58. Nikitin, S.P.; Zemtsov, V.A. *The Variability of Hydrological Parameters of Western Siberia*; Nauka: Novosibirsk, Russia, 1986; 204p. (In Russian)

59. Liss, O.L.; Abramova, L.I.; Avetov, N.A.; Berezina, N.A.; Inisheva, L.I.; Kurnishnikova, T.V.; Sluka, Z.A.; Tolpysheva, T.Yu.; Shvedchikova, N.K. *Wetland Systems of West Siberia and their Importance for Nature Conservation*; Grif, i K: Tula, Russia, 2001; 584p. (In Russian)

60. Khrenov, V.Y. *Soils of Cryolithozone of Western Siberia: Morphology, Physico-Chemical Properties and Geochemistry*; Nauka: Moscow, Russia, 2011; 214p. (In Russian)

61. Pokrovsky, O.S.; Shirokova, L.S.; Kirpotin, S.N. *Biogeochemistry of Thermokarst Lakes of Western Siberia*; Nova Science Publishers: Hauppauge, NY, USA, 2014; 163p.

62. Rikhter, G.D. *Western Siberia*; AS USSR Press: Moscow, Russia, 1963; 188p. (In Russian)

63. Ivanov, K.E.; Novikov, S.M. *Mires of Western Siberia: Their Structure and Hydrological Regime*; Hydrometeoizdat: Leningrad, Russia, 1976; 448p. (In Russian)

64. Kozlov, S.A. Evaluation of stability of the geological environment at offshore fields hydrocarbons in the Arctic. *Elektron. Nauchnii Zhurnal Neftegazov.* **2005**, *3*, 15–24. (In Russian)

65. Savchenko, N.V. Nature of lakes in subarctic of West Siberia. *Geogr. Prir. Resur.* **1992**, *1*, 85–92. (In Russian)

66. Viktorov, A.S. *Main Problems of Mathematic Morphology of Landscape*; Nauka: Moscow, Russia, 2006; 252p. (In Russian)

67. Grosse, G.; Jones, B.; Arp, C. Thermokarst lakes, drainage, and drained basins. In *Treatise on Geomorphology*, 3rd ed.; Shroder, J., Giardino, R., Harbor, J., Eds.; Academic Press: San Diego, CA, USA, 2013; Volume 8, pp. 325–353.

68. USGS Global Visualization Viewer. Available online: http://glovis.usgs.gov/ (accessed on 20 March 2017).

69. Kennedy, M.D.; Goodchild, M.F.; Dangermond, J. *Introducing Geographic Information Systems with ArcGIS: A Workbook Approach to Learning GIS*; John Wiley & Sons, Inc.: Hoboken, NJ, USA, 2013; 672p.

70. Zhu, Z.; Wang, S.; Woodcock, C.E. Improvement and expansion of the Fmask algorithm: Cloud, cloud shadow, and snow detection for Landsats 4–7, 8, and Sentinel 2 images. *Remote Sens. Environ.* **2015**, *159*, 269–277. [CrossRef]

71. Frey, K.E.; Smith, L.C. How well do we know northern land cover? Comparison of four global vegetation and wetland products with a new ground-truth database for West Siberia. *Glob. Biogeochem. Cycles* **2007**, *21*, GB1016. [CrossRef]

72. Shirokova, L.S.; Pokrovsky, O.S.; Kirpotin, S.N.; Dupré, B. Heterotrophic bacterio-plankton in thawed lakes of northern part of Western Siberia controls the CO_2 flux to the atmosphere. *Int. J. Environ. Stud.* **2009**, *66*, 433–445. [CrossRef]

73. Kirpotin, S.N.; Polishchuk, Y.M.; Zakharova, E.; Shirokova, L.; Pokrovsky, O.; Kolmakova, M.; Dupré, B. One of possible mechanisms of thermokarst lakes drainage in West-Siberian North. *Int. J. Environ. Stud.* **2008**, *65*, 631–635. [CrossRef]

74. Audry, S.; Pokrovsky, O.S.; Shirokova, L.S.; Kirpotin, S.N.; Dupré, B. Organic matter mineralization and trace element post-depositional redistribution in Western Siberia thermokarst lake sediments. *Biogeosciences* **2011**, *8*, 565–583. [CrossRef]

75. Halicki, W.; Kochanska, M.; Pokrovsky, O.S.; Kirpotin, S.N. Assessment of physical properties and pH of selected surface waters in the northern part of Western Siberia. *Int. J. Environ. Stud.* **2015**, *72*, 557–566. [CrossRef]

76. Kochanska, M.; Halicki, W.; Pokrovsky, O.S.; Kirpotin, S.N. Organic compounds in typical surface waters of the northern part of Western Siberia. *Int. J. Environ. Stud.* **2015**, *72*, 547–556. [CrossRef]

77. Pavlova, O.A.; Pokrovsky, O.S.; Manasypov, R.M.; Shirokova, L.S.; Vorobyev, S.N. Seasonal dynamics of phytoplankton in acidic and humic environment in shallow thaw ponds of western Siberia, discontinuous permafrost zone. *Ann. Limnol. Int. J. Limnol.* **2016**, *52*, 47–60. [CrossRef]

78. Shirokova, L.S.; Pokrovsky, O.S.; Viers, J.; Klimov, S.I.; Moreva, O.Y.; Zabelina, S.A.; Vorobieva, T.Y.; Dupré, B. Diurnal variations of trace elements and heterotrophic bacterioplankton concentration in a small boreal lake of the White Sea basin. *Ann. Limnol. Int. J. Limnol.* **2010**, *46*, 67–75. [CrossRef]

79. Pokrovsky, O.S.; Manasypov, R.M.; Loiko, S.; Shirokova, L.S.; Krickov, I.A.; Pokrovsky, B.G.; Kolesnichenko, L.G.; Kopysov, S.G.; Zemtzov, V.A.; Kulizhsky, S.P.; et al. Permafrost coverage, watershed area and season control of dissolved carbon and major elements in western Siberia rivers. *Biogeosciences* **2015**, *12*, 6301–6320. [CrossRef]

80. Pokrovsky, O.S.; Manasypov, R.M.; Loiko, S.; Krickov, I.A.; Kopysov, S.G.; Kolesnichenko, L.G.; Vorovyev, S.N.; Kirpotin, S.N. Trace element transport in western Siberia rivers across a permafrost gradient. *Biogeosciences* **2016**, *13*, 1877–1900. [CrossRef]

81. Cael, B.B.; Seekell, D.A. The size-distribution of Earth's lakes. *Sci. Rep.* **2016**, *6*, 29633. [CrossRef] [PubMed]

82. Baisalyamova, O.A.; Bogdanov, A.N.; Muratov, I.N.; Polishchuk, Y.M.; Snigireva, M.S. Spatial distribution of small thermokarst lakes in western Siberia using KH-7 and GeoEye-1 images. *Vestnik Ugra State Univ.* **2015**, *3*, 69–73. (In Russian)

83. Wang, Y. Advances in Remote Sensing of Flooding. *Water* **2015**, *7*, 6404–6410. [CrossRef]

84. Zhao, X.; Liu, Y. Lake Fluctuation Effectively Regulates Wetland Evapotranspiration: A Case Study of the Largest Freshwater Lake in China. *Water* **2014**, *6*, 2482–2500. [CrossRef]

85. Siewert, M.B.; Hanisch, J.; Weiss, N.; Kuhry, P.; Maximov, T.C.; Hugelius, G. Comparing carbon storage of Siberian tundra and taiga permafrost ecosystems at very high spatial resolution. *J. Geophys. Res. Biogeosci.* **2015**, *120*, 1973–1994. [CrossRef]

86. Pelletier, L.; Strachan, I.B.; Roulet, N.T.; Garneau, M. Can boreal peatlands with pools be net sinks for CO_2? *Environ. Res. Lett.* **2015**, *10*, 035002. [CrossRef]

87. Pelletier, L.; Strachan, I.B.; Roulet, N.T.; Garneau, M.; Wischnewski, K. Effect of open water pools on ecosystem scale surface-atmosphere carbon dioxide exchange in a boreal peatland. *Biogeochemistry* **2015**, *124*, 291–304. [CrossRef]

88. Tank, S.E.; Frey, K.E.; Striegl, R.G.; Raymond, P.A.; Holmes, R.M.; McClelland, J.W.; Peterson, B.J. Landscape-level controls on dissolved carbon flux from diverse catchments of the circumboreal. *Glob. Biogeochem. Cycles* **2012**, *26*, GB0E02. [CrossRef]

89. Tank, S.E.; Raymond, P.A.; Striegl, R.G.; McClelland, J.W.; Holmes, R.M.; Fiske, G.; Peterson, B.J. A land-to-ocean perspective on the magnitude, source and implication of DIC flux from major Arctic rivers to the Arctic Ocean. *Glob. Biogeochem. Cycles* **2012**, *26*, GB4018. [CrossRef]

90. Tank, S.E.; Striegl, R.G.; McClelland, J.W.; Kokelj, S.V. Multi-decadal increases in dissolved organic carbon and alkalinity flux from the Mackenzie drainage basin to the Arctic Ocean. *Environ. Res. Lett.* **2016**, *11*, 054015. [CrossRef]

91. Tank, S.E.; Lesack, L.F.W.; McQueen, D.J. Elevated pH regulates bacterial carbon cycling in lakes with high photosynthetic activity. *Ecology* **2009**, *90*, 1910–1922. [CrossRef] [PubMed]

92. Kokelj, S.V.; Burn, C.R. Geochemistry of the active layer and near-surface permafrost, Mackenzie delta region, Northwest Territories, Canada. *Can. J. Earth Sci.* **2005**, *42*, 37–48. [CrossRef]

93. Kokelj, S.V.; Jenkins, R.E.; Milburn, D.; Burn, C.R.; Snow, N. The influence of thermokarst disturbance on the water quality of small upland lakes, Mackenzie Delta Region, Northwest Territories, Canada. *Permafr. Periglac.* **2005**, *16*, 343–353. [CrossRef]

94. Kokelj, S.V.; Zajdlik, B.; Thompson, M.S. The impacts of thawing permafrost on the chemistry of lakes across the subarctic boreal-tundra transition, Mackenzie Delta region, Canada. *Permafr. Periglac.* **2009**, *20*, 185–199. [CrossRef]

95. Morgenstern, A.; Grosse, G.; Günther, F.; Fedorova, I.; Schirrmeister, L. Spatial analyses of thermokarst lakes and basins in Yedoma landscapes of the Lena Delta. *Cryosphere* **2011**, *5*, 849–867. [CrossRef]

96. Ulrich, M.; Matthes, H.; Schirrmeister, L.; Schütze, J.; Park, H.; Iijma, Y.; Fedorov, A.N. Differences in behavior and distribution of permafrost-related lakes in Central Yakutia and their response to climatic drivers. *Water Resour. Res.* **2017**. [CrossRef]

11

CryoSat-2 Altimetry Applications over Rivers and Lakes

Liguang Jiang [1,*], Raphael Schneider [1], Ole B. Andersen [2] and Peter Bauer-Gottwein [1]

[1] Department of Environmental Engineering, Technical University of Denmark, Bygningstorvet B115, 2800 Kongens Lyngby, Denmark; rasch@env.dtu.dk (R.S.); pbau@env.dtu.dk (P.B.-G.)

[2] National Space Institute, Technical University of Denmark, Elektrovej 327, 2800 Kongens Lyngby, Denmark; oa@space.dtu.dk

* Correspondence: ljia@env.dtu.dk

Academic Editor: Frédéric Frappart

Abstract: Monitoring the variation of rivers and lakes is of great importance. Satellite radar altimetry is a promising technology to do this on a regional to global scale. Satellite radar altimetry data has been used successfully to observe water levels in lakes and (large) rivers, and has also been combined with hydrologic/hydrodynamic models. Except CryoSat-2, all radar altimetry missions have been operated in conventional low resolution mode with a short repeat orbit (35 days or less). CryoSat-2, carrying a Synthetic Aperture Radar (SAR) altimeter, has a 369-day repeat and a drifting ground track pattern and provides new opportunities for hydrologic research. The narrow inter-track distance (7.5 km at the equator) makes it possible to monitor many lakes and rivers and SAR mode provides a finer along-track resolution, higher return power and speckle reduction through multi-looks. However, CryoSat-2 challenges conventional ways of dealing with satellite inland water altimetry data because virtual station time series cannot be directly derived for rivers. We review the CryoSat-2 mission characteristics, data products, and its use and perspectives for inland water applications. We discuss all the important steps in the workflow for hydrologic analysis with CryoSat-2, and conclude with a discussion of promising future research directions.

Keywords: CryoSat-2; radar altimetry; inland water altimetry; hydrology; water height

1. Introduction

Rivers and lakes are important fresh water resources. The global distribution of rivers and lakes is shown in Figure 1. They supply drinking water for many people in the world [1] and in particular for the vast majority of people in poverty. However, these people often are also vulnerable to flooding from the very same rivers. For instance, the Brahmaputra River sustains lives and livelihood along its banks, while draining through the Assam Valley. However, floods occur in monsoons every year and severe floods have happened frequently in the last decade, which caused huge losses to life and property [2]. Just like rivers, lakes serve many purposes. Not only do they provide freshwater for human use, including agriculture, but they also maintain important natural processes and ecosystems [3]. Nevertheless, many lakes around the world are shrinking and some have vanished under the influence of climate change and anthropogenic activities while in other regions, lakes are expanding (e.g., inner Tibetan Plateau) [4–8]. Therefore, global monitoring of the variation of rivers and lakes is an important research topic.

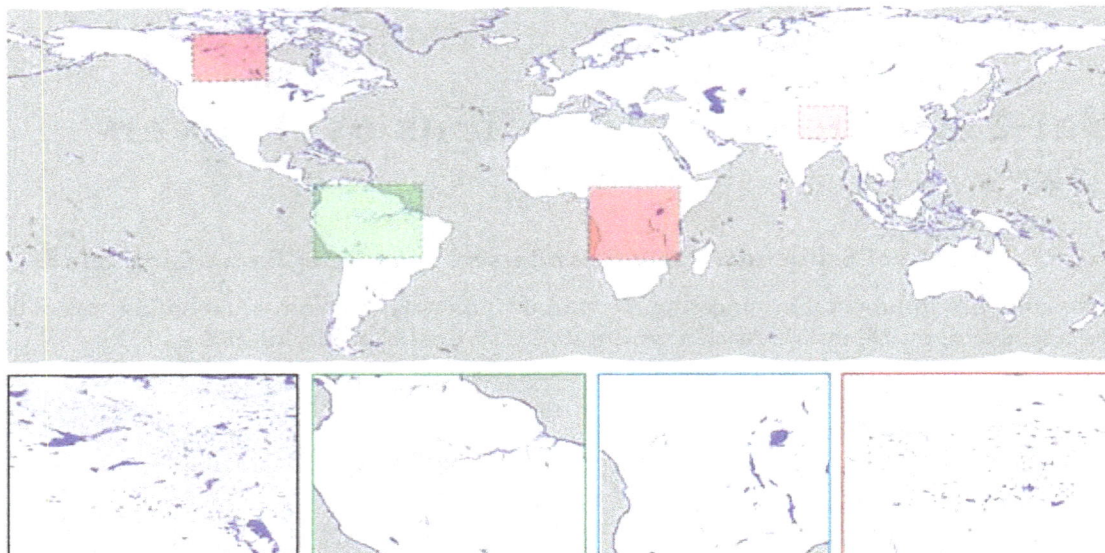

Figure 1. Global surface water distribution. (Zoom-in figures at bottom are from Kivalliq Region in Canada, Amazon River basin, source region of Congo river basin, and Tibetan Plateau, respectively. Data source: Global Surface Water Explorer [9]).

Water level is one basic and key quantity in hydrological research, which is closely related to discharge in rivers and water volume in lakes and reservoirs. Traditionally, water level observations are recorded in situ by water-level recorders or visual readings from a staff gauge. These gauging stations are normally established at point scale and often organized on a national basis. Thus, the spatial resolution is limited and the data release is slow [10]. Moreover, the number of freely accessible gauging station records in rivers is decreasing since the late 1970s [11]. Also, data sharing is a big problem, especially in transboundary river basins [12]. Here, water level observations from satellite remote sensing have advantages over traditional observations. First, remote sensing has universal spatial coverage, i.e., transboundary and inaccessible or dangerous regions are also covered. Second, data acquisition is normally free and timely, which paves the way to operational forecasting systems. With respect to observing water level, satellite altimeters make it possible to monitor water levels in lakes and sufficiently large rivers with acceptable spatio-temporal resolution. Satellite altimetry has been an important tool in inland water monitoring although the technique was initially designed for monitoring oceans [13–19]. For example, satellite altimetry makes it possible to monitor the water level and storage variation of hundreds of lakes in the Tibetan Plateau [17,20,21]. Another application is supporting river discharge modelling, in particular in transboundary areas and remote areas, such as the Brahmaputra or Amazon river basins, etc. [22–24]. An alternative technology for precise remote sensing of water heights is Global Navigation Satellite System (GNSS) Reflectometry [25–27]. This technique requires, however, ground-based or airborne GNSS receivers in the vicinity of the lake or river.

The precision of altimetry measurements has improved significantly from the first satellite altimeter (on Skylab, 1973). The list of past and current satellite altimetry missions includes GEOS-3, SeaSat, Geosat, TOPEX/Poseidon, Geosat Follow-on, Jason-1/2/3 from the National Aeronautics and Space Administration (NASA). From the European Space Agency (ESA) there are ERS-1/2, ENVISAT, CryoSat-2, and Sentinel-3. Besides that, some other missions are also in operation, such as HY-2A planned by China, or SARAL/AltiKa as a joint Indian–French project (see Table 1 for an overview). All the above mentioned missions carry radar altimeters. Besides radar altimeter missions, there has been a satellite lidar mission, ICESat, which provided similar data products for inland waters. Common to all missions except CryoSat-2 is a repeat orbit with a short repeat cycle of 10 to 35 days.

Such repeat cycles have sparse ground track patterns with an inter-track distance of at least 80 km at the equator. Precision and bias are compared across various satellite radar altimetry missions in [28].

Table 1. Summary of satellite altimetry missions.

Satellite	Agency	Period	Altitude (km)	Altimeter	Frequency Used	Repetitivity (Day)	Equatorial Inter-Track Distance (km)
Skylab	NASA	May 1973– February 1974	435	S193	Ku-band		
GEOS 3	NASA	April 1975– July 1979	845	ALT	Ku and C-band		
SeaSat	NASA	July– October 1978	800	ALT	Ku-band	17	
Geosat	US Navy	October 1985– January 1990	800		Ku-band	17	
ERS-1	ESA	July 1991– March 2000	785	RA	Ku-band	35	80
Topex/ Poseidon	NASA/ CNES	September 1992– October 2005	1336	Poseidon	Ku and C-band	10	315
ERS-2	ESA	April 1995– July 2011	785	RA	Ku-band	35	80
GFO	US Navy/ NOAA	February 1998– October 2008	800	GFO-RA	Ku-band	17	165
Jason-1	CNES/ NASA	December 2001– June 2013	1336	Poseidon-2	Ku and C-band	10	315
Envisat	ESA	March 2002– April 2012	800	RA-2	Ku and S-band	35	80
OSTM/ Jason-2	CNES/ NASA/ Eumetsat/ NOAA	Jun 2008– present	1336	Poseidon-3	Ku and C-band	10	315
CryoSat-2	ESA	April 2010– present	720	SIRAL	Ku-band	369	7.5
HY-2	China	August 2011– present	971		Ku and C-band	14, 168	
Saral	ISRO/ CNES	February 2013– present	800	AltiKa	Ka-band	35	80
Jason-3	CNES/ NASA/ Eumetsat/ NOAA	January 2016– present	1336	Poseidon-3B	Ku and C-band	10	315
Sentinel-3A	ESA	February 2016– present	814	SRAL	Ku and C-band	27	104

ESA's CryoSat-2 is distinctive due to its long repeat (369 days) and corresponding drifting ground track pattern and due to the SIRAL instrument (see details in Section 3). For short-repeat missions, one can derive water level time series at the locations where the satellite's ground track regularly intersects with the water body—the so-called virtual station. This eases many aspects of processing altimetry data and integrating it into hydrologic models (more details in Section 6.3). CryoSat-2 with its drifting ground track pattern and a repeat cycle of 369 days has an entirely different sampling pattern. For this reason, use of CryoSat-2 for inland water research, especially river modeling, has been limited so far. However, the long-repeat orbit has the advantage of short inter-track distances (larger spatial coverage, see Table 1); moreover, CryoSat-2 has other important advantages, for example, a finer along-track resolution (for SAR and SARIn modes) compared to traditional pulse-limited radar altimeters [29]. The drifting ground track pattern with a small inter-track spacing of 7.5 km at the equator enables (i) monitoring of a much larger number of lakes and (ii) derivation of high resolution water level profiles along rivers. Moreover, these dense ground tracks increase the temporal resolution for large lakes. These characteristics create new opportunities for hydrologic research: Nielsen et al. [18] showed that small lakes (~9 km²) can be observed by CryoSat-2. Schneider et al. [30]

calibrated a hydrodynamic river model with CryoSat-2 observations. Published study results show that performance of CryoSat-2 achieves root mean square error (RMSE) of a few centimeters for lakes [18,31] and between about 30 cm down to less than 5 cm using special data handling strategies for the Amazon River [32].

In this paper, we review the application of CryoSat-2 altimetry data over inland waters since its operation started in April 2010 up to today. Specifically, we first present the basic principle of satellite radar altimetry, and provide a CryoSat-2 mission overview. Then, we present data processing and data products, and review the use of CryoSat-2 altimetry data over lakes and rivers, and finally we discuss prospects of potential use of CryoSat-2 in hydrological applications.

2. Basic Principles of Radar Altimetry

In satellite radar altimetry, a microwave pulse is sent out by the altimeter, reflected by the surface, and finally part of its echo is recorded at the altimeter. The time series of returned power measured by the altimeter is commonly referred to as waveform [33] (inset in Figure 2). The principle of satellite radar altimeters obtaining surface height is to measure the two-way travel time of the microwave pulse travelling between the altimeter and the surface. This time interval can be then converted into a distance, also called range (Figure 2), by multiplying with the speed of light at which electromagnetic waves travel. With the position of the satellite, i.e., the altitude of satellite, surface height can be obtained by subtracting the range from the altitude of satellite. A Doppler Orbit and Radio Positioning Integration by Satellite (DORIS) receiver is deployed on-board for real-time measurements of satellite position, velocity and time. The measurement accuracy of the satellite position is 2–6 cm [34].

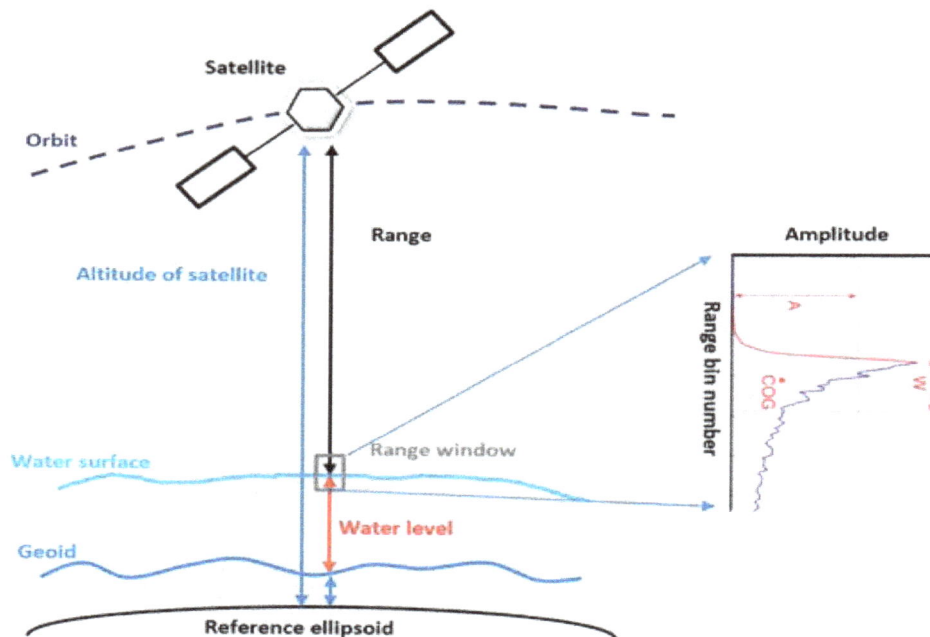

Figure 2. Altimetry principle of water level measurement (Modified from [35]).

The range window, i.e., the elevation window where the satellite altimeter is sensitive to observations, has to be adapted dynamically to the topography. This can either be done in closed loop or open loop. In open loop, the range window is positioned based on a DEM. Closed loop means that the altimetry instrument itself constantly adapts the range window based on measurements [36], which is the mode implemented on CryoSat-2.

In general, waveforms returned from small inland open water bodies have a single strong peak due to the quasi-specular scattering of the smooth water surface. Those returned from large rough water bodies usually have only one peak with a steep leading edge and a slowly decaying trailing edge (Figure 3). If the exact location of the leading edge in the return waveform can be determined, the signal travel time can be estimated, therefore the range (and ultimately the surface elevation) can be calculated [37]. This is performed by the on-board tracker.

However, due to the diversity of reflecting surfaces, waveforms change dramatically in shape and power. Waveforms returned from inhomogeneous terrain (e.g., water and land transition area) usually have a complicated shape with multiple peaks [32,38]. Therefore, the leading edge of the returned waveform deviates from the altimeter tracking position (nominal tracking point), causing an error in the measured range which is accounted for by retracking [39] (Figure 3).

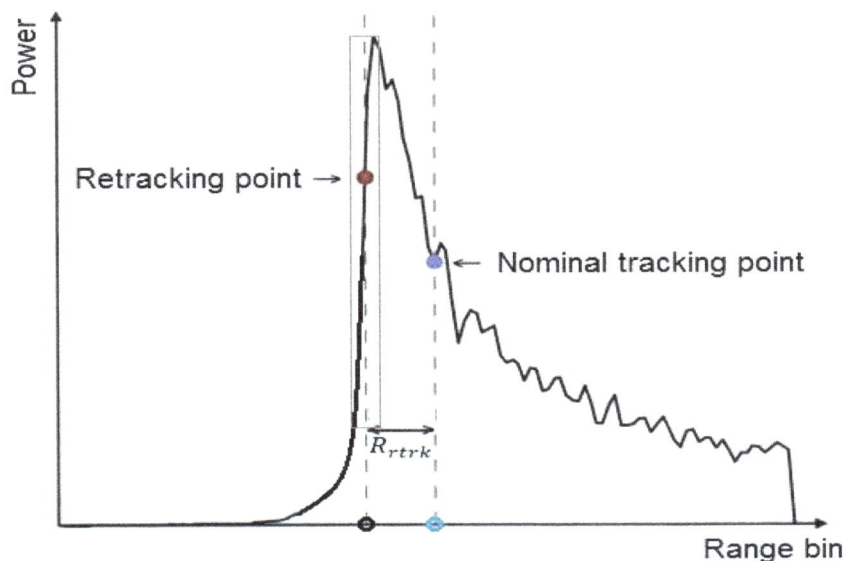

Figure 3. Illustration of waveform showing the nominal tracking point and retracking point.

The altimeter waveform is provided in a set of power signals with respect to time at a specified number of sample bins [37] (Figure 3). Waveform retracking is the process of finding the mid-bin of the leading edge (retracking point) in the return waveform to calculate the difference between the nominal tracking position and the retracking position (Figure 3), thus correcting the on-board tracker range [37,38].

$$R_{corr} = R + R_{rtrk} - \Delta R_{geo} \tag{1}$$

where R is the range computed by the on-board tracker; R_{rtrk} is the retracker correction and ΔR_{geo} is the sum of corrections including ionosphere, wet and dry troposphere, solid earth tide, ocean loading tide, and pole tide. All corrections are available in the L1b data product. Thus, the corrected range R_{corr} between the satellite and water surface can be derived.

For hydrological purpose, one can conveniently refer the surface to the geoid. Finally, the surface elevation H is obtained by subtracting the corrected range R and geoid undulation N from the satellite altitude h:

$$H = h - R_{corr} - N \tag{2}$$

3. Mission Overview

ESA's CryoSat-2 satellite was launched on 8 April 2010. The primary objectives of this mission are monitoring the Arctic sea ice thickness variation and the influence of the Antarctic and Greenland ice sheets on global sea-level [34,40]. However, like previous satellite altimetry missions, it also proved to be useful for monitoring of inland water levels [13,19,41].

3.1. Instrument

The radar altimetry instrument on CryoSat-2 is called SIRAL (Synthetic Aperture Interferometric Radar Altimeter). It is a single Ku-band radar altimeter using the full deramp range compression. It is operating in three distinct modes: Low Resolution Mode (LRM), Synthetic Aperture mode (SAR), and Synthetic Aperture Interferometric mode (SARIn) (Figure 4). Over the central regions of the ice sheets and most of the continental area, the instrument will provide the measurements as a conventional radar altimeter in LRM. SAR mode enhances the along-track spatial resolution to, for example, measure ice flows and narrow leads of open water which cannot be achieved by LRM. It is also used over some coastal regions. SARIn mode is used over the topographic surfaces of the ice-sheet margins, over mountain glaciers, and over other regions of interest, for example, large river systems such as the Danube or Congo River. In this mode, the altimeter performs synthetic aperture processing with two antennas and thus precisely determines the position of the ground surface in the return pulse. The operation mode is selected from a geographical mask [42] (see Figure 4), which is updated every two weeks to allow for changes in sea ice extent. Especially the two-antenna SARIn mode makes CryoSat-2 unique among current satellite altimetry missions. For more details, please refer to [34,43].

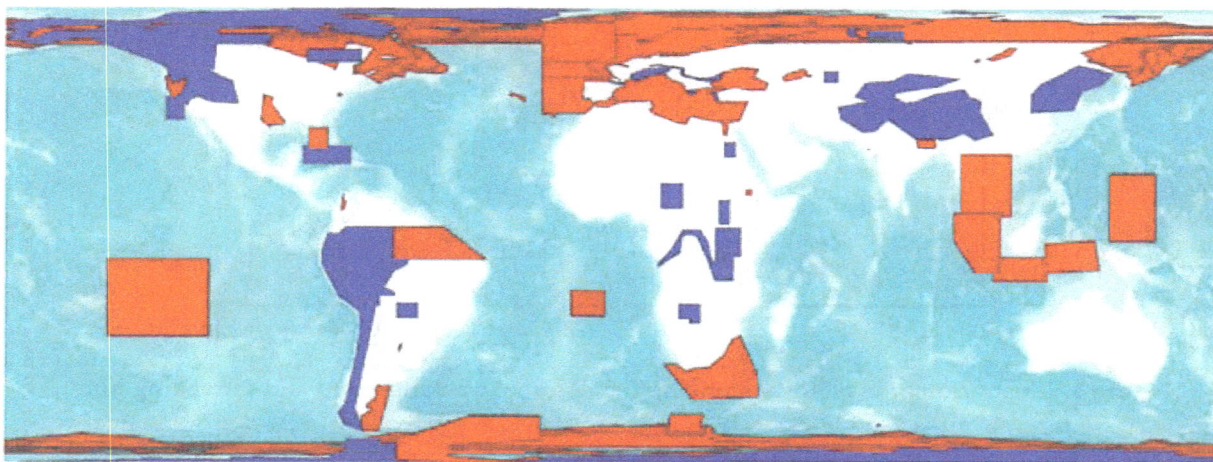

Figure 4. Geographical mode mask 3.8 (synthetic aperture radar (SAR) mode in red, SARIn mode in blue, all remaining areas in low resolution mode (LRM) [42]).

3.2. Orbit

Satellite radar altimeters sample elevation globally along the orbit ground track. Orbits are constrained by the equations of motion. The primary factors that affect the orbit geometry are the altitude, inclination and eccentricity [44]. CryoSat-2's orbit is non-sun-synchronous with a mean altitude of 717 km and a high inclination of 92°. The repeat period is 369 days or 5344 orbits. However, the orbit also has a 30-day subcycle, which encompasses the full 369-day repeat by successive shifts. In other words, the orbit shifts about 7.5 km at the equator every 30 days and returns to the same place every 369 days. More information is detailed in [43].

3.3. Ground Track

As already mentioned, CryoSat-2's orbit is long-repeat (geodetic orbit) and leads to a particular drifting ground track pattern (Figure 5). Up to today, with the exception of CryoSat-2, all satellite altimeters used to measure river water levels were on short-repeat orbits. The resulting ground tracks for the main missions are displayed in Figure 5 for the Brahmaputra River in the Assam Valley, India as an example. At locations where the conventional missions with short-repeat cycles intersect with the river (or a lake), time series of water level measurements are established. This also means that processing efforts, such as water masking, can be limited to these specific locations, which are spaced with the inter-track distance indicated in Table 1. Besides that, observations from short-repeat orbits result in water level time series with relatively high temporal resolution. The availability of dense time series eases outlier filtering and integration into hydrologic models. CryoSat-2 however, as can be seen from Figure 5 and Table 1, has a much smaller inter-track distance. This requires, amongst others, continuous water masks for processing. The challenges of dealing with the drifting ground track are described in Section 5.1 for lakes and Section 6.1 for rivers.

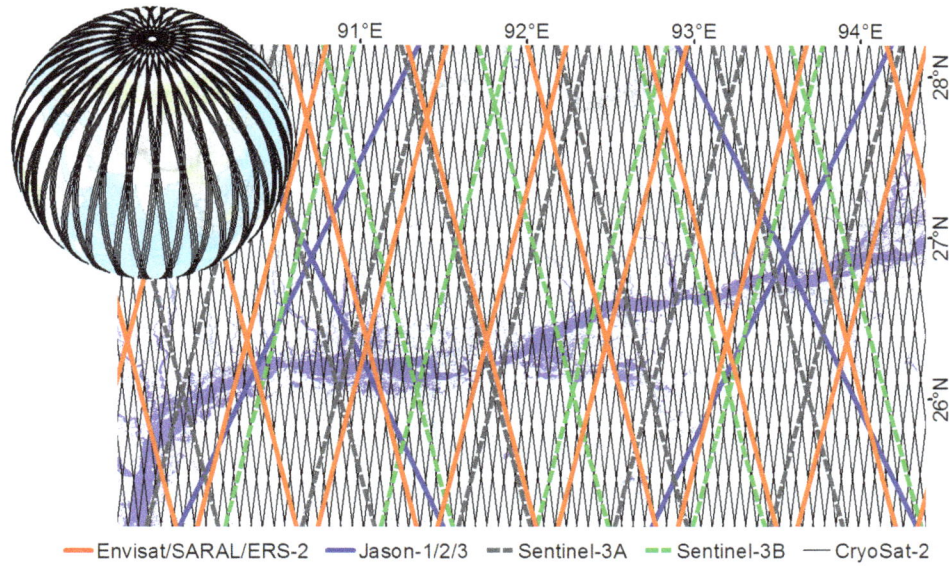

Figure 5. Ground track patterns. Inset: Ground tracks of CryoSat-2 over a period of 16 days. Main: Ground tracks of different altimetry missions over the Assam Valley with the Brahmaputra River in South Asia. Jason-1/2/3 have a repeat cycle of 10 days, Sentinel-3A and B of 27 days, and Envisat, ERS-2 and SARAL/AltiKa of 35 days. CryoSat-2 has a full repeat cycle of 369 days.

3.4. Footprint

The ground footprint size is an important characteristic that determines what the altimeter can measure [36]. The footprint is the area on the Earth's surface illuminated by the radar beam. For the pulse-limited altimeter (used by all previous radar altimetry missions), a very short duration of the pulse means that a small area is illuminated simultaneously. This is also referred to as the Pulse-limited Footprint (PLF) [45].

LRM on CryoSat-2 is the conventional pulse-limited radar altimeter mode (Figure 6). One pulse is transmitted with a very short duration (3.125 ns), so the pulse does not illuminate the whole beam width at the same time. Specifically, the illuminated area continues to grow linearly until the rear of the pulse intersects the surface at nadir [33]. Thereafter, the footprint becomes annulus with constant area for smooth surfaces. The radius r can be calculated:

$$r = \sqrt{c \cdot \tau (c \cdot \tau + h)} \tag{3}$$

where c is the speed of light, h is the altitude of the satellite and τ is the pulse length. For LRM, r is about 830 m and thus the PLF area is about 2.15 km^2. It should be noted that the true illuminated area may be discontinuous or irregular in shape due to the roughness and slope of surface [46].

In SAR/SARIn modes, the delay/Doppler beam allows the relative along-track position to be estimated relative to the position of the altimeter (Figure 6). Therefore, the illuminated area has two independent variables, i.e., along-track position and cross-track position (time delay) [47]. In the cross-track direction, the illuminated area width is the same as that in LRM. In the along-track direction, it can be seen as sharpened beam-limited. The illuminated area can be approximated by the rectangle defined by the cross-track radius and along-track width Δx (Figure 6d).

$$f_D = \frac{2v_r}{c} f_c \tag{4}$$

$$v_r = v \cdot sin\theta \tag{5}$$

$$f_c = \frac{c}{\lambda} \tag{6}$$

$$\Delta x = h \cdot sin\theta \tag{7}$$

Therefore, Δx can be expressed as:

$$\Delta x = \frac{\lambda \cdot h}{2v} \cdot f_D \tag{8}$$

where v is the velocity of the satellite; λ is the wavelength; f_D is approximately equal to the inverse of the time during which the surface is covered by the beam, i.e., PRF/64 [34].

The pulse-limited width in the cross-track direction is about 1.65 km and the sharpened beam-limited width in the along-track direction is about 300 m, thus the footprint for SAR/SARIn is about 0.5 km^2.

The SAR/SARIn modes compensate for the extra delay and thus the return waveform is much sharper than that derived from LRM [47] (Figure 6e,f). In addition, in SARIn mode, two antennas allow to precisely determine the ground position of the returned echo, because the returned echo is not necessarily from nadir point.

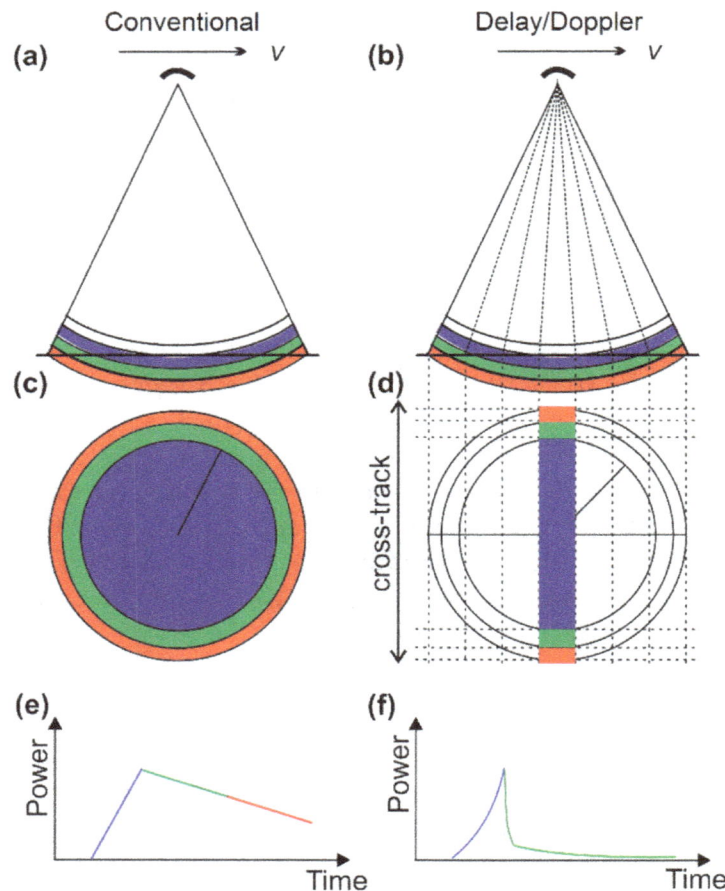

Figure 6. Comparison of a conventional pulse-limited radar altimeter and a SAR altimeter: (**a,b**) footprint side view; (**c,d**) footprint plan view; and (**e,f**) waveform. Adapted from [47].

4. Data Products

ESA provides different datasets (ftp://science-pds.cryosat.esa.int). Here, we give an introduction to Level-1b and Level-2 datasets which are used in inland water research.

4.1. Level-1b Data

Level-1b data contain the reflected waveforms and average waveforms (for LRM and SAR) along with the measurement time and geographical location. Calibration corrections are included and have been applied to the window delay computations. Signal propagation delays and other geophysical corrections are included in the data products but have not been applied to the range, therefore, the range needs to be corrected by taking these corrections into account. Data record structure is described in [34].

4.2. Level-2 GDR Data

Level-2 GDR (Geophysical Data Record) data, i.e., ground elevation, corrected for range and geo-physical effects (see Equation (2)), are produced by ESA systematically. They are the result of retracking and correcting the above discussed Level-1b data. Furthermore, other research groups produce their own Level-2 data, usually based on ESA's Level-1b data [31,32,48].

4.3 Level-3 (Along-Track) Products

Besides the ESA L2 product, very few sources provide water level data. AltWater (http://altwater.dtu.space/) from DTU space (National Space Institute, Technical University of Denmark) is the only one providing L3 water levels (along-track product) derived from CryoSat-2. While this product just covers a limited amount of lakes and reservoirs, those who aim at rivers or other inland water bodies need to process time series from scratch. Figure 7 gives a brief overview of the procedure to produce time series for lakes and water level for rivers. More details for time series construction are given in Sections 5 and 6.

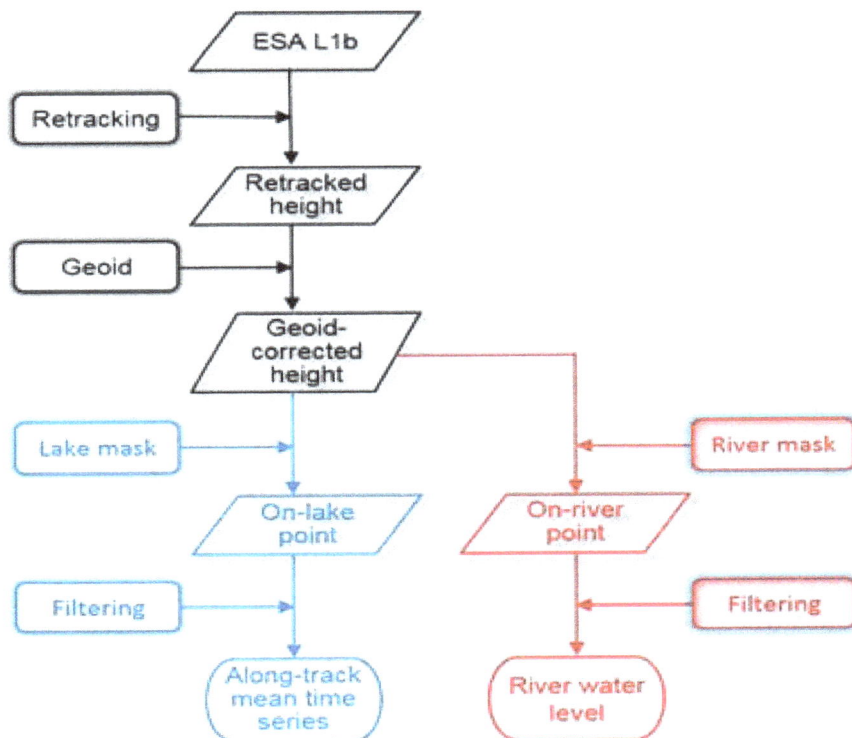

Figure 7. Flow chart of CryoSat-2 data processing.

5. Use of CryoSat-2 over Lakes

CryoSat-2, due to its special ground-track pattern, visits many lakes at global scale and provides water levels for smaller lakes than any of the previous missions. In the following sections, we will present lake level and storage variations analysis using CryoSat-2 data.

5.1. Time Series Construction

Unlike in situ hydrometric equipment which records the water level at a fixed location, altimetry provides multiple along-track measurements at different locations during a very short time slice (Figure 8). In order to investigate the variation of water level, the first step is to construct a water level time series. The most straightforward method is taking an average of all values along one track. However, outliers in the altimetry observations should be excluded. They occur, for example, near the shore because the waveforms are contaminated by land.

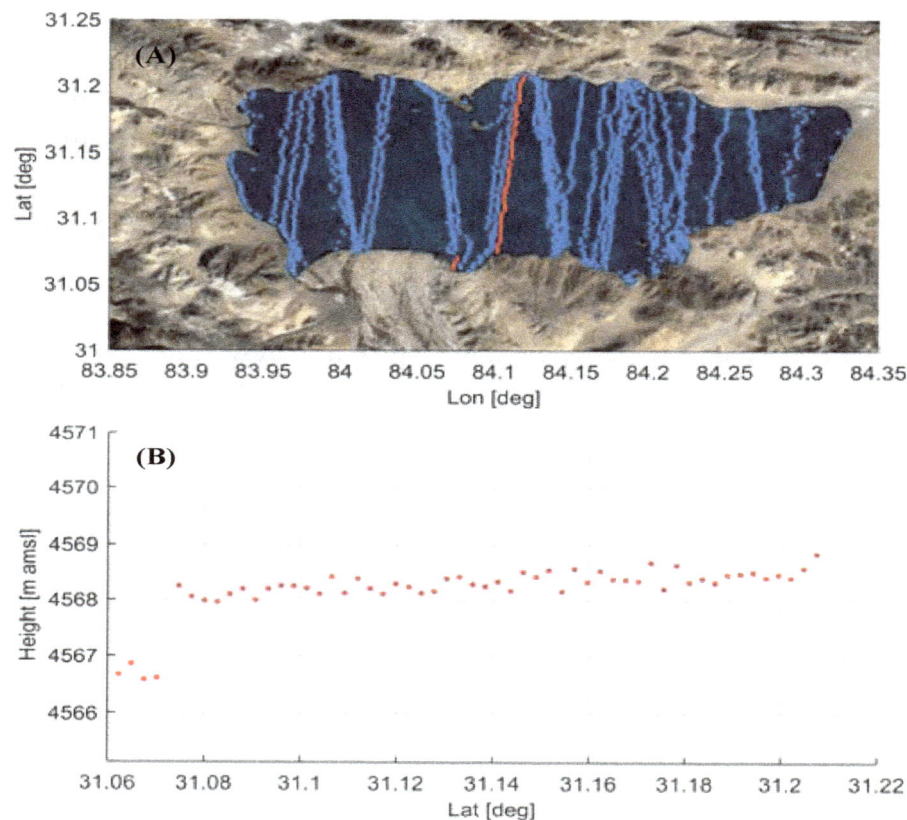

Figure 8. CryoSat-2 ground tracks over Taro Co in the Tibetan Plateau. **Upper:** ground tracks; **bottom:** measurements from track 18942, indicated by red dots on the left figure.

There exist several methods to generate along-track water levels. Kleinherenbrink et al. [48] suggested a tailored outlier removal procedure to derive the along-track mean lake level. More specifically, they utilized the mode and a threshold as the filter to identify outliers. The threshold of 1 m was chosen under the assumption that the measurements should not deviate more than 1 m from the mode. Schwatke et al. [49] employed several criteria to remove outliers, such as height error threshold, using the deviation around the median of measurements from a moving part of each track (7 km and 3 km) and support vector regression, which applies a linear regression with zero-slope constraint. However, these two approaches provide different results on outlier detection (Figure 2 in [49]). Similarly, Göttl et al. [31] excluded outliers using the moving 5-point standard deviation with a threshold of 10 cm. Nielsen et al. [18] proposed a robust method to obtain the mean water level, which assumes that the observation error follows a mixture distribution between a Gaussian and a Cauchy distribution. The advantage of this method is that the estimated mean water levels are not significantly biased by the outlying observations. The accuracy of all methods also depends on the water mask, that is, to make sure that spurious measurements are not dominant in each individual track.

The time series can be constructed by a straightforward connection of all along-track mean measurements if each individual along-track mean is accurate. Several methods are employed to estimate the water level time series, taking into account the error of measurements. In [50], besides simply connecting individual measurements, two weighted moving averaging methods were used to construct the time series. Specifically, three consecutive measurements are averaged using their error as weighting factor. Kalman filtering is used for the construction of water level time series in DAHITI [49]. The water surface was assumed to be controlled by deterministic and stochastic processes. By taking the accuracy of each track height into account, this algorithm produces an optimal estimate of water level time series. Similarly, Nielsen et al. [18] proposed a state-space model to describe the lake level variation with time under the assumption that lake level observations taken in a short time span are more strongly correlated. In this model, the true unobserved water level is described by a simple random walk. The model provides predictions of the evolution of the true lake water level. Lake levels produced with the described procedure by Nielsen et al. can be found in the aforementioned AltWater database.

5.2. Lake Level Trend Estimation

In order to investigate the characteristics of lake level changes, both inter-annual and intra-annual, generally linear or periodic non-linear regression models are used.

The Tibetan Plateau is a crucial test ground for inland water altimetry research due to the vast number of lakes, most of them unmonitored. Kleinherenbrink et al. [17] studied 30 lakes on the Tibetan Plateau and Tian Shan areas over a 2-year period. They used a simple harmonic model to determine the phases and amplitudes for both annual and semi-annual variations in addition to a long-term trend. This was successful except for some lakes, which probably exhibit water level changes with different cycles, and thus cannot be captured by this harmonic model. Jiang et al. [21] studied 70 lakes on the Tibetan Plateau over the past five years using a weighted linear regression model. Their results show that lakes are still rising at similar rate to that of 2003–2009, especially in the northern Tibetan Plateau.

Other studies focused on specific individual lakes. Song et al. [51] investigated the variation of Namco based on multiple data sources including Cryosat-2 using an iterative reweighted linear model. Similarly, Tourian et al. [52] studied the desiccation of Lake Urmia in Iran.

Lake levels, especially in endorheic lakes, like the majority of the lakes in Tibetan Plateau, are sensitive to regional climate change, which leads to changes in precipitation and evaporation. Moreover, lake level change is also controlled by regional hydrological conditions. Many studies have investigated the driving factors of significant lake level changes in order to understand and explain the mechanisms behind the change. However, lake rise on the Tibetan Plateau seems to be driven by many factors and their interactions [17,21,53–56], and a simple, process-based model remains elusive. In this context, CryoSat-2 can play an important role to further our understanding of lake response to climate change due to its dense spatial coverage.

5.3. Lake Storage Calculation

The relationship between water volume and water level is known if the bathymetry is available (Figure 9). We can calculate it by summing all small volumes as below:

$$V = \sum_{i=1}^{n} (H - H_i) A \qquad (9)$$

with V total storage, H lake surface height, H_i the bottom elevation of each volume, and A is the sectional area of volume. This can be used to establish the stage-storage curve. Thus, the instantaneous storage can be calculated using altimetry data.

However, for real-world applications, the bathymetric data is often unavailable. Instead of the total storage, storage changes can be estimated combining lake level with corresponding extent, both of which can be obtained from remote sensing datasets. Then, we can calculate the storage change under the assumption that the volume is a circular cone [57] (Figure 9):

$$S = \frac{1}{3}(H_2 - H_1)\left(A_1 + A_2 + \sqrt{A_1 \times A_2}\right) \tag{10}$$

where, S is the storage change; H_2, H_1 and A_2, A_1 are lake levels and areal extents at different dates, respectively. Sometimes, the estimation of storage change based on a constant extent gives a reasonable approximation [58]. However, with the high temporal resolution SAR imagery from Sentinel-1, dynamic lake storage changes can be estimated. For example, Baup et al. [59] estimated the volume of small lakes by combining high-resolution SAR images and altimetry.

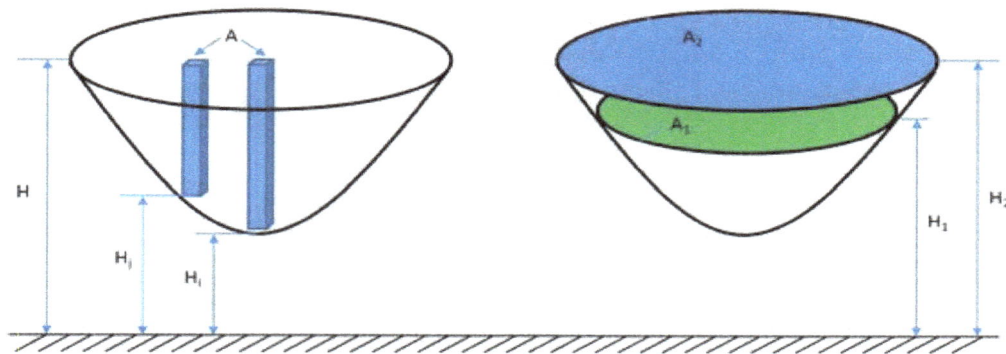

Figure 9. Illustration of lake storage and storage change calculation.

Water storage of lakes and reservoirs inferred from altimetry is of great value to regional water management and hydrologic modeling. In addition, it contributes to the understanding of total water storage changes (including ground water storage) variation inferred from GRACE. For example, the lake mass change inferred from altimetry accounts for 61% of storage increases derived from GRACE in the inner Tibetan Plateau [20].

6. Use of CryoSat-2 over Rivers

In general, the application of satellite altimetry requires the river to be of a certain minimum width around a hundred to a few hundred meters due to the footprint size of the altimeter. Otherwise, the waveform is too contaminated by the surrounding land surface. Furthermore, steep river valleys make it hard for the range window to be adapted to the river water surface if the altimeter operates in closed loop mode such as on CryoSat-2 [60]. Another important factor is the orientation of the river in relation to the ground track: All satellite altimetry missions have ground tracks with a predominantly north–south direction, which gives most regular observations over rivers flowing in the east–west direction.

6.1. Masking and Filtering

The unique drifting ground track pattern of CryoSat-2 challenges common ways of dealing with satellite altimetry data over rivers. All the conventional satellite altimetry missions have an inter-track spacing between two consecutive ground tracks of at least 80 km at the equator. This is also reflected in the processing methods of the currently active databases for inland water satellite altimetry: HydroWeb (http://www.legos.obs-mip.fr/en/soa/hydrologie/hydroweb/) applies rectangular masks at the locations of the virtual stations [61]. The DAHITI (http://dahiti.dgfi.tum.de/en/) database uses simple latitude thresholds, exploiting the fact that the altimetry satellites' ground tracks run in a predominantly north–south direction [49]. CryoSat-2 however has a much finer inter-track spacing of approximately 7.5 km (compare Figure 4 and Table 1). This means that, instead of locally limited river masks, continuous river masks are needed.

Another database, the ESA project River&Lake (http://tethys.eaprs.cse.dmu.ac.uk/RiverLake/shared/main), already used a continuous global river mask, however it is static and at low resolution [62], not accounting for potential changes in the course of the rivers over the years. High-resolution and potentially dynamic water masks, to take into account changes of the water body's extent, are derived from remote sensing imagery. Often, optical imagery, e.g., from Landsat, is used [30,63]. Such optical imagery provides high spatial and temporal resolution (30 m and 16 days, respectively, in the case of Landsat), is available freely, and easy to process. However, issues with cloud cover can severely limit the actual amount of data available, for example, making it possible to only derive one mask each year for the Brahmaputra River [30] as shown in Figure 10. Recently, readily processed, global and multi-temporal water masks from Landsat imagery have been made available, for example, the Global Surface Water Explorer [9]. Such masks could support the development of global altimetry databases. However, there also exists a weather independent alternative to derive water masks: SAR imagery. Since the start of the Sentinel-1A mission in 2014 as part of the Copernicus programme, high resolution SAR imagery is freely available. Its use could improve the river masks significantly, and hence improve the amount and quality of the extracted CryoSat-2 water level observations.

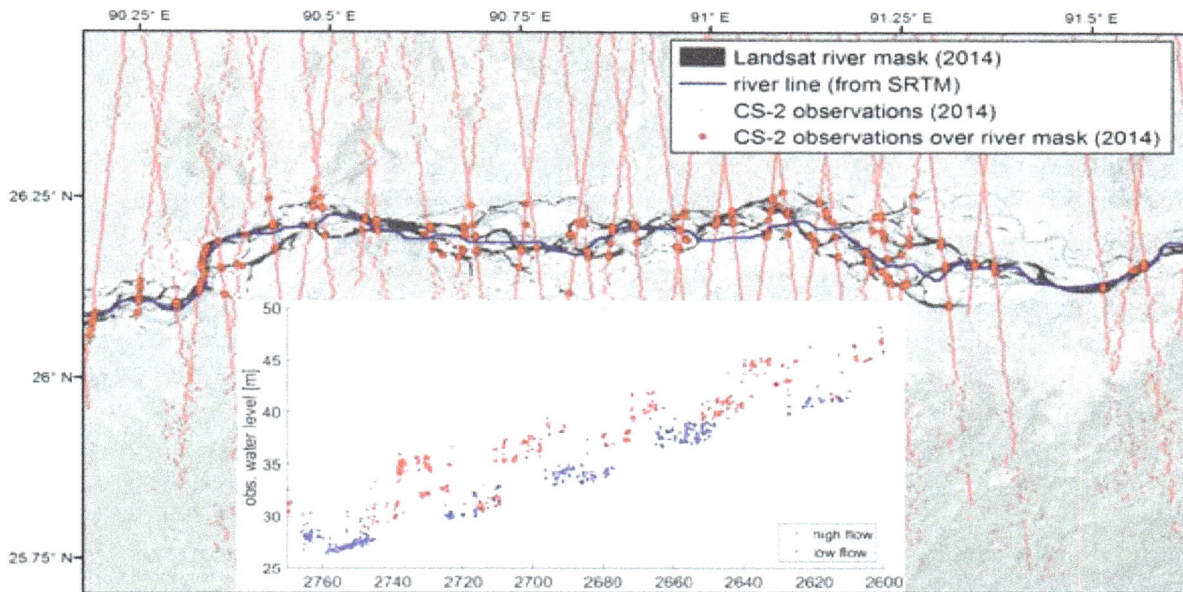

Figure 10. CryoSat-2 SARIn data over the Brahmaputra River in the Assam Valley. The map displays data for 2014, while the graph shows all data from the river stretch in the map view for 2010 to 2015, displaying the river water level profile observed by CryoSat-2.

Besides the challenges to river masking, the absence of a (short) repeat cycle also means that time series analysis for outlier filtering and uncertainty estimation of the observations cannot be applied for CryoSat data over rivers, at least not directly: In the case of virtual station time series, water level amplitudes can be observed during the course of a year, potentially detecting outliers. Both DAHITI and HydroWeb use some outlier filtering based on the whole time series for their virtual station products.

Due to these challenges, neither DAHITI nor HydroWeb (nor the now inactive River&Lake database) currently offer readily available water level measurements over rivers from CryoSat-2, but only from the repeat orbit missions.

6.2. Densification

For various purposes, such as the creation of time series and the validation of CryoSat-2 observations against in situ station observations, it can be necessary to interpolate the spatio-temporally distributed observations of CryoSat-2 onto certain points (e.g., virtual stations). To achieve this, there exist, in principle, three methods:

First, the interpolation of water level observations along the river, using the water level slope. This water level slope can, for example, be estimated from the very same data, as done by Villadsen et al. [19] with CryoSat-2 data over the Ganges and Brahmaputra rivers. They interpolated the CryoSat-2 data to Envisat virtual stations along the Brahmaputra River and found a reasonable agreement between Envisat and CryoSat-2 data. However, such simple interpolation introduces errors, for example, by not taking into account changing water level slopes between a high- and low-flow season or approaching or receding flood waves.

Another suggestion was made by Tourian et al. [50]. They used quantiles of relative water levels at virtual stations, and transferred those to defined stations using simple time lags along the river network. By this, they can densify multi-mission altimetry datasets. They also included CryoSat-2 data in their study. However, the application of this method to CryoSat-2 still is challenging as the proposed method derives the quantiles from empirical cumulative distribution functions. With the proposed method, these distributions can only be derived from virtual station time series with a sufficient number of observations at the same location. For CryoSat-2, the quantile values were interpolated from downstream and upstream virtual stations with a short repeat cycle. This interpolation potentially introduces errors: In a direct comparison of in situ and CryoSat-2 data close to the in situ stations, Tourian et al. found that CryoSat-2 performs better than most other missions. Its inclusion in the entire dataset of repeat orbit missions however slightly deteriorated the performance of the multi-mission dataset.

The third possibility is the spatio-temporal interpolation of the scattered CryoSat-2 observations to obtain continuous water level surfaces. Such an interpolation should be aware of the correlations of the water levels in time and in space, due to the river network and the physics of river flow. Kriging methods have been suggested for such tasks and applied, for example, with synthetic SWOT data over the Ganges-Brahmaputra-Meghna Delta [64] or the Tennessee River [65]. A multi-mission dataset has been interpolated with spatio-temporal kriging by Boergens et al. [66], however also here CryoSat-2 was not included. This is due to the fact that the method requires the altimetry data to be in the form of virtual station time series. Simple spatio-temporal interpolation of CryoSat-2 water level observations can be successful, at least for simple river networks as shown by Bercher et al. for CryoSat-2 data over a tributary of the Amazon River [67]. Their results, however, were not validated against in situ data.

6.3. Merging with Hydrodynamic Models

River discharge is one key component of the water cycle and is one of the most important quantities for the hydrology community. Although discharge can be inferred from altimetry data based on Manning's equation if cross-sectional geometry and roughness are available or can be estimated, the estimates of discharge are only produced at the time of altimetry overpasses [68]. As discussed above, data based densification approaches also have their shortcomings and inaccuracies. The best estimators of water levels, continuous in time and space, are hydrodynamic models. Such models then can be informed by altimetry data in different ways.

On the one hand, calibration of model parameters such as channel shape and channel roughness is possible. This was performed with data from conventional altimetry missions such as Envisat and ERS-2 [69,70], TOPEX/Poseidon [71]. Schneider et al. [30] exploited the fine spatial resolution along the course of a river (compare Figure 10). Here, CryoSat-2 data was used over the Brahmaputra River in combination with Envisat virtual station data to calibrate cross section shapes and datums of a hydrodynamic model of the river. The calibrated model is able to reproduce water level–discharge relationships with an accuracy that hardly can be obtained by using globally available DEMs such as SRTM for model parameterization only.

On the other hand, satellite altimetry data also can be used to update the states of a hydrodynamic river model, i.e., its water levels via data assimilation. This has, for example, been successfully demonstrated with Envisat data over the Brahmaputra River [22], the Amazon River [72] or over the Zambezi River [16]. Jason-2 altimetry has successfully been assimilated to a hydrodynamic model of the Ganges-Brahmaputra-Meghna river system in a real-time flood forecasting system [73]. As for many of the aforementioned methods, all these studies, however, do use altimetry data in the form of virtual station time series. To assimilate CryoSat-2 data, a flexible modelling-data assimilation approach is needed, being able to handle altimetry measurements arbitrarily distributed in time and space. One such framework was proposed by Schneider et al. [74], and has been tested with a 1D hydrodynamic model of the Brahmaputra River. With this modelling and data assimilation approach, any kind of altimetry data, including multi-mission datasets, can be ingested.

7. Discussion and Perspectives

CryoSat-2, with its SIRAL altimetry instrument, began a new era of SAR altimetry, which outperforms conventional altimeters by providing a finer along-track resolution [32]. However, to date, applications of CryoSat-2 data for hydrologic studies are scarce. This is due to the new challenges posed by CryoSat-2 compared to previous missions which originate from the drifting ground track pattern. These issues, mentioned above, are reflected in the limited availability of CryoSat-2 water level data in inland water altimetry databases.

However, besides the new altimetry instrument, CryoSat-2 has other advantages over the previous short-repeat missions. The small inter-track distance is beneficial over lakes: For example, the study of Kleinherenbrink et al. [17] shows that CryoSat visited 125 lakes with at least four passes over the period February 2012 to January 2014 in the Tibetan Plateau and Tian Shan area. Actually, CryoSat-2 sampled more than 400 lakes with at least ten passes over 6 years in the Tibetan Plateau (Figure 11), including practically all lakes with surface areas exceeding 5 km². This is unachievable using previous altimetry missions. Moreover, the study of Nielsen et al. [18] indicates that CryoSat-2 has better precision than Envisat with regard to mean water level. Therefore, CryoSat-2 shows great advantages over other missions in lake level monitoring.

CryoSat-2 also can be used beneficially over rivers. One important novelty and advantage of CryoSat-2 is that its ground track pattern allows deriving high-resolution water level profiles [26] (Figure 10). Similarly, data from ICESat with slightly lower along-river resolution than CryoSat-2 has been used for a hydraulic characterization of the Congo River [75], allowing to derive water surface slopes with greater detail than those previously derived from virtual station altimetry. Such hydraulic characteristics of rivers can be used to parameterize or calibrate hydrodynamic models of the same rivers, with higher spatial resolution than previously possible. Furthermore, with a flexible data assimilation approach, CryoSat-2 data can be used to update hydrodynamic models [74].

Moreover, the higher resolution and the better signal-to-noise ratio of SAR/SARIn data allow monitoring water levels of narrow rivers [30,76], for which the application of satellite altimetry for water level measurements has been restricted in the past [77]. Given that most major rivers are less than 1 km wide [78,79], SAR altimetry will be of great value for river water level monitoring due to its higher spatial resolution and higher precision. With the operation of Sentinel-3A and upcoming Sentinel-3B, SAR altimetry is likely to get more attention in river monitoring with the increase in both accuracy and coverage (inter-track distance at the equator 52 km of the constellation).

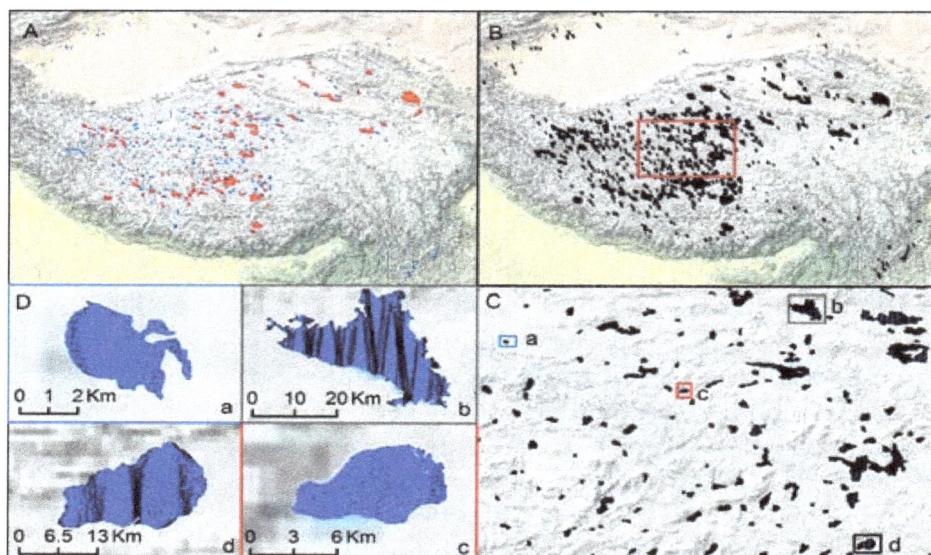

Figure 11. Map of lakes in the Tibetan Plateau seen by Envisat/SARAL ((**A**), in red) and CryoSat-2 (SARIn mode from 2010 to 2016 in black); (**C**) Zoom-in of the red rectangle in (**B**); (**D**) CryoSat-2 measurements of four lakes which have surface areas between 10 to 420 km^2.

Some of the strategies developed for CryoSat-2, such as new approaches to densification of altimetry measurements and combination with hydrodynamic models will support processing and application of data from the upcoming SWOT mission. SWOT, expected to be launched in 2020 by NASA, will be the first satellite altimetry mission to provide images of water heights (instead of points at nadir) [80]. Hence, its capabilities and potential applications over inland water bodies have been assessed in many studies [64,81–84]. The mission's new sampling pattern will require some of the techniques developed for CryoSat-2 data, such as flexible data assimilation approaches or interpolation methods. Also, SWOT will provide instantaneous water level profiles along rivers within single measurements (images). This is different from CryoSat-2, which provides water level profiles sampled in a series of consecutive overflights. Combination of all the different missions and data types will greatly enhance our understanding of water storage and flow characteristics and support model calibration and operational forecasting.

8. Conclusions

Being an ongoing mission, CryoSat-2 will continue to build up its dataset starting in 2010. In this review, promising applications of CryoSat-2 over lakes and rivers have been summarized. CryoSat-2 can be used beneficially over inland water. If exploited correctly, it features some significant advantages over previous short-repeat cycle missions mostly related to its drifting ground track pattern and resulting shorter inter-track distance.

Using CryoSat-2 data requires moving beyond the concept of virtual stations, or station data in general. Some of the data processing developed for CryoSat-2 inland water altimetry data will be useful for handling data delivered by the state-of-the-art Sentinel-3 altimeter. In general, the technology push created by CryoSat-2 can ultimately lead to data processing and model integration techniques being able to effectively handle altimetry datasets with arbitrary spatio-temporal distribution, or multi-mission datasets. This could also include new, unconventional data types, such as data acquired from UAVs (Unmanned Aerial Vehicle).

The new generation of SAR altimeters (Sentinel-3, Jason-CS/Sentinel-6 and SWOT) is expected to provide higher resolution data at unprecedented spatial coverage. This will further improve our abilities to monitor surface water variations, especially in data-sparse regions, and also help to advance forecasting applications.

Acknowledgments: Liguang Jiang is supported by China Scholarship Council, which is gratefully acknowledged.

Author Contributions: All authors contributed extensively to this paper.

References

1.   Vorosmarty, C.J. Global Water Resources: Vulnerability from Climate Change and Population Growth. *Science* **2000**, *289*, 284–288. [CrossRef] [PubMed]
2.   Flood Fury: Why Brahmaputra's Trail of Destruction Has Become Annual Ritual in Assam. The Indian Express. Available online: http://indianexpress.com/article/india/india-news-india/flood-fury-why-brahmapurtas-trail-of-destruction-has-become-annual-ritual-in-assam-2958587/ (accessed on 16 January 2017).
3.   United States Geological Survey (USGS). The Water Cycle. Available online: http://water.usgs.gov/edu/watercyclefreshstorage.html (accessed on 10 March 2017).
4.   Ma, R.; Duan, H.; Hu, C.; Feng, X.; Li, A.; Ju, W.; Jiang, J.; Yang, G. A half-century of changes in China's lakes: Global warming or human influence? *Geophys. Res. Lett.* **2010**, *37*. [CrossRef]
5.   Oren, A.; Plotnikov, I.S.; Sokolov, S.S.; Aladin, N.V. The Aral Sea and the Dead Sea: Disparate lakes with similar histories. *Lakes Reserv. Res. Manag.* **2010**, *15*, 223–236. [CrossRef]
6.   Carroll, M.L.; Townshend, J.R.G.; Dimiceli, C.M.; Loboda, T.; Sohlberg, R.A. Shrinking lakes of the Arctic: Spatial relationships and trajectory of change. *Geophys. Res. Lett.* **2011**, *38*, 1–5. [CrossRef]
7.   Gao, H.; Bohn, T.J.; Podest, E.; McDonald, K.C.; Lettenmaier, D.P. On the causes of the shrinking of Lake Chad. *Environ. Res. Lett.* **2011**, *6*, 34021. [CrossRef]
8.   Tao, S.; Fang, J.; Zhao, X.; Zhao, S.; Shen, H.; Hu, H.; Tang, Z.; Wang, Z.; Guo, Q. Rapid loss of lakes on the Mongolian Plateau. *Proc. Natl. Acad. Sci. USA* **2015**, *112*, 2281–2286. [CrossRef] [PubMed]
9.   Pekel, J.-F.; Cottam, A.; Gorelick, N.; Belward, A.S. High-resolution mapping of global surface water and its long-term changes. *Nature* **2016**, *540*, 418–422. [CrossRef] [PubMed]
10. Da Silva, J.S.; Calmant, S.; Seyler, F.; Moreira, D.M.; Oliveira, D.; Monteiro, A. Radar Altimetry Aids Managing Gauge Networks. *Water Resour. Manag.* **2014**, *28*, 587–603. [CrossRef]
11. Global Runoff Data Center (GRDC). Global Runoff Database 2015. Available online: http://www.bafg.de/GRDC/EN/Home/homepage_node.html (accessed on 10 March 2017).
12. Biancamaria, S.; Hossain, F.; Lettenmaier, D.P. Forecasting transboundary river water elevations from space. *Geophys. Res. Lett.* **2011**, *38*, 1–5. [CrossRef]
13. Birkett, C.M. The contribution of TOPEX/POSEIDON to the global monitoring of climatically sensitive lakes. *J. Geophys. Res.* **1995**, *100*, 25179. [CrossRef]
14. Berry, P.A.M.; Garlick, J.D.; Freeman, J.A.; Mathers, E.L. Global inland water monitoring from multi-mission altimetry. *Geophys. Res. Lett.* **2005**, *32*, 1–4. [CrossRef]
15. Crétaux, J.F.; Birkett, C. Lake studies from satellite radar altimetry. *C. R. Geosci.* **2006**, *338*, 1098–1112. [CrossRef]
16. Michailovsky, C.I.; Bauer-Gottwein, P. Operational reservoir inflow forecasting with radar altimetry: The Zambezi case study. *Hydrol. Earth Syst. Sci.* **2014**, *18*, 997–1007. [CrossRef]
17. Kleinherenbrink, M.; Lindenbergh, R.C.; Ditmar, P.G. Monitoring of lake level changes on the Tibetan Plateau and Tian Shan by retracking Cryosat SARIn waveforms. *J. Hydrol.* **2015**, *521*, 119–131. [CrossRef]
18. Nielsen, K.; Stenseng, L.; Andersen, O.B.; Villadsen, H.; Knudsen, P. Validation of CryoSat-2 SAR mode based lake levels. *Remote Sens. Environ.* **2015**, *171*, 162–170. [CrossRef]
19. Villadsen, H.; Andersen, O.B.; Stenseng, L.; Nielsen, K.; Knudsen, P. CryoSat-2 altimetry for river level monitoring—Evaluation in the Ganges–Brahmaputra River basin. *Remote Sens. Environ.* **2015**, *168*, 80–89. [CrossRef]
20. Zhang, G.; Yao, T.; Xie, H.; Kang, S.; Lei, Y. Increased mass over the Tibetan Plateau: From lakes or glaciers? *Geophys. Res. Lett.* **2013**, *40*, 2125–2130. [CrossRef]
21. Jiang, L.; Nielsen, K.; Andersen, O.B.; Bauer-Gottwein, P. Monitoring recent lake level variations on the Tibetan Plateau using CryoSat-2 SARIn mode data. *J. Hydrol.* **2017**, *544*, 109–124. [CrossRef]

22. Michailovsky, C.I.; Milzow, C.; Bauer-Gottwein, P. Assimilation of radar altimetry to a routing model of the Brahmaputra River. *Water Resour. Res.* **2013**, *49*, 4807–4816. [CrossRef]

23. Maswood, M.; Hossain, F. Advancing River Modeling in Ungauged Basins using Satellite Remote Sensing: The case of the Ganges-Brahmaputra-Meghna basins. *Int. J. River Basin Manag.* **2016**, *14*, 103–117. [CrossRef]

24. Paris, A.; de Paiva, R.D.; da Silva, J.S.; Moreira, D.M.; Calmant, S.; Garambois, P.-A.; Collischonn, W.; Bonnet, M.-P.; Seyler, F. Stage-discharge rating curves based on satellite altimetry andmodeled discharge in the Amazon basin. *Water Resour. Res.* **2016**, *52*, 3787–3814. [CrossRef]

25. Semmling, M.; Beyerle, G.; Beckheinrich, J.; Ge, M.; Wickert, J. Airborne GNSS reflectometry using crossover reference points for carrier phase altimetry. In Proceedings of the 2014 IEEE International Geoscience and Remote Sensing Symposium, Quebec City, QC, Canada, 13–18 July 2014; pp. 3786–3789.

26. Jin, S.; Komjathy, A. GNSS reflectometry and remote sensing: New objectives and results. *Adv. Space Res.* **2010**, *46*, 111–117. [CrossRef]

27. Richter, A.; Popov, S.V.; Fritsche, M.; Lukin, V.V.; Matveev, A.Y.; Ekaykin, A.A.; Lipenkov, V.Y.; Fedorov, D.V.; Eberlein, L.; Schröder, L.; et al. Height changes over subglacial Lake Vostok, East Antarctica: Insights from GNSS observations. *J. Geophys. Res. Earth Surf.* **2014**, *119*, 2460–2480. [CrossRef]

28. Asadzadeh Jarihani, A.; Callow, J.N.; Johansen, K.; Gouweleeuw, B. Evaluation of multiple satellite altimetry data for studying inland water bodies and river floods. *J. Hydrol.* **2013**, *505*, 78–90. [CrossRef]

29. Ricker, R.; Hendricks, S.; Helm, V.; Gerdes, R. Classification of CryoSat-2 radar echoes. In *Towards an Interdisciplinary Approach in Earth System Science: Advances of a Helmholtz Graduate Research School*; Springer: Cham, Switzerland, 2015; pp. 149–158.

30. Schneider, R.; Godiksen, P.N.; Villadsen, H.; Madsen, H.; Bauer-Gottwein, P. Application of CryoSat-2 altimetry data for river analysis and modelling. *Hydrol. Earth Syst. Sci.* **2017**, *21*, 751–764. [CrossRef]

31. Göttl, F.; Dettmering, D.; Müller, F.; Schwatke, C. Lake Level Estimation Based on CryoSat-2 SAR Altimetry and Multi-Looked Waveform Classification. *Remote Sens.* **2016**, *8*, 885. [CrossRef]

32. Villadsen, H.; Deng, X.; Andersen, O.B.; Stenseng, L.; Nielsen, K.; Knudsen, P. Improved inland water levels from SAR altimetry using novel empirical and physical retrackers. *J. Hydrol.* **2016**, *537*, 234–247. [CrossRef]

33. Chelton, D.B.; Ries, J.C.; Haines, B.J.; Fu, L.-L.; Callahan, P.S. Satellite altimetry. In *Satellite Altimetry and Earth Sciences: A Handbook of Techniques and Applications*; Academic Press: Cambridge, MA, USA, 2001; Volume 69, pp. 2504–2510.

34. European Space Agency. *Mullar Space Science Laboratory CryoSat Product Handbook*; European Space Agency: Paris, France, 2012; Volume DLFE-3605. Available online: https://earth.esa.int/c/document_library/get_file?folderId=125272&name=DLFE-3605.pdf (accessed on 27 January 2017).

35. Jain, M.; Andersen, O.B.; Dall, J. *Improved Sea Level Determination in the Arctic Regions through Development of Tolerant Altimetry Retracking*; Technical University of Denmark: Kgs. Lyngby, Denmark, 2015.

36. Rosmorduc, V.; Benveniste, J.; Lauret, O.; Maheu, C.; Milagro, M.; Picot, N. *Radar Altimetry Tutorial*; Benveniste, J., Picot, N., Eds.; European Space Agency: Paris, France, 2011.

37. Bao, L.; Lu, Y.; Wang, Y. Improved retracking algorithm for oceanic altimeter waveforms. *Prog. Nat. Sci.* **2009**, *19*, 195–203. [CrossRef]

38. Jain, M.; Andersen, O.B.; Dall, J.; Stenseng, L. Sea surface height determination in the Arctic using Cryosat-2 SAR data from primary peak empirical retrackers. *Adv. Space Res.* **2015**, *55*, 40–50. [CrossRef]

39. Davis, C.H. Growth of the Greenland ice sheet: A performance assessment of altimeter retracking algorithms. *IEEE Trans. Geosci. Remote Sens.* **1995**, *33*, 1108–1116. [CrossRef]

40. European Space Agency. CryoSat Level-2 Product Evolutions and Quality Improvements in Baseline C. 2015. Available online: https://earth.esa.int/documents/10174/1773005/C2-Evolution-BaselineC-Level2-V3 (accessed on 27 January 2017).

41. Birkett, C.M.; Beckley, B.D. Investigating the Performance of the Jason-2/OSTM Radar Altimeter over Lakes and Reservoirs. *Mar. Geod.* **2010**, *33*, 204–238. [CrossRef]

42. Geographical Mode Mask. Content—Earth Online—ESA. Available online: https://earth.esa.int/web/guest/-/geographical-mode-mask-7107 (accessed on 16 January 2017).

43. Wingham, D.J.; Francis, C.R.; Baker, S.; Bouzinac, C.; Brockley, D.; Cullen, R.; de Chateau-Thierry, P.; Laxon, S.W.; Mallow, U.; Mavrocordatos, C.; et al. CryoSat: A mission to determine the fluctuations in Earth's land and marine ice fields. *Adv. Space Res.* **2006**, *37*, 841–871. [CrossRef]

44. Ulaby, F.T.; Long, D.G.; Blackwell, W.J.; Elachi, C.; Fung, A.K.; Ruf, C.; Sarabandi, K.; Zebker, H.A.; van Zyl, J. *Microwave Radar and Radiometric Remote Sensing*; University of Michigan Press: Ann Arbor, MI, USA, 2014.

45. Soussi, B. ENVISAT ALTIMETRY Level 2 User Manual. Available online: https://earth.esa.int/pub/ESA_DOC/ENVISAT/RA2-MWR/PH_light_1rev4_ESA.pdf (accessed on 27 January 2017).

46. Fetterer, F.M.; Drinkwater, M.R.; Jezek, K.C.; Laxon, S.W.C.; Onstott, R.G.; Ulander, L.M.H. Sea ice altimetry. In *Microwave Remote Sensing of Sea Ice*; Carsey, F.D., Ed.; Geophysical Monograph Series; American Geophysical Union (AGU): Washington, DC, USA, 1992; Volume 68, pp. 111–135.

47. Keith Raney, R. The delay/doppler radar altimeter. *IEEE Trans. Geosci. Remote Sens.* **1998**, *36*, 1578–1588. [CrossRef]

48. Kleinherenbrink, M.; Ditmar, P.G.; Lindenbergh, R.C. Retracking Cryosat data in the SARIn mode and robust lake level extraction. *Remote Sens. Environ.* **2014**, *152*, 38–50. [CrossRef]

49. Schwatke, C.; Dettmering, D.; Bosch, W.; Seitz, F. DAHITI—An innovative approach for estimating water level time series over inland waters using multi-mission satellite altimetry. *Hydrol. Earth Syst. Sci.* **2015**, *19*, 4345–4364. [CrossRef]

50. Tourian, M.J.; Tarpanelli, A.; Elmi, O.; Qin, T.; Brocca, L.; Moramarco, T.; Sneeuw, N. Spatiotemporal densification of river water level time series by multimission satellite altimetry. *Water Resour. Res.* **2016**, *52*, 1140–1159. [CrossRef]

51. Song, C.; Ye, Q.; Cheng, X. Shifts in water-level variation of Namco in the central Tibetan Plateau from ICESat and CryoSat-2 altimetry and station observations. *Sci. Bull.* **2015**, *60*, 1287–1297. [CrossRef]

52. Tourian, M.J.; Elmi, O.; Chen, Q.; Devaraju, B.; Roohi, S.; Sneeuw, N. A spaceborne multisensor approach to monitor the desiccation of Lake Urmia in Iran. *Remote Sens. Environ.* **2015**, *156*, 349–360. [CrossRef]

53. Zhang, G.; Xie, H.; Kang, S.; Yi, D.; Ackley, S.F. Monitoring lake level changes on the Tibetan Plateau using ICESat altimetry data (2003–2009). *Remote Sens. Environ.* **2011**, *115*, 1733–1742. [CrossRef]

54. Gao, L.; Liao, J.; Shen, G. Monitoring lake-level changes in the Qinghai–Tibetan Plateau using radar altimeter data (2002–2012). *J. Appl. Remote Sens.* **2013**, *7*, 73470. [CrossRef]

55. Li, Y.; Liao, J.; Guo, H.; Liu, Z.; Shen, G. Patterns and potential drivers of dramatic changes in Tibetan lakes, 1972–2010. *PLoS ONE* **2014**, *9*, e111890. [CrossRef] [PubMed]

56. Song, C.; Huang, B.; Ke, L. Heterogeneous change patterns of water level for inland lakes in High Mountain Asia derived from multi-mission satellite altimetry. *Hydrol. Process.* **2015**, *29*, 2769–2781. [CrossRef]

57. Taube, C.M. Chapter 12: Three methods for computing the volume of a lake. In *Manual of Fisheries Survey Mehtods II: With Periodic Updates*; Schneider, J.C., Ed.; Michigan Department of Natural Resources: Lansing, MI, USA, 2000.

58. Phan, V.H.; Lindenbergh, R.; Menenti, M. ICESat derived elevation changes of Tibetan lakes between 2003 and 2009. *Int. J. Appl. Earth Obs. Geoinf.* **2012**, *17*, 12–22. [CrossRef]

59. Baup, F.; Frappart, F.; Maubant, J. Combining high-resolution satellite images and altimetry to estimate the volume of small lakes. *Hydrol. Earth Syst. Sci.* **2014**, *18*, 2007–2020. [CrossRef]

60. Dehecq, A.; Gourmelen, N.; Shepherd, A.; Cullen, R.; Trouvé, E. Evaluation of CryoSat-2 for height retrieval over the Himalayan range. In Proceedings of the CryoSat-2 Third User Workshop, Dresden, Germany, 12–14 March 2013.

61. Rosmorduc, V. *Hydroweb Product User Manual Version 1.0*; Theia—Land Data Centre: Ramonville-Saint-Agne, France, 2016.

62. Berry, P.A.M.; Wheeler, J.L. *JASON2-ENVISAT Exploitation—Development of Algorithms for the Exploitation of JASON2-ENVISAT Altimetry for the Generation of a River and Lake Product*; Product Handbook v3.5; De Montfort University: Leicester, UK, 2009.

63. Michailovsky, C.I.; McEnnis, S.; Berry, P.A.M.; Smith, R.; Bauer-Gottwein, P. River monitoring from satellite radar altimetry in the Zambezi River basin. *Hydrol. Earth Syst. Sci.* **2012**, *16*, 2181–2192. [CrossRef]

64. Paiva, R.C.D.; Durand, M.T.; Hossain, F. Spatiotemporal interpolation of discharge across a river network by using synthetic SWOT satellite data. *Water Resour. Res.* **2015**, *51*, 430–449. [CrossRef]

65. Yoon, Y.; Durand, M.; Merry, C.J.; Rodriguez, E. Improving temporal coverage of the SWOT mission using spatiotemporal kriging. *IEEE J. Sel. Top. Appl. Earth Obs. Remote Sens.* **2013**, *6*, 1719–1729. [CrossRef]

66. Boergens, E.; Buhl, S.; Dettmering, D.; Klüppelberg, C.; Seitz, F. Combination of multi-mission altimetry data along the Mekong River with spatio-temporal kriging. *J. Geod.* **2016**. [CrossRef]

67. Bercher, N.; Dinardo, S.; Lucas, B.M.; Fleury, S.; Calmant, S.; Crétaux, J.-F.; Femenias, P.; Boy, F.; Picot, N.; Benveniste, J. Applications of CryoSat-2 SAR & SARIn modes for the monitoring of river water levels. In Proceedings of the CryoSat Third User Workshop, Dresden, Germany, 12–14 March 2013; pp. 1–7.

68. Lettenmaier, D.P.; Alsdorf, D.; Dozier, J.; Huffman, G.J.; Pan, M.; Wood, E.F. Inroads of remote sensing into hydrologic science during the WRR era. *Water Resour. Res.* **2015**, *51*, 7309–7342. [CrossRef]

69. Domeneghetti, A.; Tarpanelli, A.; Brocca, L.; Barbetta, S.; Moramarco, T.; Castellarin, A.; Brath, A. The use of remote sensing-derived water surface data for hydraulic model calibration. *Remote Sens. Environ.* **2014**, *149*, 130–141. [CrossRef]

70. Yan, K.; Tarpanelli, A.; Balint, G.; Moramarco, T.; Di Baldassarre, G. Exploring the Potential of SRTM Topography and Radar Altimetry to Support Flood Propagation Modeling: Danube Case Study. *J. Hydrol. Eng.* **2014**, *20*, 04014048. [CrossRef]

71. Biancamaria, S.; Bates, P.D.; Boone, A.; Mognard, N.M. Large-scale coupled hydrologic and hydraulic modelling of the Ob River in Siberia. *J. Hydrol.* **2009**, *379*, 136–150. [CrossRef]

72. Paiva, R.C.D.; Collischonn, W.; Bonnet, M.-P.; de Gonçalves, L.G.G.; Calmant, S.; Getirana, A.; da Santos Silva, J. Assimilating in situ and radar altimetry data into a large-scale hydrologic-hydrodynamic model for streamflow forecast in the Amazon. *Hydrol. Earth Syst. Sci.* **2013**, *10*, 2879–2925.

73. Hossain, F.; Maswood, M.; Siddique-E-Akbor, A.H.; Yigzaw, W.; Mazumdar, L.C.; Ahmed, T.; Hossain, M.; Shah-Newaz, S.M.; Limaye, A.; Lee, H.; et al. A promising radar altimetry satellite system for operational flood forecasting in flood-prone Bangladesh. *IEEE Geosci. Remote Sens. Mag.* **2014**, *2*, 27–36. [CrossRef]

74. Schneider, R.; Godiksen, P.N.; Ridler, M.-E.; Villadsen, H.; Madsen, H.; Bauer-Gottwein, P. Combining Envisat type and CryoSat-2 altimetry to inform hydrodynamic models. In *Proceedings Living Planet Symposium 2016*; Ouwehand, L., Ed.; ESA Special Publications SP-740; European Space Agency: Paris, France, 2016.

75. O'Loughlin, F.; Trigg, M.A.; Schumann, G.J.P.; Bates, P.D. Hydraulic characterization of the middle reach of the Congo River. *Water Resour. Res.* **2013**, *49*, 5059–5070. [CrossRef]

76. Cotton, P.D.; Andersen, O.; Stenseng, L.; Boy, F.; Cancet, M.; Cipollini, P.; Gommenginger, C.; Dinardo, S.; Egido, A.; Fernandes, M.J.; et al. *Improved Oceanographic Measurements from SAR Altimetry: Results and Scientific Roadmap from ESA Cryosat Plus for Oceans Project*; Special Publication ESA SP 2016; European Space Agency: Paris, France, 2016.

77. Maillard, P.; Bercher, N.; Calmant, S. New processing approaches on the retrieval of water levels in Envisat and SARAL radar altimetry over rivers: A case study of the São Francisco River, Brazil. *Remote Sens. Environ.* **2015**, *156*, 226–241. [CrossRef]

78. Yamazaki, D.; O'Loughlin, F.; Trigg, M.A.; Miller, Z.F.; Pavelsky, T.M.; Bates, P.D. Development of the Global Width Database for Large Rivers. *Water Resour. Res.* **2014**, *50*, 3467–3480. [CrossRef]

79. Andreadis, K.M.; Schumann, G.J.P.; Pavelsky, T. A simple global river bankfull width and depth database. *Water Resour. Res.* **2013**, *49*, 7164–7168. [CrossRef]

80. Biancamaria, S.; Lettenmaier, D.P.; Pavelsky, T.M. The SWOT Mission and Its Capabilities for Land Hydrology. *Surv. Geophys.* **2016**, *37*, 307–337. [CrossRef]

81. Biancamaria, S.; Durand, M.; Andreadis, K.M.; Bates, P.D.; Boone, A.; Mognard, N.M.; Rodríguez, E.; Alsdorf, D.E.; Lettenmaier, D.P.; Clark, E.A. Assimilation of virtual wide swath altimetry to improve Arctic river modeling. *Remote Sens. Environ.* **2011**, *115*, 373–381. [CrossRef]

82. Yoon, Y.; Durand, M.; Merry, C.J.; Clark, E.A.; Andreadis, K.M.; Alsdorf, D.E. Estimating river bathymetry from data assimilation of synthetic SWOT measurements. *J. Hydrol.* **2012**, *464–465*, 363–375. [CrossRef]

83. Durand, M.; Neal, J.; Rodríguez, E.; Andreadis, K.M.; Smith, L.C.; Yoon, Y. Estimating reach-averaged discharge for the River Severn from measurements of river water surface elevation and slope. *J. Hydrol.* **2014**, *511*, 92–104. [CrossRef]

84. Garambois, P.-A.; Monnier, J. Inference of effective river properties from remotely sensed observations of water surface. *Adv. Water Resour.* **2015**, *79*, 103–120. [CrossRef]

12

European Rice Cropland Mapping with Sentinel-1 Data: The Mediterranean Region Case Study

Duy Ba Nguyen [1,2,*] **and Wolfgang Wagner** [2]

[1]   Department of Geodesy and Geoinformation, Vienna University of Technology, Gusshausstrasse 27-29, A-1040 Vienna, Austria

[2]   Department of Photogrammetry and Remote Sensing, Hanoi University of Mining and Geology, Hanoi 10000, Vietnam; wolfgang.wagner@geo.tuwien.ac.at

*   Correspondence: nguyenbaduy@humg.edu.vn

Academic Editors: Arjen Y. Hoekstra, Frédéric Frappart and Luc Bourrel

Abstract: Rice farming is one of the most important activities in the agriculture sector, producing staple food for the majority of the world's growing population. Accurate and up-to-date assessment of the spatial distribution of rice cultivated area is a key information requirement of all stakeholders including policy makers, rice farmers and consumers. Timely assessment with high precision is, e.g., crucial for water resource management, market prices control and during humanitarian food crisis. Recently, two Sentinel-1 (S-1) satellites carrying a C-band Synthetic Aperture Radar (SAR) sensor were launched by the European Space Agency (ESA) within the homework of the Copernicus program. The advanced data acquisition capabilities of S-1 provide a unique opportunity to monitor different land cover types at high spatial (20 m) and temporal (twice-weekly to biweekly) resolution. The objective of this research is to evaluate the applicability of an existing phenology-based classification method for continental-scale rice cropland mapping using S-1 backscatter time series. In this study, the S-1 images were collected during the rice growing season of 2015 covering eight selected European test sites situated in six Mediterranean countries. Due to the better rice classification capabilities of SAR cross-polarized measurement as compared to co-polarized data, S-1 cross-polarized (VH) data were used. Phenological parameters derived from the S-1 VH backscatter time series were used as an input to a knowledge-based decision-rule classifier in order to classify the input data into rice and non-rice areas. The classification results were evaluated using multiple regions of interest (ROIs) drawn from high-resolution optical remote sensing (SPOT 5) data and the European CORINE land cover (CLC 2012) product. An overall accuracy of more than 70% for all eight study sites was achieved. The S-1 based classification maps reveal much more details compared to the rice field class contained in the CLC 2012 product. These findings demonstrate the potential and feasibility of using S-1 VH data to develop an operational rice crop monitoring framework at the continental scale.

Keywords: rice mapping; Sentinel-1 A; SAR time series; remote sensing

---

1. Introduction

Europe is the fourth largest importer of rice in the world (Figure 1a). Over the last five years, Europe's annual rice imports were on average about 1 million tons (milled basis), ranging between 873 million tons in 2011 and 1190 million tons in 2015 [1]. Meanwhile, the size of rice cultivated areas has either reduced or remained stable around 420,000 hectares (Figure 1b). Information about how the area of rice croplands varies from year to year is an important piece of information for the European economy being relevant for: risk management for the insurance industry [2], environmental reporting, contributions to greenhouse gases [3], life cycle inventory [4,5], life cycle assessment [6,7], water cycle analysis [8], crop forecasting [9], and others.

According to the FAO Technical Guidelines, the standard procedure to derive rice cropland layers is to manually digitize optical (near-infrared) satellite imagery [10,11]. This process provides a data product with an internal and external quality assurance. However, the process is time-consuming, expensive, labor-intensive and best suited for small scale applications, whereas over large regions classification results may not always be directly comparable because the results achieved by different experts may differ. One rice cropland layer produced in this way is part of the CORINE land cover (CLC) product that has a resolution of 100 m and will be updated after every 4–6 years. This does not meet the requirements of many users who prefer having annual rice cropland maps with higher spatial resolution (e.g., 20 m).

Day–night and weather independent data acquisition capabilities of SAR sensors have made them attractive for monitoring land surface dynamics. With the launch of Sentinel-1A/B (a C-band space-borne SAR sensor), data with much improved spatio-temporal sampling characterizations have become available (over Europe 20 m spatial resolution images are acquired every three days). Furthermore, the free and open data policy adopted by the Copernicus program will make the data accessible to a large user community, and will help accelerating the progress in geophysical research in general and paddy rice cropland mapping in specific.

There is a long history of rice cropland area mapping with SAR sensor imagery [12–16]. Multi-temporal SAR data can be used to retrieve the rice growing cycle based on the temporal variations in the SAR backscatter ($\sigma°$(dB)) signal [12,16,17]. The annual variation in $\sigma°$(dB) from rice fields is higher than any other agricultural crop [13]. In this particular application domain, a substantial number of studies have already been reported in the literature. Most of them used medium resolution SAR data over test sites, mostly located in India [18], Bangladesh [19], Thailand [20], China [21–23], Vietnam [14–16,24] and also a few in Europe [25,26]. High spatial resolution SAR data ($\leq$20 m) have been investigated over a fewer study sites because of the difficulty of collecting dense and long-enough image time series [27–30].

Time series analysis of SAR backscatter values is the most common data analysis approach used for paddy rice identification. Both single and multi-polarization SAR data have been used for rice monitoring and the discrimination of different growth stages [14,31]. In this approach, C-band SAR sensor have been the most attractive data source for rice mapping at regional or continental scale because data from other SAR sensors is hampered either by limited spatial coverage (e.g., TerraSAR-X) or longer revisit time (e.g., ALOS PALSAR).

Several investigations [14,32] demonstrated that the C-band like-polarized ratio (HH/VV) is a useful parameter for mapping and monitoring rice cropland. Wu et al. (2011) reported that the HH/VV ratio was best for discriminating rice from bananas, forest, and water [33]. However, Schmitt and Brisco (2013) reported that the HH/VV combination produced a significantly lower accuracy than the combination of HH/HV and VV/VH [34]. The backscatter coefficients of cross-polarized data have a significant correlation with the development of rice plants [33,35]. Schmitt and Brisco (2013) also found that the cross-polarized data gave the best relationship with rice age after transplantation. Due to the improved spatio-temporal resolution and ability to acquire data in different polarizations (VV, VH), S-1 is expected to improve the accuracy of rice cropland monitoring and mapping applications.

Literature review suggests that most of the studies have been limited to map paddy rice from C-band SAR data either by using single polarization (HH) or a combination of different polarizations (HH/VV, HH/HV or VV/VH). However, only a few of them have investigated dense VH backscatter time series in order to have a better understanding of the SAR response to growth stages of the rice fields. Furthermore, to our knowledge, no work using real S-1 data for paddy rice mapping in the Mediterranean region has been published yet.

The goal of this research is to evaluate the potential and transferability of a phenology-based classification strategy developed by Nguyen et al. (2016) [36] over a regional test site in the Mekong

Delta to a continental scale (Mediterranean region). To achieve this objective, we used a dense time series stack of S-1 backscatter data as input to map rice paddy area at fine spatial scale over eight study sites in six European countries.

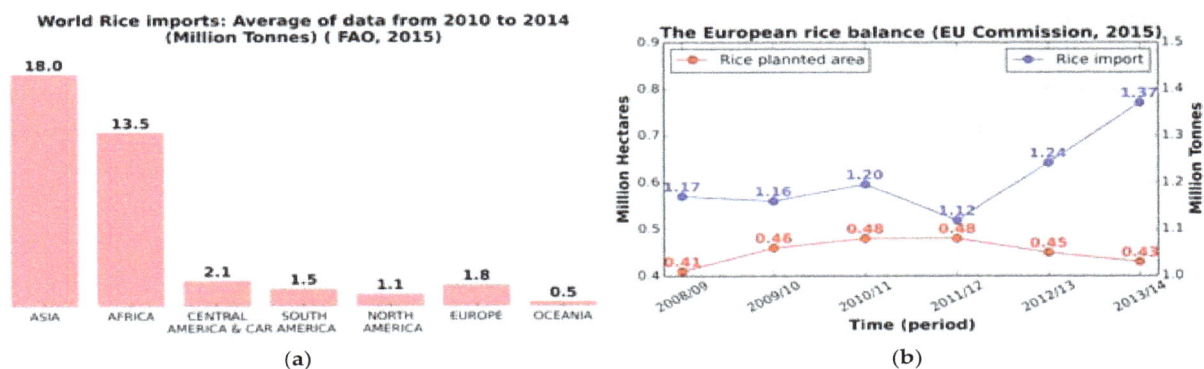

Figure 1. (**a**) Global rice imports, average of data for 2010 to 2014 [1]; and (**b**) the EU rice balance for 2008 to 2013 [37].

2. Study Area and Materials

2.1. Study Sites Characteristics

In the European Union (EU) the total area of cultivated rice is about 430,000 hectares. The growing areas are mostly located in the Mediterranean countries [1] (Figure 2), where the summer seasons are warm and dry. The normal growing season is either from April/May to September/October or from May to October/November depending on the temperature. Italy and Spain are the top rice producer EU member countries followed by Greece, Portugal and France [37].

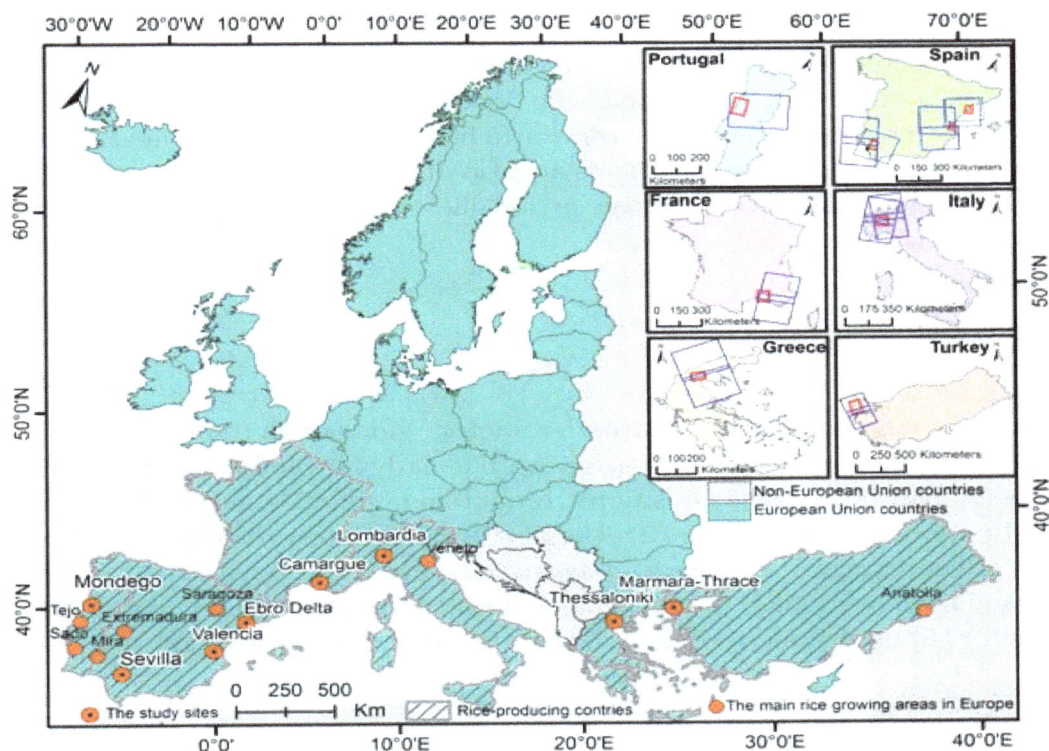

Figure 2. Study sites (red) and spatial extent of the used S-1 scenes (blue).

Rice varieties grown in Europe mostly belong to the Japonica (70%) and Indica (30%) species group. The recent evolution of Japonica and Indica areas (in hectares) in the EU member States is shown in Figure 3. Rice is mostly grown in congregated areas such as in the Po valley in Italy, the Rhône delta in France, and the Thessaloniki area in Greece. In Spain, rice cultivation is more scattered; rice growing areas are found in the Aragon region, the Guadalquivir valley, the Ebro delta and Valencia Albufera [38]. In Portugal, rice cultivation area is concentrated mainly in three regions: the Tagus and Sorraia valleys, the Mondego, and the Sado and Caia river valleys [39,40]. In Turkey, the most productive regions are Thrace and Marmara, which are producing 10–15% of the total national rice production.

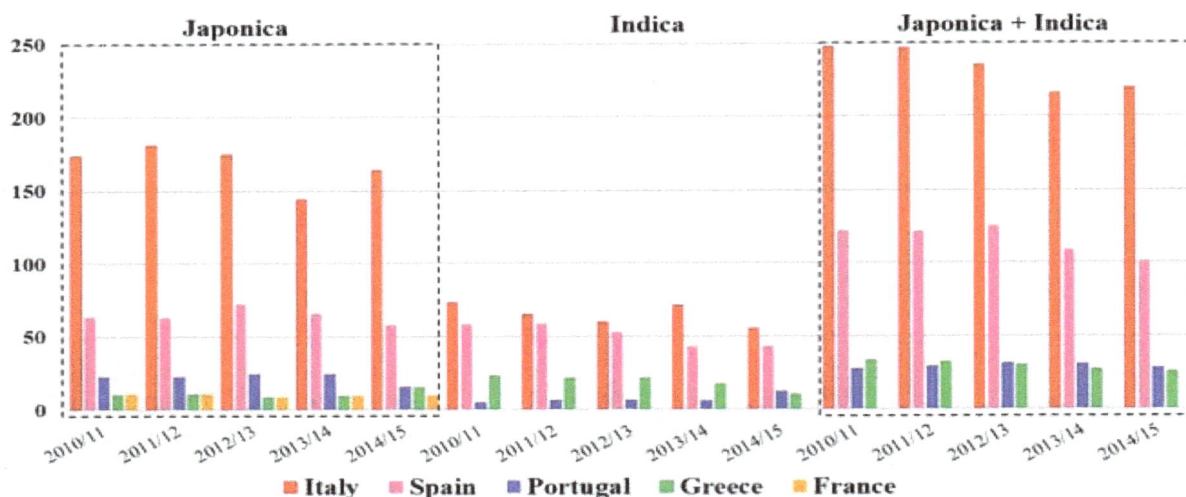

Figure 3. Rice varieties areas in the EU member states [37].

Rice planting in Europe involves direct seeding into flooded soil (water seeding) or dry soil (dry seeding). With both the farming methods, the floodwater is maintained until the harvest season. Water is drained several times from the field prior to harvesting so that the fields can be dried and harvesting equipment can pass through. Fields are also drained for early foliar herbicide treatments, then re-flooded within few days. The drainage period allows certain weed species to germinate in the aerobic environment (Figure 4).

2.2. Materials

2.2.1. Sentinel 1A Data

For this study, we accessed archived Interferometric Wide Swath (IW) mode S-1 data acquired during the rice crop growing cycle in 2015, from April to early November to completely cover every test site. The S-1A SAR IW mode acquisitions come from eight different tracks, whereas each test site is covered by the same track (Figure 2). The details for each test site regarding data acquisition, location, date, path and range of incidence angle are shown in Figure 4. All the images were obtained from ESA as standard Level 1 GRD (ground-range detected) high resolution images. Only the S-1 IW acquisitions with VH polarization were selected for rice cropland classification.

2.2.2. Optical Data and Ancillary Data

For the study sites in France, and Italy, multi-temporal Spot 5 (10 m spatial resolution) for the year 2015 were downloaded from the SPOT website [41]. These optical datasets were used to produce reference classification maps for validating the S-1 rice maps. Rice cropland area—a vector dataset with a minimum mapping unit (MMU) of 1 hectare—over Seville in Spain for the year 2015 has been obtained from the Institute of Statistics and Cartography of Andalusia's website [42] Two others vector dataset of rice cropland area—over Valencia in Spain and Thessaloniki in Greece for the year 2015 were retrieved from the ERMES (An Earth Observation Model Based Rice Information Service)'s website [43] Google Earth imagery and Sentinel-2 [44] optical data were used to produce reference data for the rest of study areas. In addition to this, the European CORINE land cover (CLC 2012) product was also used for comparing classification results across all European test sites in a consistent manner. The data set is not as detailed and accurate as the other two datasets. However, it is the only available reference datasets covering all our study regions. Figure 5 shows the land cover composition over the eight selected test sites in 2012. Rice cropland is the most common land cover in Lombardia, Italy with 33.6%, followed in Ebro, Spain (18.8%) in Valencia, Spain (12.4%), and only 2.3% of rice cropland in Mondego, Portugal. Inland water makes up over 10% of the total land covers in Camargue, France and below 1.5% for the others. It is the same for wetland class, e.g., 7% in Camargue, France, and below 2% for the others.

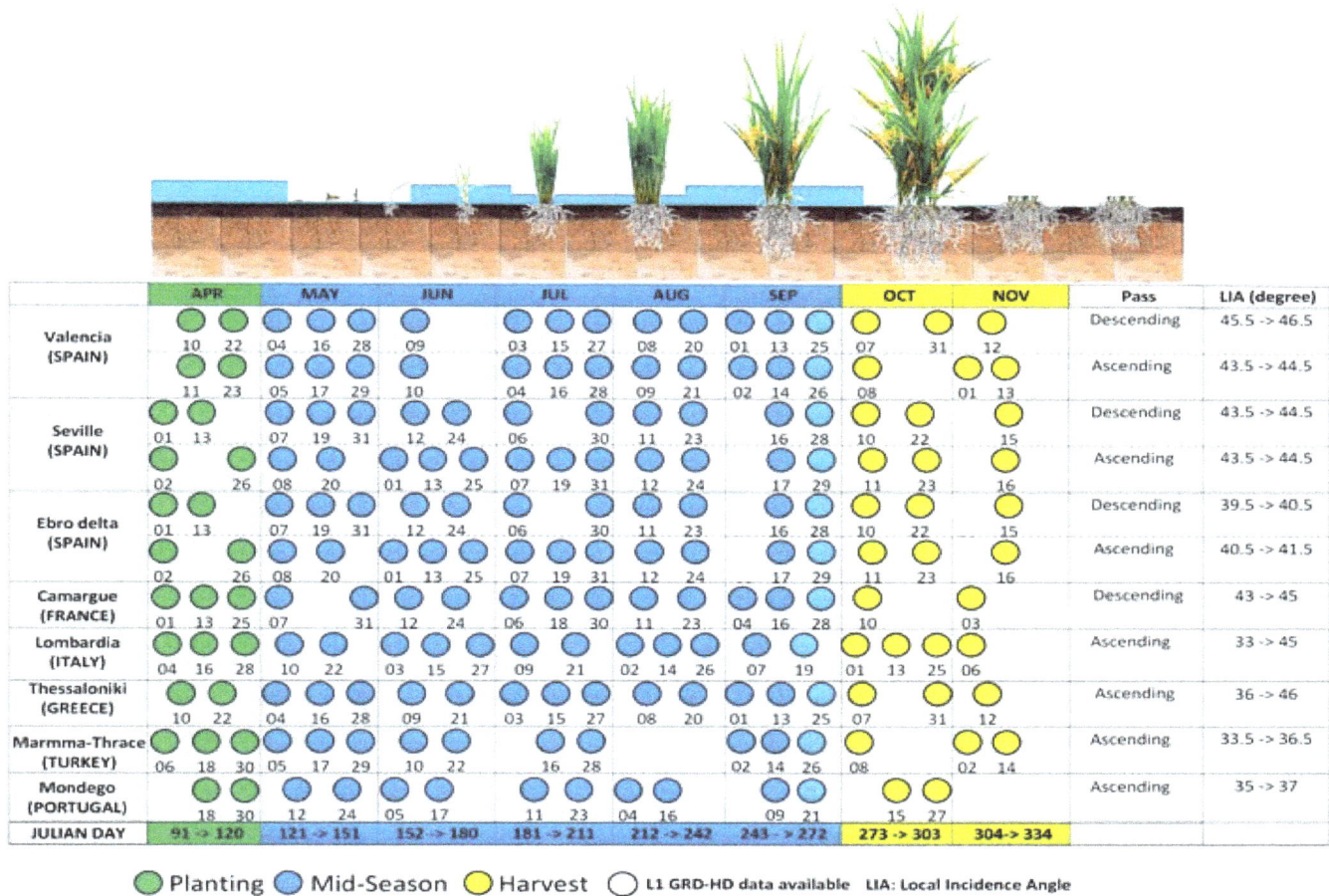

Figure 4. Growing season of paddy rice crop and the available S-1A SAR scenes over eight selected test sites.

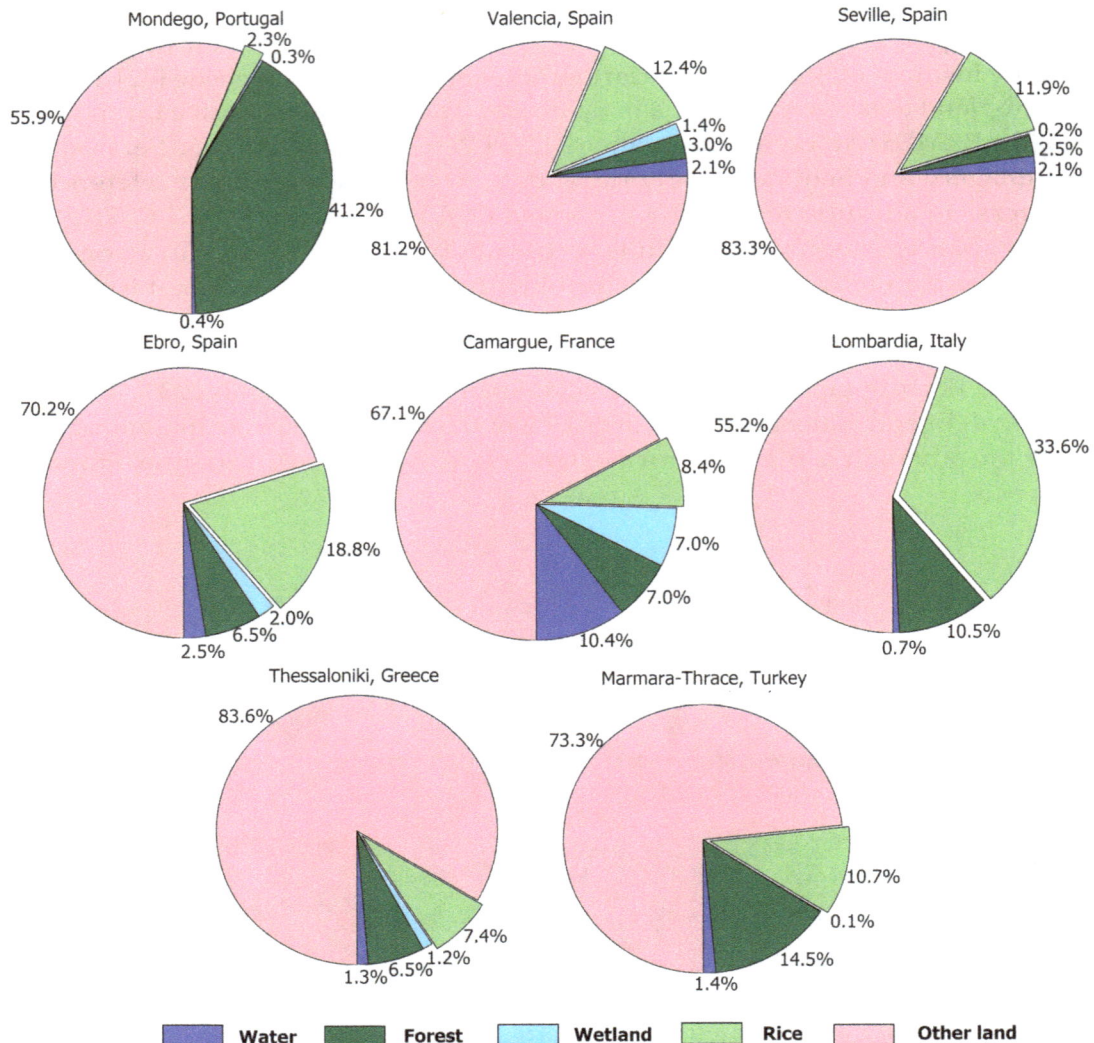

Figure 5. Relative contribution of harmonized land-cover categories in eight selected test sites.

3. Methodology

The time series algorithm used in this study was introduced by Nguyen et al. [36]. It consists of six steps as illustrated in Figure 6: (1) S-1A pre-processing; (2) segmentation to extract the potential rice areas ($\sigma^{\circ} \rightarrow \sigma^{\circ}_{potential}$); (3) time-series smoothing with a Gaussian moving window filter ($\sigma^{\circ}_{potential} \rightarrow \sigma^{\circ}_{potential\_smooth}$); (4) vegetation phenology parameters extraction ($\sigma^{\circ}_{potential\_smooth} \rightarrow$ DoS (Date of Start Season), DoM (Date of Maximum backscatter), LoS (Length of Season)); (5) classification using a knowledge-based decision-tree approach; and (6) accuracy assessment based on reference data.

3.1. Pre-Processing

All the selected S-1 SAR IW images were pre-processed (orbit correction, radiometric calibration, resampling and geocoding) using ESA's Sentinel-1 Toolbox. The geocoding step involved a Range Doppler Terrain correction algorithm that uses the elevation data from the 1 arc-second DEM product from the Shuttle Radar Topography Mission (SRTM) and POD orbit state vectors provided by ESA. Notice that precise orbits files (POD) are produced few weeks after the acquisition and they are automatically downloaded from ESA website [45]. In this process, data are resampled and geo-coded to a grid of 10 m spacing preserving the 20 m spatial resolution.

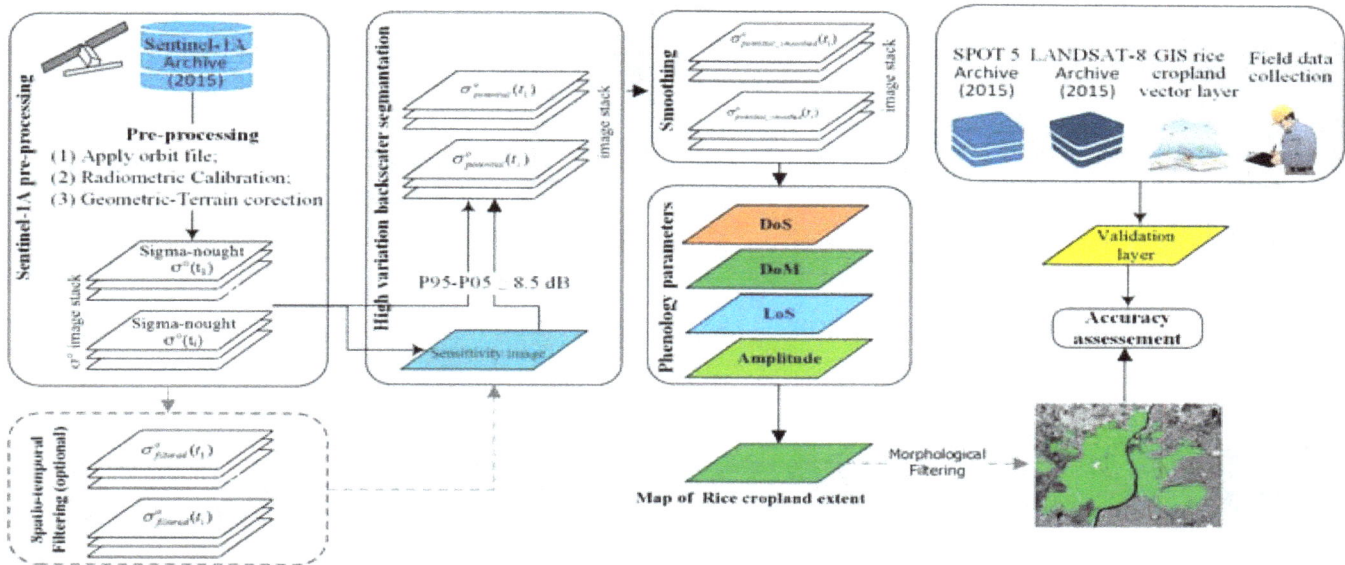

Figure 6. Systematic workflow of S-1 SAR (VH polarization) data processing, rice area classification and validation.

3.2. Identification of Potential Rice Pixels

For the identification of potential rice growing areas, our approach is to threshold the dynamic range backscatter image to identify image pixels that change more than the defined threshold value (dB). Threshold value selection depends on the nature and expected changes in the magnitude of VH backscatter and the SAR geometry (e.g., incidence angle). A generalized threshold for the rice fields can only be determined if the optimum SAR data acquisition is guaranteed (e.g., SAR observations are available in the flooded and vegetative stages). Otherwise, the threshold must be optimized considering the data acquisition and the constraints of local crop calendar. S-1A provides at least one acquisition after every 12 days over the selected study sites (see Figure 4); after the launch of S-1B the temporal sampling was reduced to 5–6 days. Based on the visual interpretation of optical imagery for the selected period and expert knowledge acquired from the ancillary data, a threshold of 8.5 dB was used to extract the potential rice pixels referred to as $\sigma^o_{VH\_potential}$.

Figure 7 (first column) shows sensitivity images based on dry reference (P05) and wet reference (P95) images for all study areas, and the results of applying thresholds of 8, 8.5 and 9 dB to these images are illustrated in following columns of the Figure 7. The CLC 2012 map is used to indicate the compartment boundaries, and pixels exhibiting change below and upper the threshold (8.5 dB) are classified as potential rice cropland areas. Combining Figure 6 with Figure 7 suggests a threshold on sensitivity of 8.5 dB to extract the potential rice cropland pixels. Lowering the threshold to 8.0 dB increases area outside the potential rice cropland areas in all the study sites leads to computation time increases. Raising the threshold to 9.0 dB decreases the potential rice cropland areas may leads to numerous rice pixels are omitted. Thus 8.5 dB seems the better choice as long as we accept that physically, processing time, as well as classification precision. With this threshold, the majority of the rice cultivated areas are correctly picked out in all data sets. From the basic concept of the approach, it is clear that classification error will occur for the rice pixels whose variation is less than the selected threshold. Different of farming activities during the growing season (rice varieties, water level in the fields, density of rice plants in the fields), SAR acquisitions period may give rise to this type of misclassification.

Mondego (Portugal)

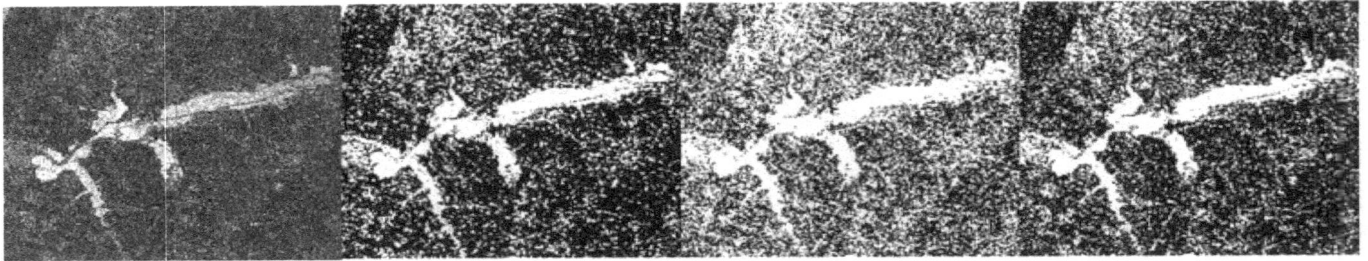

Sensitivity image   ≥8.0 dB   ≥8.5 dB   ≥9.0 dB

Seville (Spain)

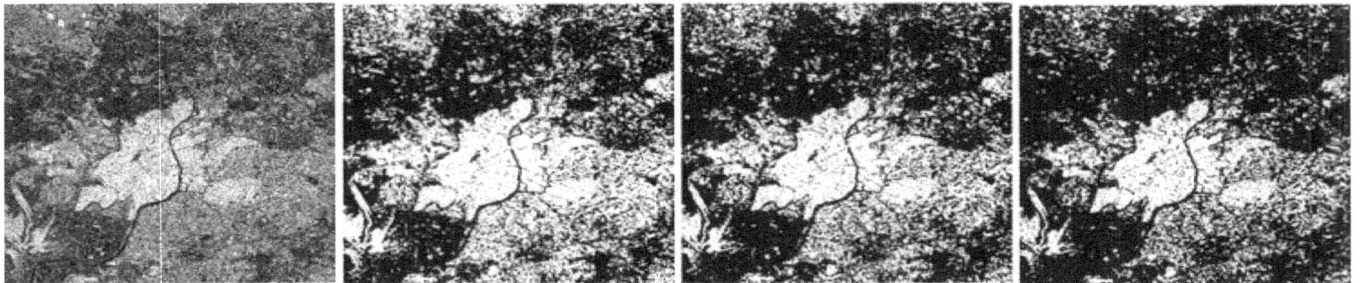

Sensitivity image   ≥8.0 dB   ≥8.5 dB   ≥9.0 dB

Valencia (Spain)

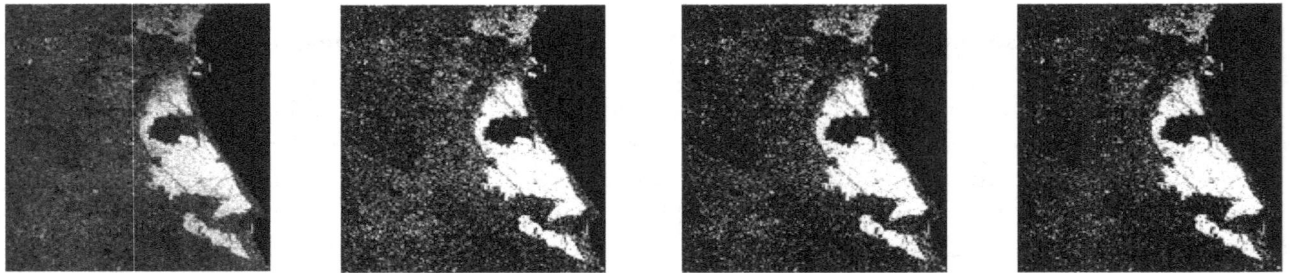

Sensitivity image   ≥8.0 dB   ≥8.5 dB   ≥9.0 dB

Ebro delta (Spain)

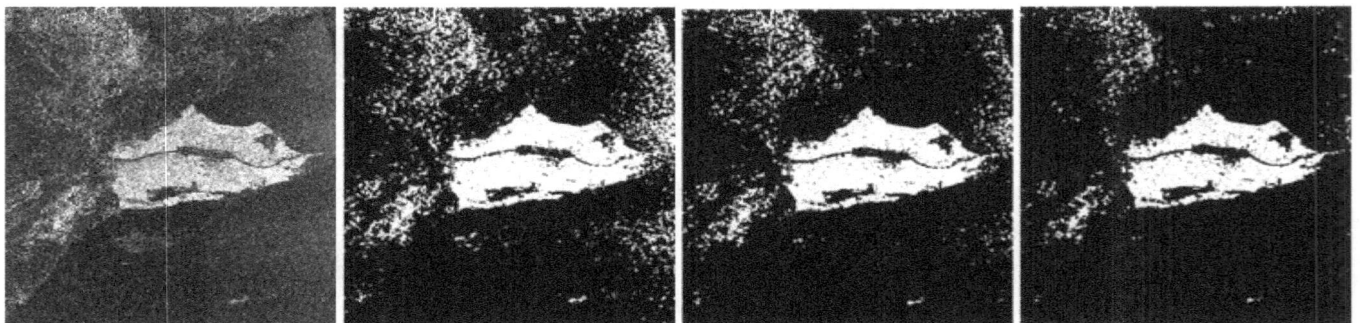

Sensitivity image   ≥8.0 dB   ≥8.5 dB   ≥9.0 dB

Figure 7. *Cont.*

Camargue (France)

| Sensitivity image | ≥8.0 dB | ≥8.5 dB | ≥9.0 dB |

Lombaria (Italy)

| Sensitivity image | ≥8.0 dB | ≥8.5 dB | ≥9.0 dB |

Thessaloniki (Greece)

| Sensitivity image | ≥8.0 dB | ≥8.5 dB | ≥9.0 dB |

Marmma-Thrace (Turkey)

| Sensitivity image | ≥8.0 dB | ≥8.5 dB | ≥9.0 dB |

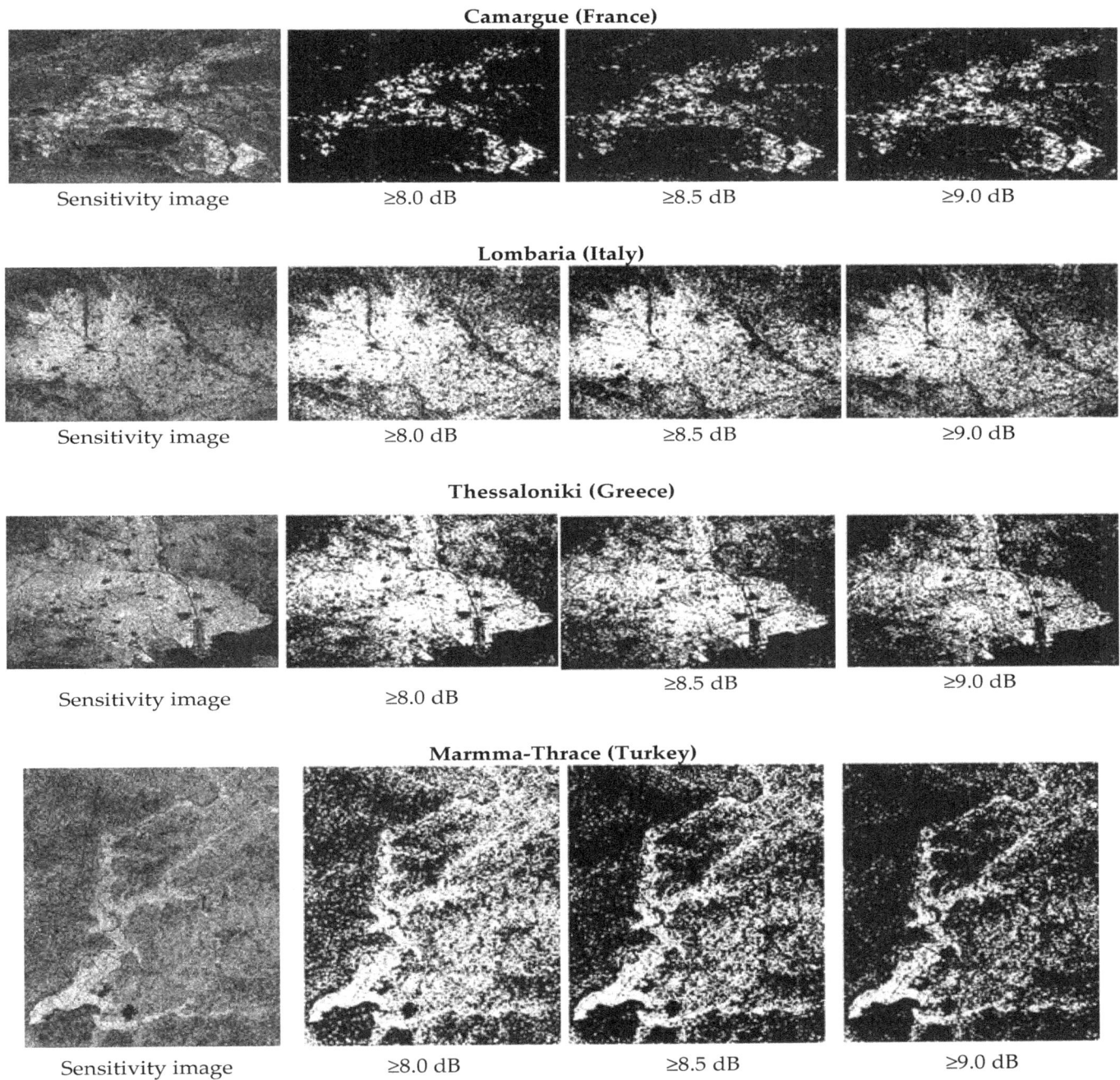

Figure 7. Sensitivity images and potential rice cropland area extent results using different thresholds.

3.3. Time Series Filtering

Filtering the backscatter time series has the purpose of reducing the short-term influence of environmental conditions and noise inherent in the S-1 data due to speckle and other noise-like influences. The processed output is a smoothed backscatter signal ($\sigma^0_{VH_{potential\_smooth}}$), which will be used for the extraction of different phenological stages of rice (e.g., a start of the season, heading time, and length of season). For the temporal filtering of the $\sigma^0_{VH\_potential}$ time series a Gaussian smoothing filter (with the standard deviation of 3 dB for the kernel) was used. A detailed investigation on the selection of kernels with different standard deviation was already reported by Nguyen, Wagner et al. 2015 [16]. Moreover, in order to discriminate the rice pixels from the other land cover classes we have empirically defined a list of three more static thresholds based on the $\sigma^0_{VH_{potential\_smooth}}$ values.

3.4. Extraction of Vegetation Phenology Parameters

For this step, we calculated three phenological indicators, namely, date of beginning of season (DoS), date of maximum backscatter (DoM), and Length of the season (LoS). This allows to delineating the areas as "rice paddy", and all other areas were placed in a generalized "non-rice" class. These parameters, introduced by [37], are summarized in Table 1.

Table 1. Phenological parameters for the rule-based classification.

Parameters	Definition and Explanation
DoS	During the growing season, the date of the beginning of season is defined as the first local minima in $\sigma^0_{VH_{potentia\_smooth}}$ time-series.
DoM	During the growing season, the date when backscatter reaches a maximum value is defined as the local maxima in $\sigma^0_{VH_{potentia\_smooth}}$ time-series. This date must come after the date of the beginning of season, where it reaches its local minimum backscatter value.
LoS	The length of the season is defined as the number of days difference between DoM and DoS.

3.5. Rice Paddy Identification

Due to the high temporal variability in the SAR backscatter signal across the different study sites, the raw output from the thresholding of phenological profiles contained some noisy pixels. This implies that most fields were not fully classified as rice and non-rice class at the pixel level. Therefore, we constrained the minimum mapping unit to the average farm size in the Mediterranean region, which is 1/4 hectares. This means that no polygon was composed of fewer than 25 S-1 pixels (20 m spatial resolution, 10 m pixel spacing). We implemented this through a post-classification processing step, whereas a majority/minority analysis with the window size of 5 × 5 pixels was applied to remove the small pixel groups in order to obtain refined classification results.

3.6. Accuracy Assessment

For validation and evaluation of classification results, standard accuracy assessment measures were used, i.e., kappa coefficient, overall accuracy, omission error, and commission error. Rice cropland vector layer 2015 for study site in Spain (Seville), rice cropland maps created through interpretation of SPOT-5 data for three study sites in Spain (Valencia), Italy, and France; rice cropland raster layers 2006 from CLC 2006 for Marmara-Thrace, Turkey; and rice cropland raster layers from CLC 2012 for all study sites were used.

4. Results and Discussion

4.1. Temporal Rice Backscatter Signature from Sentinel-1 SAR Data

Seasonal VH time series are shown in Figure 8a–h, for Mondego (Portugal), Seville (Spain), Valencia (Spain), Ebro delta (Spain), Camargue (France), Lombaria (Italy), Thessaloniki (Greece) and Marmma-Thrace (Turkey), respectively. In addition, Figure 8a–h also shows the results of smooth backscatter profiles (magenta color), and the date of maximum and minimum backscatter values for the selected rice fields (represented by the red and blue dots, respectively). To complement this analysis, Figure 9 shows the VH backscattering coefficients and false color composites of the study sites.

Figure 9 (column 4) shows the color composites, which are created by using the multi-temporal SAR acquisitions in order to highlight the temporal characteristics within the rice fields. The red, blue, and green colors in these figures (Figure 9, column 4) correspond to the images acquired during the flooded/seeding (April/May), heading (August/September) and post harvested (October/November) period, respectively. The blue and dark green color regions in these figures (Figure 9, column 4) indicate the rice cropland areas. At the beginning of the growing season (April–May) the σ^0_{VH} values from the rice fields were very low due to flooding after sowing. Thus, in the early stage of rice crop growth the fields appeared as dark areas in SAR images. This condition corresponds to those scenes that were acquired during the months April and May (see black areas in Figure 9, the first column). In general, during this period the rice fields show low backscatter values, e.g., less than -20 dB (for reference, see Figure 8). However, there are several conditions related to the soil preparation and wind speed/direction that must also be considered. Some rice fields require a special tillage, such as furrows, or the presence of low water level in the field for a shorter period due to the diversity in farming activities among different parts of the region. If these furrows are shallow and under light wind conditions, the surface is smooth irrespective of the furrow direction.

However, if the furrows are deep, the furrow direction becomes an important factor regarding SAR image acquisition geometry. In case that the deep furrows are nearly parallel to the radar viewing direction, the surface seems smooth in radar image. On the other hand, if the deep furrows are perpendicular to the radar viewing direction, the surface becomes strongly rough, and the signal backscatter becomes very strong. These results are in accordance with Brisco et al. (1991) when evaluating the effect of tillage row direction in relation to the radar's look direction using radar backscattering coefficient from three different radar frequencies [45]. The Sentinel-1 data are also influenced by the incidence angle, whereas the strength of the radar's backscatter signal gradually decreases with the increase of incidence angle. Therefore, backscatter is in general higher over the Marmma-Thrace (Turkey) and Mondego (Portugal) sites, where the incidence angles were 35° and 36° compared to the other regions where the incidence angle was about 40° (Figures 4 and 8).

In the second stage (vegetative stage), backscatter value increases as the vegetation grows (e.g., plant size increases), and eventually the SAR images show no significant difference between rice fields and other agricultural fields or vegetated areas (see bright areas in Figure 9, second column). One month to 45 days after the start date of the growing season it reaches the first peak in June/July. Specifically, a simple visual inspection to the $\sigma^o{}_{VH}$ images which were acquired at the end of July (where DoY is around 210, Figure 8a–h)) reveals that the backscatter value has dropped suddenly, despite the fact that vegetation is fully developed. This anomaly is observed for all study sites, except Mondego (Portugal) and Lombaria (Italy). It is clearly illustrated in the case of Seville (Spain) and Valencia (Spain) in the Figure 8. The scale of these changes is quite different for different study sites due to different agricultural practices. Like in other European regions, rice crop in Seville and Valencia (Spain) begins in mid-April with the deep placement of fertilizers under dry conditions. Flooding starts during the first week of May and then the seeds are sown. At the end of June there is a short dry period of ten days [46].

Furthermore, this period is characterized by the increase of the rice plant height and the number and size of tillers that lead to make free space among narrowed or blocked rice stems. As a result, the absence of water from the rice fields will minimize the double bounce effect of SAR signals and this explains the decrease in backscattering values at this stage. For the other regions, throughout the rice cultivation period, water is commonly kept at a depth of 4–8 cm, and drained away 2–3 times during the growing season to improve the crop rooting, reduce the algae growth and to allow application of herbicides. For the reproductive stage (August/September), backscatter values continuously keep increasing until they achieve the maximum value in September. During this period the values of backscattering coefficient vary between the range of -17 dB and -13 dB, which might get influenced by the variations in incidence angle, water level in the fields, cultivation activities or the rice varieties. However, during the ripening phase a slight decrease in SAR backscatter signal is observed. One potential reason for this could be due to the fact that the plants will dry before the harvesting. Normally, rice fields are drained towards the end of August to allow harvesting [47].

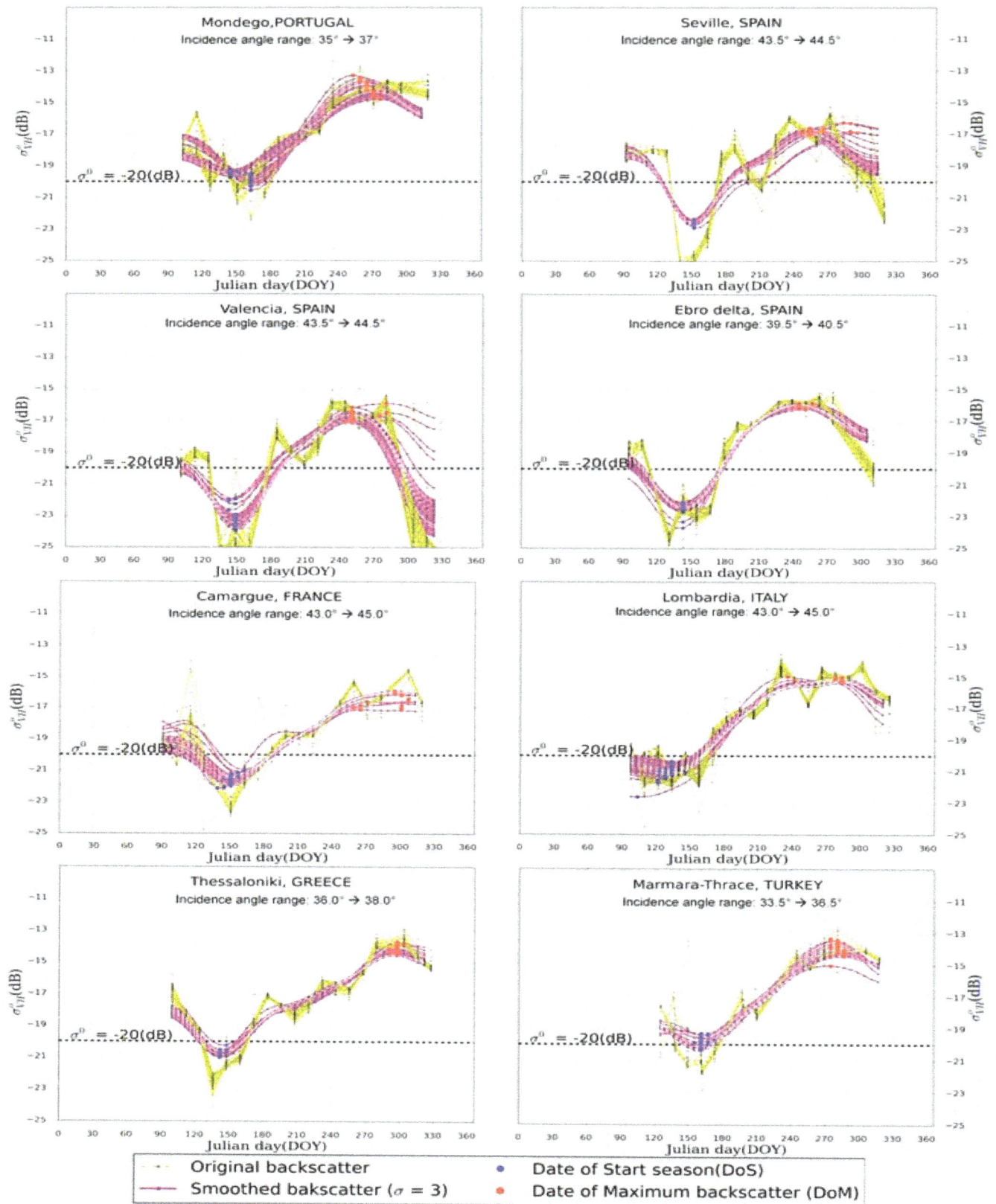

Figure 8. Temporal evolution of the backscattering coefficients derived from VH polarization (where, $\sigma^o_{VH} = -20$ (dB) is base line).

Mondego (Portugal)

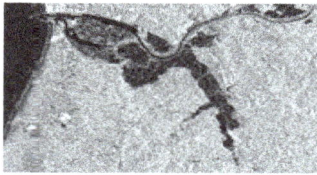

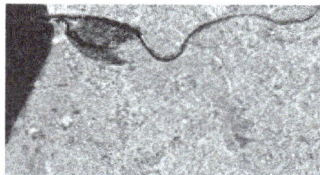

6 May 2015 (R)   22 August 2015 (G)   2 November 2015 (B)   RGB

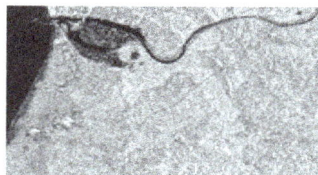

Valencia (Spain)

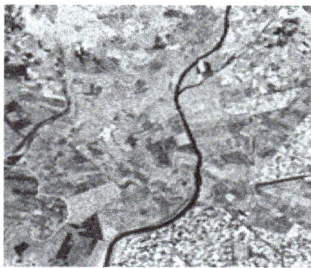

19 May 2015 (R)   23 August 2015 (G)   15 November 2015 (B)   RGB

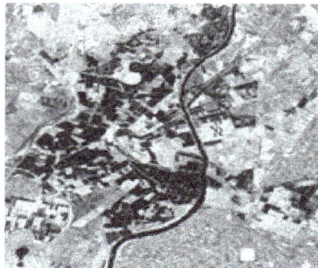

Seville (Spain)

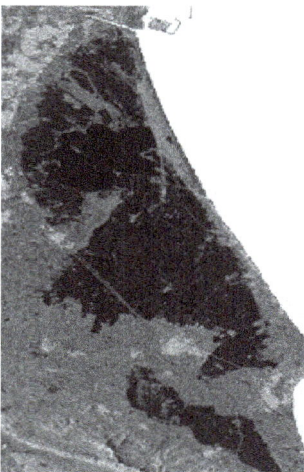

16 May 2015 (R)   20 August 2015 (G)   12 November 2015 (B)   RGB

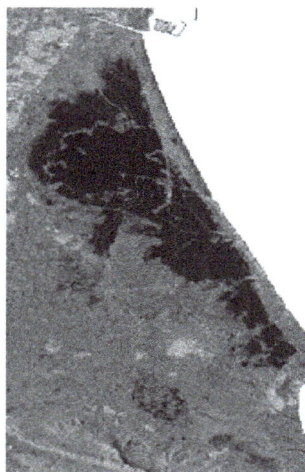

Ebro delta (Spain)

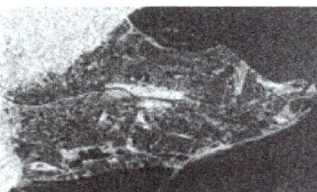

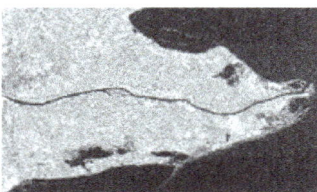

11 May 2015 (R)   27 August 2015 (G)   7 November 2015 (B)   RGB

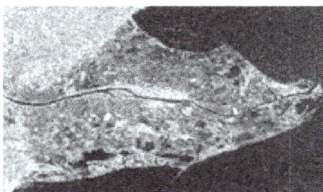

Figure 9. *Cont.*

Camargue (France)

7 May 2015 (R)   4 September 2015 (G)   3 November 2015 (B)   RGB

Lombaria (Italy)

10 May 2015 (R)   26 August 2015 (G)   6 November 2015 (B)   RGB

Thessaloniki (Greece)

16 May 2015 (R)   20 August 2015 (G)   31 October 2015 (B)   RGB

Marmma-Thrace (Turkey)

17 May 2015 (R)   2 September 2015 (G)   14 November 2015 (B)   RGB

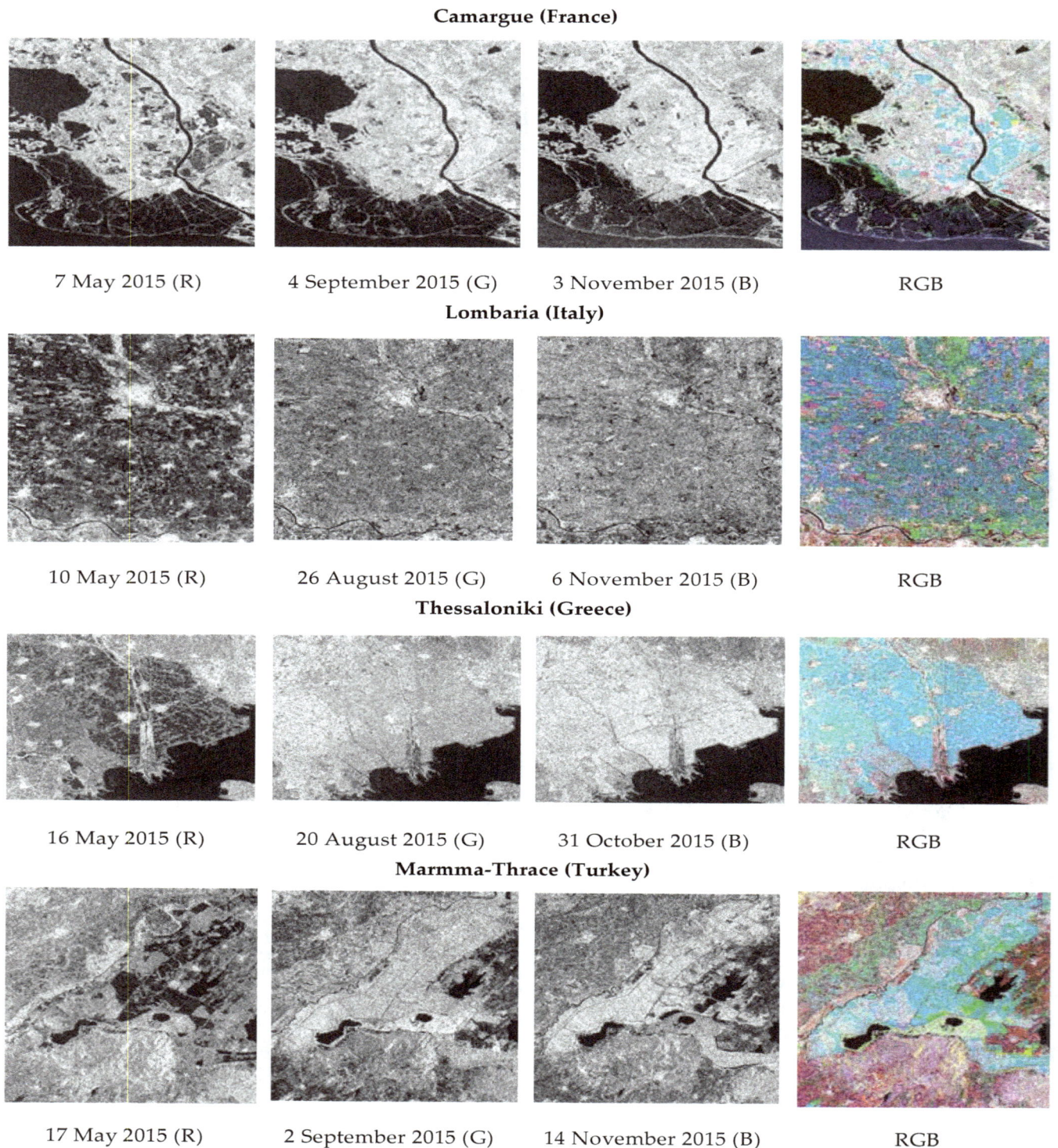

Figure 9. Columns 1–3: images of σ^0_{VH} acquired at three dates (from left to right) and; Column 4: images of false color composite over the part of study sites.

After harvest, the fields can have diverse conditions, either bare and dry fields or covered with weeds in wet conditions. Fields may also be flooded due to local farming activities, e.g., in some areas fields are flooded again until January for duck hunting [46]. This event is clearly visible at the study sites in Spain (Figures 8b–d and 9). However, the levels of change are quite different among the study sites due to the differences in farming activities. Meanwhile, in other regions, some small crops or weeds cover is possible right after the rice crop is cultivated. The radar backscatter is thus variable and in most cases can have high backscatter values. Moreover, as a result of rain which is typically high in the winter season in the Mediterranean region, the water content in bare soil increases, and this may also explain the increase of backscattering. Consequently, there is little possibility to interpret this last stage of the cultivation.

4.2. Thresholds Selection

Optical data sources (Spot 5, Sentinel-2, and Google Earth imagery), vector dataset and expert knowledge were used to create polygons for threshold selection. These polygons were carefully selected and digitized across eight study sites. Training areas were considered in terms of distribution and variety of rice fields (e.g., size, density, rice variety) and SAR geometry characteristic (e.g., incidence angle) to build robust training data set. These polygons used for threshold determination were then excluded from the set of polygons used for accuracy assessment. Time series analysis was carried out using all available S-1 IW Mode data between April and November 2015 (Figure 8 and Section 4.1). The overall behavior of σ^o_{VH} is comparable to our previous investigations [37]. However, the dynamic range over different regions varies due to different incidence angles and farming activities. In Figure 8, it is important to note that the response signal patterns from all sample parcels are very consistent. Therefore, we can conclude that these temporal signatures are associated with the rice crop type, and the empirical thresholds of three phenological parameters based on the $\sigma^o_{VH_{potentia\_smooth}}$ values can be used for the identification of rice cropland. This study, we followed the approach of Nguyen et al. [15] who used phenological parameters (derived from σ^o_{VH} profile) to map the rice areas of eight sites in the Mediterranean region.

For rice areas mapping, the first two key parameters, σ^o_{DoM}—the peak (maximum) of VH backscatter and $\Delta\sigma^o$ the amplitude backscatter (the difference between σ^o_{DoM}—backscatter at date of maximum backscatter and σ^o_{DoS}—backscatter at the begin of the growing season) are very critical. The peak and valley of VH backscatter (i.e., σ^o_{DoM}, σ^o_{DoS}) within the one rice-growing cycle can be identified by a local extrema algorithm. To eliminate unrealistic peaks, a threshold for VH backscatter is required. These two parameter thresholds (i.e., $\sigma^o_{DoM} \geq -19$ dB and $\Delta\sigma^o \geq 2.5$ dB) were selected based on a conservative deduction from training data, limited census statistics, and expert knowledge [36]. The corresponding positions (date) and backscatter values of these two points showed in the Figure 8 where big-red dots are DoS(s) and big-blue dots are DoM(s) respectively. All areas that practice rice crops start the growing season between May–June where the lowest backscatter is observed. Rice flowering period then comes after more than one and a haft month (July–August) where S-1 respond signal in the temporal signature reaches the peak.

The third parameter, length of the season (LoS) which is defined as the temporal distance (number of days difference) between the date of beginning of season and the date when maximum backscatter value is recorded. This temporal distance has to be greater than the shortest possible rice growing cycle and smaller than the longest possible rice growing cycle. From crop calendar and our expert knowledge about rice growing season in the study areas, the threshold of temporal distance which is about 50 and 120 days, respectively. If all three conditions are met then the pixel is classified as rice, otherwise as non-rice area.

4.3. Spatial Distribution and Comparison of S-1 Derived Rice Area with Reference Data

The temporal backscatter signatures is shown in Figure 8 (and Figure S1 in the supplementary material) served to define specific thresholds which were applied to the S-1 time series in order to generate rice area maps for the growing season of 2015. The classification results of rice fields over eight selected sites in the Mediterranean region are shown in Figure 10. The classification accuracy has been assessed by comparing the classification results to the 2015 rice cropland vector layer for four study sites in Seville (Spain), Valencia (Spain), Lombardia (Italy), and Camargue (France); and CLC (2012) for all study sites. The confusion matrices are presented in Tables 1 and 2, respectively.

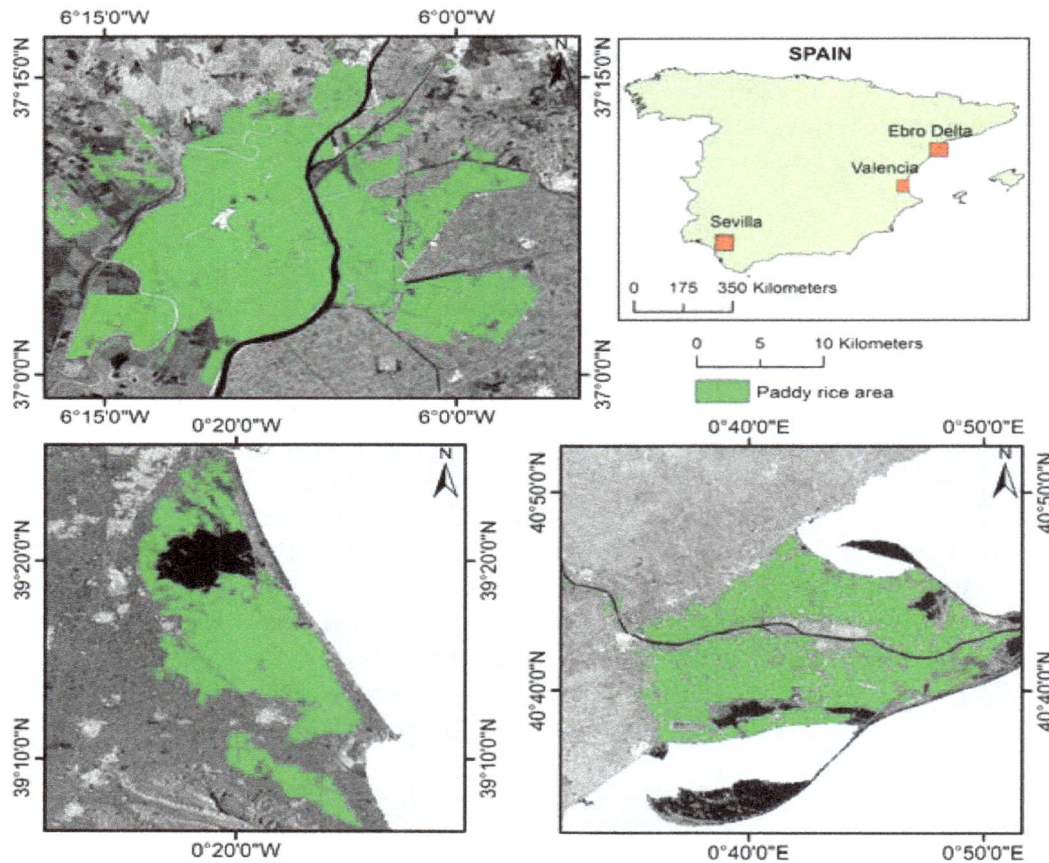

Figure 10. Rice cropland mapping in Spain.

To provide direct comparison, the classification results in all study sites: Seville and Valencia in Spain, and Thessaloniki in Greece were evaluated by using the reference vector dataset from the same year 2015; Lombardia (Italy), and Camargue (France) were evaluated by using higher spatial resolution (SPOT 5, 10 m); and Earth Imagery and Sentinel-2 data were used to produce reference data to validate the classification results for the rest of study areas (Table 2). The Kappa coefficients, rice user accuracy (commission error), and rice producer accuracy (omission error) were, respectively, 0.76, 70.2%, and 82.7% for Mondego, Portugal; 0.87, 86.8%, and 89.2% for Seville, Spain; 0.85, 81.7%, and 93.3% for Valencia, Spain; 0.79, 70.3 and 95.3% for Ebro delta, Spain ; 0.85, 80.2%, and 94.0% for Camargue, France; 0.82, 79.8%, and 86.2% for Lombardia, Italy; 0.79, 74.1%, and 94.0% for Thessaloniki, Greece; and 0.76, 71.1%, and 84.1% for Marmara-Thrace, Turkey (Table 1). The lower accuracy level was observed in Mondego, Portugal, Ebro delta, Spain, Thessaloniki, Greece and Marmara-Thrace, Turkey, because of most rice fields in these regions associated with the small farms, and scattered throughout the mixed agricultural landscape and, thus, were easily omitted with non-rice classes. One potential source of error lies in the time difference between the reference data and the Sentinel-1 image acquisition. For example, the reference data were produced by digitizing homogenous sites of rice fields based on the image visualization and interpretation of existing high-resolution Google Earth imagery and Sentinel-2 optical data, while the classification maps were produced from 2015 S-1A data. Moreover, a number of different of farming activities during the growing season (rice varieties, water level in the fields, and density of rice plants in the fields) and SAR acquisitions period could also cause an increase in mapping errors. Although the proposed approach can properly detect most rice areas, some land cover types with high variation in backscatter values (e.g., wetland or seasonal water bodies areas) can cause commission errors.

Table 2. Confusion matrices of the accuracy assessment.

SAR Data					
		Mondego, Portugal			
	Class	**Non-Rice**	**Rice**	**Total**	**Producer Accuracy**
	non-rice	4251	71	4322	98.4%
	rice	35	167	202	82.7%
	total	4286	238	4524	-
	user accuracy	99.2%	70.2%	-	Kappa = 0.76
		Seville, Spain			
	Class	**Non-Rice**	**Rice**	**Total**	**Producer Accuracy**
	non-rice	10,057	140	10,197	98.6%
	rice	111	919	1030	89.2%
	total	10,168	1059	11,227	-
	user accuracy	98.9%	86.8%	-	Kappa = 0.87
		Valencia, Spain			
	Class	**Non-Rice**	**Rice**	**Total**	**Producer Accuracy**
	non-rice	6344	176	6520	97.3%
	rice	56	786	842	93.3%
	total	6400	962	7362	-
	user accuracy	99.1%	81.7%	-	Kappa = 0.85
		Ebro delta, Spain			
	Class	**Non-Rice**	**Rice**	**Total**	**Producer Accuracy**
	non-rice	5637	256	5893	95.6%
	rice	30	606	636	95.3%
	total	5667	862	6529	-
S-1	user accuracy	99.5%	70.3%	-	Kappa = 0.79
		Camargue, France			
	Class	Non-Rice	Rice	Total	Producer Accuracy
	non-rice	8915	252	9167	97.3%
	rice	65	1023	1088	94.0%
	total	8980	1275	10,255	-
	user accuracy	99.3%	80.2%		Kappa = 0.85
		Lombardia, Italy			
	Class	**Non-Rice**	**Rice**	**Total**	**Producer Accuracy**
	non-rice	3392	356	3748	90.5%
	rice	56	1404	1460	86.2%
	total	3448	1760	5208	-
	user accuracy	98.4%	79.8%	-	Kappa= 0.82
		Thessaloniki, Greece			
	Class	**Non-Rice**	**Rice**	**Total**	**Producer Accuracy**
	non-rice	8915	357	9272	96.2%
	rice	65	1023	1088	94.0%
	total	8980	1380	10,360	-
	user accuracy	99.3%	74.1%	-	Kappa = 0.79
		Marmara-Thrace, Turkey			
	Class	**Non-Rice**	**Rice**	**Total**	**Producer Accuracy**
	non-rice	22,975	453	23,428	98.1%
	rice	210	1112	1322	84.1%
	total	23,185	1565	24,750	-
	user accuracy	99.1%	71.1%	-	Kappa = 0.76

For all the study sites, CLC from 2012 were also used for comparison and checking the consistency of classification results over all study areas. Despite the limitations of the CLC data (e.g., there was a three years difference between CLC and S-1 datasets) the results suggest that the application of S-1 time series data for rice area mapping has produced consistent results for all the test sites with overall accuracies (in quotation {}) ranging from 77.7% to 98.9% (kappa average at 0.53). The average accuracies at Camargue (France), Thessaloniki (Greece) and Mondego (Portugal) are lower than that of Seville and Valencia (Spain). The results showed relatively high error rate for both commission and omission measure (see Table 3 for details). The best results were achieved over Seville and Valencia (Spain) with kappa of 0.70 (Rice omission 31.9%) and 0.82, respectively (Rice omission 22.6%), which is significantly higher than over the other regions. The worst performance (Kappa 0.37, overall accuracy: 79.9%) was reported from Thessaloniki (Greece) with the highest commission errors (63.0%) and omission error (nearly 30.7%). In order to better match the spatial scale between the classification map produced and the CLC 2012 validation layer, the classification results were smoothed by gap filling using a 10 pixels × 10 pixels window. The reason for this step is to compensate the smoothness level between the classified map and the CLC 2012 validation layer. The CLC 2012 product is consist of large homogeneous polygons which includes the small roads and fields boundaries, which are major source of error in the validation process. After smoothing the classified rice map a significant improvement in the overall accuracy and the kappa value is observed (Table 3). The comparison results in the Table 3 show that the poor results of accuracy comparison are not due to the errors classification output, but due to the coarser resolution of the CLC 2012 product.

Table 3. Comparisons of rice crop extraction accuracies from time series S-1 data with Corine Land cover products (CLC 2012).

Sites, Country	Kappa	Overall Accuracy %	Rice Commission %	Rice Omission %
	Comparison with CLC 2012 after Smoothing and with Original CLC 2012 (in {})			
Seville, Spain	0.85 {0.70}	93.4 {88.1}	11.8 {9.3}	9.3 {31.9}
Valencia, Spain	0.93 {0.82}	98.9 {92.7}	3.8 {2.0}	9.5 {22.6}
Ebro delta, Spain	0.88 {0.52}	94.3 {77.7}	2.6 {2.8}	10.7 {49.2}
Camargue, France	0.46 {0.37}	87.8 {86.5}	17.7 {18.0}	62.3 {71.2}
Lombardia, Italy	0.60 {0.50}	83.9 {81.3}	12.6 {12.6}	37.3 {53.3}
Thessaloniki, Greece	0.57 {0.37}	83.5 {79.9}	29.8 {30.7}	34.1 {63.0}
Marmara-Thrace, Turkey	0.74 {0.52}	97.3 {94.5}	16.4 {11.3}	31.21 {60.5}
Mondego, Portugal	0.65 {0.43}	95.4 {98.9}	31.2 {40.7}	34.3 {65.7}

5. Conclusions

In this paper, we investigated a phenology-based approach to map rice crop at a continental scale by using space-borne C-band SAR data. Time series of S-1A IW mode with 20 m spatial resolution and VH polarization covering eight sites in the Mediterranean region were used. The results show that the proposed approach is efficient and operationally feasible for extracting rice cropland areas with high accuracy (above 70%) at 20 m spatial resolution (single-polarization) by using S-1A time series.

Our results show that dense SAR time series are critical for monitoring rice areas, which also gave an insight into the farmers' management practices within each rice cropping system.

This study emphasizes the use of annual SAR time series to generate a timely, accurate and high-resolution rice cropland spatial extent. It will be very helpful to assist decision making in the identification of rice areas for intensification, and areas for the development of irrigation as one of the necessary steps in dealing with food security related issues. Despite some current limitations of gaps in S-1A data acquisitions, this study suggests that the current and future SAR systems such as S-1B can be used complementarily to provide valuable spatial, thematic and temporal information about rice crop areas in the Mediterranean region and worldwide.

Information regarding the time of land preparation and water supply would be necessary to improve the classification in areas where the temporal signature for rice is unusual. Acquiring to such data would also help to improve the accuracy of rice cropland monitoring at the continental scale using S-1 time series.

Acknowledgments: The authors would like to thank VIED (Vietnam International Education Development) and OeAD (Österreichischer Austauschdienst) for their support during the studying period. We also thank Ali Iftikhar for the valuable comments on the paper in its early version.

Author Contributions: Duy Ba Nguyen and Wolfgang Wagner conceived and designed the study. The algorithm was developed by Duy Ba Nguyen and the experiments were implemented by Duy Ba Nguyen with guiding comments by Wolfgang Wagner. SAR data pre-processing and geophysical parameters retrieval were performed by Duy Ba Nguyen. Duy Ba Nguyen wrote a first draft version of the manuscript and produced all graphics. Wolfgang Wagner suggested detailed improvements. All authors provided assistance in editing and organizing the manuscript.

References

1. Food and Agriculture Organization of the United Nations (FAO). *The State of Food and Agriculture*; FAO: Rome, Italy, 2015.

2. Van der Sande, C.J.; De Jong, S.M.; De Roo, A.P.J. A segmentation and classification approach of IKONOS-2 imagery for land cover mapping to assist flood risk and flood damage assessment. *Int. J. Appl. Earth Obs. Geoinform.* **2003**, *4*, 217–229. [CrossRef]

3. Koellner, T.; Scholz, R.W. Assessment of land use impacts on the natural environment. *Int. J. Life Cycle Assess.* **2006**, *13*, 32–48.

4. Koellner, T.; Baan, L.; Beck, T.; Brandão, M.; Civit, B.; Goedkoop, M.; Margni, M.; Canals, L.M.; Müller-Wenk, R.; Weidema, B.; et al. Principles for life cycle inventories of land use on a global scale. *Int. J. Life Cycle Assess.* **2012**, *18*, 1203–1215. [CrossRef]

5. Roy, P.; Ijiri, T.; Nei, D.; Orikasa, T.; Okadome, H.; Nakamura, N.; Shiina, T. Life cycle inventory (LCI) of different forms of rice consumed in households in Japan. *J. Food Eng.* **2009**, *91*, 49–55. [CrossRef]

6. Blengini, G.A.; Busto, M. The life cycle of rice: LCA of alternative agri-food chain management systems in Vercelli (Italy). *J. Environ. Manag.* **2009**, *90*, 1512–1522. [CrossRef] [PubMed]

7. Roy, P.; Nei, D.; Orikasa, T.; Xu, Q.; Okadome, H.; Nakamura, N.; Shiina, T. A review of life cycle assessment (LCA) on some food products. *J. Food Eng.* **2009**, *90*, 1–10. [CrossRef]

8. Weng, X.-Y.; Zheng, C.-J.; Xu, H.-X.; Sun, J.-Y. Characteristics of photosynthesis and functions of the water-water cycle in rice (*Oryza sativa*) leaves in response to potassium deficiency. *Physiol. Plant.* **2007**, *131*, 614–621. [CrossRef] [PubMed]

9. Genovese, G.; Vignolles, C.; Negre, T.; Passera, G. A methodology for a combined use of normalised difference vegetation index and CORINE land cover data for crop yield monitoring and forecasting. A case study on Spain. *Agronomie* **2001**, *21*, 91–111. [CrossRef]

10. Agency, E.E. *Clc2006 Technical Guidelines*; European Environment Agency: Copenhagen, Denmark, 2007.

11. Agency, E.E. *Clc2012: Addendum to clc2006 Technical Guidelines*; European Environment Agency: Copenhagen, Denmark, 2012.

12. Kurosu, T.; Fujita, M.; Chiba, K. Monitoring of rice crop growth from space using the ERS-1 C-band SAR. *IEEE Trans. Geosci. Remote Sens.* **1995**, *33*, 1092–1096. [CrossRef]

13. Toan, T.L.; Ribbes, F.; Wang, L.F.; Floury, N.; Ding, K.H.; Kong, J.A.; Fujita, M.; Kurosu, T. Rice crop mapping and monitoring using ERS-1 data based on experiment and modeling results. *IEEE Trans. Geosci. Remote Sens.* **1997**, *35*, 41–56. [CrossRef]

14. Bouvet, A.; Le Toan, T.; Lam-Dao, N. Monitoring of the rice cropping system in the Mekong Delta using ENVISAT/ASAR dual polarization data. *IEEE Trans. Geosci. Remote Sens.* **2009**, *47*, 517–526. [CrossRef]

15. Bouvet, A.; Le Toan, T. Use of ENVISAT/ASAR wide-swath data for timely rice fields mapping in the Mekong River Delta. *Remote Sens. Environ.* **2011**, *115*, 1090–1101. [CrossRef]

16. Nguyen, D.; Clauss, K.; Cao, S.; Naeimi, V.; Kuenzer, C.; Wagner, W. Mapping rice seasonality in the Mekong Delta with multi-year Envisat Asar Wsm data. *Remote Sens.* **2015**, *7*, 15868–15893. [CrossRef]

17. Choudhury, I.; Chakraborty, M. SAR signature investigation of rice crop using RADARSAT data. *Int. J. Remote Sens.* **2006**, *27*, 519–534. [CrossRef]

18. Panigrahy, S.; Manjunath, K.R.; Chakraborty, M.; Kundu, N.; Parihar, J.S. Evaluation of RADARSAT standard beam data for identification of potato and rice crops in india. *ISPRS J. Photogramm. Remote Sens.* **1999**, *54*, 254–262. [CrossRef]

19. Panigrahy, S.; Jain, V.; Patnaik, C.; Parihar, J.S. Identification of Aman Rice Crop in Bangladesh using temporal C-band SAR—A feasibility study. *J. Indian Soc. Remote Sens.* **2012**, *40*, 599–606. [CrossRef]

20. Kaojarern, S.-A.; Delsol, J.P.; Toan, T.L.; Kam, S.P. Assessment of multi-temporal radar imagery in mapping land system for rainfed lowland rice in Northeast Thailand. In Proceedings of the Map Asia 2002, Bangkok, Thailand, 7–9 September 2002.

21. Li, Z.; Sun, G.; Wooding, M.; Pang, Y.; Dong, Y.; Chen, E.; Tan, B. Rice monitoring using envisat Asar data in China. In Proceedings of the 2004 Envisat & ERS Symposium, Salzburg, Austria, 6–10 September 2004.

22. Chen, J.; Lin, H.; Pei, Z. Application of ENVISAT ASAR data in mapping rice crop growth in Southern China. *IEEE Geosci. Remote Sens. Lett.* **2007**, *4*, 431–435. [CrossRef]

23. Yu, G.; Yang, S.; Zhao, X.; Shen, S. Application of ENVISAT ASAR data to rice monitoring in Jiangsu Province, China. In Proceedings of the 2010 International Conference on Multimedia Technology, Ningbo, China, 29–31 October 2010.

24. Liew, S.C.; Kam, S.-P.; Tuong, T.-P. Application of multitemporal ERS-2 synthetic aperture radar in delineating rice cropping systems in the Mekong River Delta, Vietnam. *IEEE Trans. Geosci. Remote Sens.* **1998**, *36*, 1412–1420. [CrossRef]

25. Lopez-Sanchez, J.M.; Ballester-Berman, J.D.; Hajnsek, I. First results of rice monitoring practices in Spain by means of time series of TerraSAR-X dual-pol images. *IEEE J. Sel. Topics Appl. Earth Obs. Remote Sens.* **2011**, *4*, 412–422. [CrossRef]

26. Özküralpli, İ.; Sunar, F. Monitoring crop growth in rice padies in the thrace-meriç basin with multitemporal RADARSAT-1 satellite images. In Proceedings of the ISPRS Commission VII—Conference on Information Extraction From SAR and Optical Data, with Emphasis on Developing Countries, Istanbul, Turkey, 15–17 May 2007.

27. Rosenqvist, A. Temporal and spatial characteristics of irrigated rice in JERS-1 L-band SAR data. *Int. J. Remote Sens.* **1999**, *20*, 1567–1587. [CrossRef]

28. Chakraborty, M.; Panigrahy, S. A processing and software system for rice crop inventory using multi-date RADARSAT ScanSAR data. *ISPRS J. Photogramm. Remote Sens.* **2000**, *55*, 119–128. [CrossRef]

29. Tan, Q.; Hu, J.; Bi, S.; Liu, Z. A study on rice field edge extraction in RADARSAT SAR images. In Proceedings of the 2004 IEEE Interbational Geoscience and Remote Sensing Symposium, Anchorage, AK, USA, 20–24 September 2004.

30. Zhang, Y.; Wang, C.; Wu, J.; Qi, J.; Salas, W.A. Mapping paddy rice with multitemporal ALOS/PALSAR imagery in Southeast China. *Int. J. Remote Sens.* **2009**, *30*, 6301–6315. [CrossRef]

31. Dao, N.L. Rice Crop Monitoring Using New Generation Synthetic Aperture Radar (SAR) Imagery. Ph.D. Thesis, University of Southern Queensland, Darling Heights, QLD, Australia, 2009.

32. Lam-Dao, N.; Apan, A.; Le-Toan, T.; Young, F.; Le-Van, T.; Bouvet, A. Towards an operational system for rice crop inventory in the Mekong River Delta, Vietnam using ENVISAT-ASAR data. In Proceedings of the 7th FIG Regional Conference, Spatial Data Serving People: Land Governance and the Environment—Building the Capacity, Hanoi, Vietnam, 19–22 October 2009.

33. Fan, W.; Chao, W.; Hong, Z.; Bo, Z.; Yixian, T. Rice crop monitoring in South China with RADARSAT-2 quad-polarization SAR data. *IEEE Geosci. Remote Sens. Lett.* **2011**, *8*, 196–200.

34. Schmitt, A.; Brisco, B. Wetland monitoring using the curvelet-based change detection method on polarimetric SAR imagery. *Water* **2013**, *5*, 1036–1051. [CrossRef]

35. Yang, S.B.; Shen, S.H.; Zhao, X.Y.; Gao, W.; Jackson, T.J.; Wang, J.; Chang, N.B. Assessment of RADARSAT-2 quad-polarization SAR data in rice crop mapping and yield estimation. *Remote Sens. Model. Ecosyst. Sustain. IX* **2012**. [CrossRef]

36. Nguyen, D.B.; Gruber, A.; Wagner, W. Mapping rice extent and cropping scheme in the Mekong Delta using Sentinel-1A data. *Remote Sens. Lett.* **2016**, *7*, 1209–1218. [CrossRef]

37. European Commission; DG AGRI. EU Rice Economic Fact Sheet. Available online: http://ec.europa.eu/agriculture/cereals/trade/rice/economic-fact-sheet_en.pdf (accessed on 31 May 2017).

38. Ferrero, A.; Tinarelli, A. Rice cultivation in the EU ecological conditions and agronomical practices. In *Pesticide Risk Assessment in Rice Paddies: Theory and Pratice*; Ettore, C., Dimitrios, K., Eds.; Elsevier: Amsterdam, The Netherlands, 2007; pp. 1–24.

39. Calha, I.M.; Machado, C.; Rocha, F. Resistance of alisma plantago aquatica to sulfonylurea herbicides in portuguese rice fields. In *Biology, Ecology and Management of Aquatic Plants: Proceedings of the 10th International Symposium on Aquatic Weeds, European Weed Research Society*; Caffrey, J., Barrett, P.R.F., Ferreira, M.T., Moreira, I.S., Murphy, K.J., Wade, P.M., Eds.; Springer: Dordrecht, The Netherlands, 1999; pp. 289–293.

40. Silva, E.; Batista, S.; Viana, P.; Antunes, P.; Serôdio, L.; Cardoso, A.T.; Cerejeira, M.J. Pesticides and nitrates in groundwater from oriziculture areas of the 'Baixo Sado' region (Portugal). *Int. J. Environ. Anal. Chem.* **2006**, *86*, 955–972. [CrossRef]

41. Spot Take5. Available online: https://spot-take5.org/client/#/products/ (accessed on 31 May 2017).

42. The Institute of Statistics and Cartography of Andalusia. Available online: http://www.juntadeandalucia.es/institutodeestadisticaycartografia/sima/ficha.htm?mun=41902 (accessed on 31 May 2017).

43. The ERMES (An Earth Observation Model Based Rice Information Service). Available online: http://ermes.dlsi.uji.es/prototype/ (accessed on 31 May 2017).

44. The EOX: Maps. Available online: https://s2maps.eu/ (accessed on 31 May 2017).

45. Brisco, B.; Brown, R.J.; Snider, B.; Sofko, G.J.; Koehler, J.A.; Wacker, A.G. Tillage effects on the radar backscattering coefficient of grain stubble fields. *Int. J. Remote Sens.* **1991**, *12*, 2283–2298. [CrossRef]

46. Rodríguez Díaz, J.A.; Weatherhead, E.K.; Knox, J.W.; Camacho, E. Climate change impacts on irrigation water requirements in the guadalquivir river basin in Spain. *Reg. Environ. Chang.* **2007**, *7*, 149–159. [CrossRef]

47. Sharma, S.D. *Rice Origin Antiquity and History*; Taykor & Francis: New York, NY, USA, 2010; ISBN 978-1-4398-4056-6.

PERMISSIONS

LIST OF CONTRIBUTORS

Wenzhao Li, Thomas Piechota and Daniele Struppa
Schmid College of Science and Technology, Chapman University, Orange, CA 92866, USA

Hesham El-Askary
Schmid College of Science and Technology, Chapman University, Orange, CA 92866, USA
Center of Excellence in Earth Systems Modeling and Observations, Chapman University, CA 92866, USA
Department of Environmental Sciences, Faculty of Science, Alexandria University, Moharem Bek, Alexandria 21522, Egypt

Rejoice Thomas
Computational and Data Sciences Graduate Program, Schmid College of Science and Technology, Chapman University, Orange, CA 92866, USA

Surya Prakash Tiwari and Karuppasamy P. Manikandan
Center for Environment and Water, The Research Institute, King Fahd University of Petroleum & Minerals (KFUPM), Dhahran 31261, Saudi Arabia

Marie Parrens, Ahmad Al Bitar and Yann Kerr
Centre d'Etudes Spatiales de la BIOsphère (CESBIO — Université de Toulouse, CNES, CNRS, IRD), UMR5126, BPI 2801, 31401 Toulouse CEDEX 9, France

Stephane Calmant and Jean-François Crétaux
Laboratoire d'Etudes en Géophysique et Océanographie Spatiales (LEGOS), UMR5566, Université de Toulouse, CNES, CNRS, IRD, Observatoire Midi-Pyrénées (OMP), 14 Avenue Edouard Belin, 31400 Toulouse, France

Frédéric Frappart
Laboratoire d'Etudes en Géophysique et Océanographie Spatiales (LEGOS), UMR5566, Université de Toulouse, CNES, CNRS, IRD, Observatoire Midi-Pyrénées (OMP), 14 Avenue Edouard Belin, 31400 Toulouse, France
Géosciences Environnement Toulouse (GET), UMR5563, Université de Toulouse, CNES, CNRS, IRD, Observatoire Midi-Pyrénées (OMP),14 Avenue Edouard Belin, 31400 Toulouse, France

Fabrice Papa
Laboratoire d'Etudes en Géophysique et Océanographie Spatiales (LEGOS), UMR5566, Université de Toulouse, CNES, CNRS, IRD, Observatoire Midi-Pyrénées (OMP), 14 Avenue Edouard Belin, 31400 Toulouse, France
Indo-French Cell for Water Sciences (IFCWS), IRD-IISc-NIO-IITM Joint International Laboratory, Bangalore 560012, India

Jean-Pierre Wigneron
INRA, UMR 1391 ISPA, F-33140 Villenave d'Ornon, Bordeaux, France

Edward Salameh
Laboratoire d'Etudes en Géophysique et Océanographie Spatiales (LEGOS), Université de Toulouse, IRD, CNES, CNRS, UPS, Toulouse 31400, France
Université de Rouen, UMR CNRS 6143, Mont-Saint-Aignan 76821, France

Benoît Laignel
Université de Rouen, UMR CNRS 6143, Mont-Saint-Aignan 76821, France

David Labat
Géosciences Environnement Toulouse (GET), Université de Toulouse, IRD, CNES, CNRS, UPS, Toulouse 31400, France

Andreas Güntner
GFZ German Research Centre for Geosciences, Potsdam 14473, Germany

Vuruputur Venugopal
Centre for Atmospheric and Oceanic Sciences (CAOS), Indian Institute of Science, IISc, Bangalore 560012, India

Augusto Getirana
Hydrological Sciences Laboratory, NASA Goddard Space Flight Center, Greenbelt, MD 20771, USA
Earth System Science Interdisciplinary Center, University of Maryland, College Park, College Park, MD 20742, USA

Claudia Ferrara and Giannetta Fusco
Department of Science and Technologies, University of Naples Parthenope, Centro Direzionale di Napoli, Isola C4, 80143 Napoli, Italy
CINFAI, (Consorzio Interuniversitario Nazionale per la Fisica delle Atmosfere e delle Idrosfere), 62029 Tolentino (MC), Italy

Massimiliano Lega
Department of Engineering, University of Naples Parthenope, Centro Direzionale di Napoli, Isola C4, 80143 Napoli, Italy

Paul Bishop
College of Engineering, University of Rhode Island, 102 Bliss Hall 1 Lippitt Rd, Kingston, RI 02881, USA

Theodore Endreny
Department of Environmental Resources Engineering, College of Environmental Science and Forestry, SUNY, 402 Baker Labs, 1 Forestry Drive, Syracuse, NY 13244, USA

Philippe Paillou
Laboratoire d'Astrophysique de Bordeaux, Université de Bordeaux, UMR 5804-CNRS, 33600 Pessac, France

Anika Bettge, Jürgen Kusche and Anne Springer
Institute of Geodesy and Geoinformation, Bonn University, 53115 Bonn, Germany

Annette Eicker
Institute of Geodesy and Geoinformation, Bonn University, 53115 Bonn, Germany
Hafen-City University, 20457 Hamburg, Germany

Andreas Hense
Meteorological Institute, Bonn University, 52121 Bonn, Germany

Catherine Prigent and Filipe Aires
Laboratoire d'Etudes du Rayonnement et de la Matiére en Astrophysique et Atmosphéres, UMP 8112, l'Observatoire de Paris, 61 Avenue de l'Observatoire, 75014 Paris, France

Binh Pham-Duc
Laboratoire d'Etudes du Rayonnement et de la Matiére en Astrophysique et Atmosphéres, UMP 8112, l'Observatoire de Paris, 61 Avenue de l'Observatoire, 75014 Paris, France
Space and Aeronautics Department, University of Science and Technology of Hanoi, 18 Hoang Quoc Viet, Cau Giay, 10000 Hanoi, Vietnam

Karina Nielsen, Lars Stenseng, Ole Baltazar Andersen and Per Knudsen
Department of Geodesy, DTU Space, National Space Institute, 2800 Kgs. Lyngby, Denmark

Damien O'Grady
Centre for Tropical Water Research & Aquatic Ecosystem Research (TropWATER), James Cook University, 4870 Cairns, Australia

Hong Shen
Centre for Tropical Water Research & Aquatic Ecosystem Research (TropWATER), James Cook University, 4870 Cairns, Australia
State Key Laboratory of Hydro-Science and Engineering, Department of Hydraulic Engineering, Tsinghua University, Beijing 100084, China

Sarah Tweed
Research Institute for the Development, UMR G-EAU, 34000 Montpellier, France

Marc Leblanc
Centre for Tropical Water Research & Aquatic Ecosystem Research (TropWATER), James Cook University, 4870 Cairns, Australia
Research Institute for the Development, UMR G-EAU, 34000 Montpellier, France
Hydrogeology Laboratory, UMR EMMAH, University of Avignon, 84000 Avignon, France

Lucia Seoane
Géosciences Environnement Toulouse (GET)— UMR5563, CNRS, IRD, Université de Toulouse UPS, OMP-GRGS, 14 Avenue E. Belin, 31400 Toulouse, France

Albert Olioso
UMR EMMAH, INRA—University of Avignon, 84000 Avignon, France

Yury M. Polishchuk
Ugra Research Institute of Information Technologies, Mira str., 151, Khanty-Mansiysk 628011, Russia
Institute of Petroleum Chemistry SB RAS, 4 Akademichesky av., Tomsk 634021, Russia

Alexander N. Bogdanov
Ugra Research Institute of Information Technologies, Mira str., 151, Khanty-Mansiysk 628011, Russia

Vladimir Yu. Polishchuk
Institute of Monitoring of Climate and Ecological Systems SB RAS, 10/3 Akademichesky av., Tomsk 634021, Russia
Tomsk Polytechnic University, Lenina av., 30, Tomsk 634004, Russia

Sergey N. Kirpotin
BIO-GEO-CLIM Laboratory, Tomsk State University, Lenina av., 36, Tomsk 634004, Russia

Rinat M. Manasypov
BIO-GEO-CLIM Laboratory, Tomsk State University, Lenina av., 36, Tomsk 634004, Russia
Federal Center for Integrated Arctic research, Institute of Ecological Problem of the North, 23 Nab Severnoi Dviny, Arkhangelsk 163000, Russia

Liudmila S. Shirokova
Federal Center for Integrated Arctic research, Institute of Ecological Problem of the North, 23 Nab Severnoi Dviny, Arkhangelsk 163000, Russia
GET UMR 5563 CNRS University of Toulouse (France), 14 Avenue Edouard Belin, 31400 Toulouse, France

Oleg S. Pokrovsky
GET UMR 5563 CNRS University of Toulouse (France), 14 Avenue Edouard Belin, 31400 Toulouse, France

Liguang Jiang, Raphael Schneider and Peter Bauer-Gottwein
Department of Environmental Engineering, Technical University of Denmark, Bygningstorvet B115, 2800 Kongens Lyngby, Denmark

Ole B. Andersen
National Space Institute, Technical University of Denmark, Elektrovej 327, 2800 Kongens Lyngby, Denmark

Wolfgang Wagner
Department of Photogrammetry and Remote Sensing, Hanoi University of Mining and Geology, Hanoi 10000, Vietnam

Duy Ba Nguyen
Department of Geodesy and Geoinformation, Vienna University of Technology, Gusshausstrasse 27-29, A-1040 Vienna, Austria
Department of Photogrammetry and Remote Sensing, Hanoi University of Mining and Geology, Hanoi 10000, Vietnam

Index

9 781647 404369